After the Famine examines the recovery in Irish agriculture in the wake of the disastrous potato famine of the 1840s, and presents for the first time an annual agricultural output series for Ireland from 1850 to 1914. Michael Turner's detailed study is in three parts: he analyses the changing structure of agriculture in terms of land use and peasant occupancy; he presents estimates of the annual value of Irish output between 1850 and 1914; and he assesses Irish agricultural performance in terms of several measures of productivity. These analyses are placed in the context of British and European agricultural development, and suggest that, contrary to prevailing orthodoxies, landlords rather than tenants were the main income beneficiaries in the decades before the Land War and land reforms. This interpretation could help to explain the emergence of the land reform movement from the late 1870s, which culminated in peasant independence by 1914. *After the Famine* is an important contribution to an extremely controversial area of Irish and economic history.

After the Famine

After the Famine

Irish agriculture, 1850–1914

Michael Turner
University of Hull

CAMBRIDGE
UNIVERSITY PRESS

Published by the Press Syndicate of the University of Cambridge
The Pitt Building, Trumpington Street, Cambridge CB2 1RP
40 West 20th Street, New York, NY 10011–4211, USA
10 Stamford Road, Oakleigh, Melbourne 3166, Australia

First published 1996

Printed in Great Britain at the University Press, Cambridge

A catalogue record for this book is available from the British Library

Library of Congress cataloguing in publication data

Turner, Michael Edward.
  After the famine: Irish agriculture, 1850–1914/Michael Turner.
    p.   cm.
  Includes bibliographical references (p.   ).
  ISBN 0 521 55388 1
  1. Agriculture – Ireland – History – 19th century. 2. Agriculture – Ireland –
History – 20th century. 3. Agriculture – Economic aspects – Ireland – History –
19th century. 4. Agriculture – Economic aspects – Ireland – History –
20th century. I. Title.
S461.T87 1996
338.1'009415'09034–dc20   95–24534   CIP

ISBN 0 521 55388 1 hardback

CE

For Alison
Kate, James and Jessie

Contents

Figures

Tables

Appendix tables

Appendix 4

Preface and acknowledgements

Through personal and domestic circumstances I have made many visits to Belfast in the last twenty years or so. I want to take this early opportunity to record my thanks to my mother and father-in-law, and indeed to all my contacts in Northern Ireland for making me feel so welcome. Whilst in Belfast I often took the opportunity to use Queen's University Library. I can commend the marvellous collection of nineteenth-century parliamentary papers held there. This was my initial entry into Irish agricultural history, along with my extraction of material from the Public Record Office of Northern Ireland which was concerned with the County Down livestock census of 1803.[1] I shall always be grateful to Peter Roebuck for alerting me to the existence of this material. This research also allowed me to become familiar with the abundant annual agricultural census material which was collected from the late 1840s which now form the main data on which this present study is based.

A more deeply rooted connection with Ireland is literally a matter of roots. My mother came from Achill Island, a remote Atlantic island connected to its administrative county of Mayo by a causeway. She was born into an impoverished family in 1913, one of eight children. This study of Irish history inevitably took on a personal meaning.

The whole exercise has illuminated the value of inter-disciplinary co-operation. I have relied upon the expertise and good advice from many disparate areas – geography, economics, history, computing, economic history, and even geology. At a minimum, the size of the data set on which this study is based – upwards of 40 variables for 70 years at the national level, and the same 40 variables for 32 counties for each of seven specific census years – are about 12,000 separate numbers. In the text they have usually been presented as some kind of transformation of the originals. This work has only been possible with the aid of modern

---

[1] This was part of the data gathering exercises related to the invasion scares of the French Revolutionary and Napoleonic Wars. See M.E. Turner, 'Livestock in the agrarian economy of Counties Down and Antrim from 1803 to the Famine', *Irish Economic and Social History*, 11, 1984, 19–43.

technology and the grant of a year away from university teaching and administration through the generosity of the Nuffield Foundation. I am happy to thank the Nuffield for the award of a Social Science Fellowship during the year 1985–6. Much of that year was occupied in data entry and transformation, and only in a small measure to data analysis. All of the data were made adaptable to a table top computer. My thanks to Norman Davidson of the Department of Geography firstly for educating me in the use of data management systems, and secondly for introducing me to the statistical and mapping packages which were in current use at the time. He also taught me the intricacies of digitising (inputing the boundaries of the Irish counties into the computer for automatic mapping). Good students always strive for independence but there is often a battle with technological constraints. Although I acquired a computer I remained dependent on Norman and his department for the use of their digitisers and pen plotters. During this, the initial phase of development, my final thanks are to Richard Middleton, formerly of Geology, for constant encouragement and the use of his packages for some of the original transformations of the data.

Immediately the Fellowship expired it was back to other duties, and in my case this involved the acquisition of the role of Senior Tutor for Social Science students at Hull University, as well as a full teaching load. In the three years while I held office the project lay largely dormant, but there occurred an immense technological change. The geographers changed systems for a start, and in general the schools and departments throughout the university acquired PC computers. My new-found skills were not wasted because the adaptation to an alternative computing system involved a steep learning curve. I ended up completing this study on a PC using the most modern of packages. For his assistance during this phase I particularly thank Chris Hammond of the Department of Economics for his interest and work on my behalf, especially in terms of mapping and allied computer skills. I also thank John Palmer of the Department of History for the use of his digitising equipment and to George Slater of the Computer Centre for his developmental work on the mapping package. For constant encouragement, often of a non-specific nature, I thank my colleagues Donald Woodward and David Richardson of the Department of Economic History, Gerry Makepeace and Stephen Trotter in Economics, and Mahes Visvalingam of the Cartographic Information Systems Research Group. A good library is a must for any substantial research, and we have a good library in Hull. Within it we have marvellous support services, but I must single out John Morris for special mention. No inquiry was too large, no search too tedious.

Away from Hull I owe debts of thanks to Cormac Ó Gráda and Peter

Solar for constant encouragement and good advice. Peter Solar has been generous with his time commenting on earlier drafts of specific chapters and making valuable suggestions for improvements. A small army of anonymous readers, and some not so anonymous, have had their say and I am grateful to them all. In particular I would like to thank Cormac Ó Gráda a second time for his help with earlier drafts, and also Patrick O'Brien and Charles Feinstein for their help in the final tidying up process. I always took advice in the spirit in which it was given, though I may not have always applied it in the correct manner. Richard Fisher at Cambridge University Press has been a model of patience.

While the book was developing there were invitations to give papers, to test ideas, and to contribute to edited volumes. The bare data in chapter 4 and two related but separate arguments in chapter 5 appeared in a journal article in 1990 and an edited volume in 1991, but they have now been completely revised; a long general essay on post-Famine agriculture has appeared in a volume of essays, though again much of this has been revised in this book.[2]

Finally, my love and thanks go to Alison, Kate, James and Jessie for their patience over many years, for putting up with the countless late nights in the office and even the missed holidays.

[2] M.E. Turner, 'Output and productivity in Irish agriculture from the Famine to the Great War', *Irish Economic and Social History*, 17 (1990), 62-78, and, 'Agricultural output and productivity in post-Famine Ireland', in B.M.S. Campbell and M. Overton (eds.), *Land, Labour, and Livestock: Historical Studies in European Agricultural Productivity* (Manchester, 1991), pp. 410–38, and, 'Rural economies in post-Famine Ireland, c. 1850–1914', in B.J. Graham and L. Proudfoot (eds.), *An Historical Geography of Ireland* (London, 1993), pp. 293–337.

1 Introduction: Ireland and Irish agricultural history in context

The European context

It is not always remembered that the Irish economy in the first half of the nineteenth century, as well as thereafter, was inextricably bound to the rest of Britain. As Mokyr and Ó Gráda have put it 'Ireland and Britain were part of the same economy', and indeed, nearly one-third of the population of the British Isles came from Ireland. The connection was a very direct one: constitutionally in the very existence of the United Kingdom from 1800 (though legally and therefore effectively from 1 January 1801); and materially in the food stuffs and manpower which Ireland provided for industrialising Britain.[1] In addition, the country may not have been as backward as legend would have it. It had begun the process of industrialisation in the eighteenth century, especially in textiles, and later in shipbuilding and allied industries. Although the decline of the textiles industry came early, other industries remained and did not decline until much later in the nineteenth century, or indeed like shipbuilding they actually grew. Measured against some other European countries Ireland will look less the industrial laggard, or the 'peasant' economy of popular myth. Even though by the second quarter of the twentieth century something near to 50 per cent of her employed population was still engaged in agriculture, and the agricultural sector provided over 30 per cent of GDP, this profile was by no means unusual in Europe as a whole. It may not be flattering to link Ireland (in fact the Free State for this particular period) with the economies of eastern and southern Europe in this way, but unlike them it might be true to say that in relative terms she became more agricultural from a once more mixed economy in the previous century.[2] It was her nineteenth-century partner,

---

[1] J. Mokyr and C. Ó Gráda, 'Poor and getting poorer? Living standards in Ireland before the Famine', *Economic History Review*, 41, (1988), 209.

[2] See R. Munting and B.A. Holderness, *Crisis, Recovery and War: An Economic History of Continental Europe, 1918–1945* (London, 1991), pp. 48 and 51. Given the constraints of the data the summary of European data relates to c. 1930 for employment, and to the

the remains of the larger original UK, which of all European countries had deserted agriculture in the most profound way since the nineteenth century.

Let us not be misunderstood, this is not to say that Ireland was industrialised or experiencing an industrial revolution in the nineteenth century. She was not best endowed with the principal industrial raw materials of fuel (meaning coal rather than peat, which was the mainstay of the domestic hearth), and minerals, though by the late nineteenth century she had a railway network and was not short of capital. If we regard Ireland as a region of a larger economy, the United Kingdom, it should not cause surprise and alarm to see a process of specialisation take place which emphasised agriculture in Ireland. Of more importance perhaps, was not the absence of industrialisation on any scale, but rather the fact of specialisation within Ireland itself and the emphasis on industrialisation in Ulster.[3] Eventually, in her own right therefore, Ireland became, rather than remained, more essentially an agricultural economy, and a highly commercialised one at that. Although its people often existed on or not much above the margin of self-sufficiency, providing their own food, fuel and clothing,[4] the country also generated a very sizeable surplus. This is hardly the prescription for, or the traditional image of, a backward economy.[5] Ó Gráda's estimates suggest that the bulk of the livestock, and perhaps one quarter of the grain was exported as early as the 1830s and early 1840s. Furthermore, total Irish exports in 1845 could have fed over 2 million people.[6] The shock of the Famine therefore was felt all the more, not just in the decline in food for the population, but also because specialisation meant that the agricultural sector was vital to the economy at large through its exports.

Britain in the mid-nineteenth century could rely on agricultural imports from Ireland – they were on tap, but they were not of a scale to suggest a

broad period 1938–50 for GDP statistics. See also J.J. Lee, *Ireland 1912–1985* (Cambridge, 1989), pp. 69–71.

[3] The strength of Irish industrialisation is subject to much debate. For example see the review of deindustrialisation in the decades leading to the Famine in J. Mokyr, *Why Ireland Starved: A Quantitative and Analytical History of the Irish Economy, 1800–1850* (London, 1985 edn.), pp. 13–15. In this section I have followed the line taken by K.A. Kennedy, T. Giblin and D. McHugh, *The Economic Development of Ireland in the Twentieth Century*, (London, 1988), pp. 7–12 who follow J.J. Lee, *The Modernisation of Irish Society 1848–1918* (Dublin, 1973), especially pp. 11–13. For industrialisation in Ulster in the period 1820–1914 see P. Ollerenshaw, 'Industry, 1820–1914', chapter 2 of L. Kennedy and P. Ollerenshaw (eds.), *An Economic History of Ulster 1820–1939* (Manchester, 1985), pp. 62–108.

[4] Mokyr and Ó Gráda, 'Poor and getting poorer?', 211.

[5] Lee, *The Modernisation*, pp. 9–10.

[6] C. Ó Gráda, *Ireland Before and After the Famine: Explorations in Economic History, 1800–1925* (Manchester, 1988), pp. 51, 57–8.

vital dependency for Britain. In addition, they offered mutual benefits since Ireland enjoyed a protected extension of the British home market. However, when the Famine struck the Irish economy it had but marginal effects on Britain – especially given the near coincidental repeal of protection – but for Ireland, the demographic effects were stupendous, as were the immediate agricultural disruptions. Without doubt, dependency was a one-way ticket. If then the reaction of the British authorities to the Famine was 'too little too late', as the popular parlance often has it, then for the purposes of this book this is not the important point. Rather it is more generally significant that agriculture had been and then remained the mainstay of the Irish economy whereas in Britain and much of Western Europe the Industrial Revolution was in full flow to varying degrees.

Moreover, the fact that agriculture was that mainstay does not have to imply that it was an impoverished economy. We have already indicated the rumblings of an industrial revolution, and while poverty could be found, and without too much difficulty, nevertheless, it is now thought that average incomes in Ireland may actually have increased in the decades leading to the Famine, even if the base from which this progress advanced was very low and the poles of inequality had widened. Contemporary accounts of Ireland in the early to mid-nineteenth century are a damning indictment, but they are now thought to be based on different, or shall we say *foreign* appreciations and expectations regarding consumption. If the Irish were badly clothed and housed, they were, in the strictly nutritional sense better fed.[7] The potato diet may have been monotonous, though even that is subject to debate, but it was certainly body building. After the Famine those average incomes certainly did rise. If real GNP per capita in Ireland in 1841 was 40 per cent of that attained in the UK by 1830, then by 1913 it was 60 per cent. If in 1841 her real GNP per capita placed her in the bottom third of twenty-three European countries when ranked against a UK base, then by 1913 she was in the top half and not far short of France, Austria and Sweden, all of which were within five percentage points above her. On the downside, and perhaps more significantly however, was that economies with similar emphasis on agriculture, such as Denmark and the Netherlands, and to a lesser extent Belgium, were all and always above her in such an exercise. In contrast the Mediterranean and eastern European economies, certainly by 1913, were all below her.[8]

[7] See Kennedy *et al.*, *The Economic Development*, p. 17; J. Mokyr and C. Ó Gráda, 'Emigration and poverty in prefamine Ireland', *Explorations in Economic History*, 19 (1982), especially 360–62; and in greater detail see Mokyr, *Why Ireland Starved*, chapter 2.

[8] Kennedy *et al.*, *The Economic Development*, p. 14.

Table 1.1 *Population growth in selected European countries in the nineteenth century* (in 000 and inter-censal percentage changes)

	Belgium	Denmark	France	Germany	IRELAND	Italy	N'lands	GB
1801		929	27,349			17,237		10,501
			6.4			6.6		14.0
1811	4,166		29,107	22,377		18,381	2,047	11,970
			4.7			7.3		17.7
1821			30,462		6,802	19,727		14,092
			6.9		14.2	7.5		15.4
1831	4,090	1,231	32,569	28,237	7,767	21,212	2,613	16,261
	6.0	4.7	5.1	7.6	5.3	8.1	9.5	14.0
1841	4,337	1,289	34,230	30,382	8,175	22,936	2,861	18,534
	4.5	9.8	4.5	10.0	-19.9	6.2	6.9	12.3
1851	4,530	1,415	35,783	33,413	6,552	24,351	3,057	20,817
	6.6	13.6	4.5	6.4	-11.5	2.7	8.2	11.1
1861	4,828	1,608	37,386	35,567	5,799	25,017	3,309	23,128
		11.0	-3.4	15.4	- 6.7	7.1	8.2	12.7
1871		1,785	36,103	41,059	5,412	26,801	3,580	26,072
		10.3	3.6	10.2	- 4.4	6.2	12.1	14.0
1881	5,520	1,969	37,406	45,234	5,175	28,460	4,013	29,710
	9.9	10.3	1.9	9.3	- 9.1		12.4	11.2
1891	6,069	2,172	38,133	49,428	4,705		4,511	33,029
	10.3	12.8	0.8	14.0	- 5.2		13.1	12.0
1901	6,694	2,450	38,451	56,367	4,459	32,475	5,104	37,000
	10.9	12.5	1.9	15.2	- 1.5	6.8	14.8	10.4
1911	7,424	2,757	39,192	64,926	4,390	34,671	5,858	40,831

Note: The dates conform to the UK census. For other countries the population figures relate to dates which may be up to five years adrift of the dates indicated.
Source: B.R. Mitchell, *European Historical Statistics 1750-1975* (Cambridge, 1981), pp. 29–34.

This general improvement took place in spite of a rise of population that was greater than occurred in Britain and most of the rest of Western Europe, though that rise had clearly slowed down in the decades immediately preceding the Famine to a level which was nothing very special compared with other countries (for which see table 1.1). By then the nutritional advantages of a potato diet had built up a head of steam. The evidence on nutrition as displayed in the mean height of men in the mid-century suggests that only Norway produced taller men than Ireland.[9] This must say something for some kind of efficiency within its

[9] On population growth in the century before c. 1850 see C. Ó Gráda, *The Great Irish Famine* (London, 1989), p. 13; on the changing population after this see B.R. Mitchell, *European Historical Statistics 1750–1975* (Cambridge, 1981 edn.), p. 31; and on the issue of stature and height see Ó Gráda, *Ireland Before and After the Famine*, p. 17 and

economic system, though it does rather stress the nutritional advantages of that potato diet. The efficiency perhaps was in the transformation of a crop into energy, and there were also good agricultural by-products from potato cultivation into livestock fodder. Therefore it was a particularly cruel irony that the population which gained such strength from its dietary mainstay was dealt such a crushing blow when the potato crop failed.

It is not always appreciated that the potato blight which heralded the Famine was not confined to Ireland, nor was it by any means the first time that there was a failure of the potato crop. It has been suggested that the death rate resulting from the subsistence crisis of 1740–41, for example, was equal to that of the 1840s.[10] The importance of the crisis of the 1840s therefore might more importantly be explored not in the innate damage it caused, but in the different response it invoked from the economy and society of Ireland. The earliest indications of the profound damage the potato blight could do came from the Continent, and this had been preceded by its visitation on North America in the previous two years. The disease first struck the crop in Belgium in June 1845, and within a month or so had spread over the whole of Flanders and into neighbouring France and the Netherlands. It proceeded in August to stretch into more central parts of Europe, to Switzerland and generally along the Rhine and Rhone river systems, and then across the Channel into England. By early September 1845 it was into Ireland, but it was not until the harvest of October that the extent of the damage was realised and made an impact on the consciousness. Crop losses were greatest in areas where the disease first appeared, therefore not in Ireland but in Belgium, Holland and north-east France. This east to west diffusion of the blight was reversed in subsequent years.

It was the succession of bad crops, and also the relative susceptibility of the very potato varieties in more general use in Ireland than elsewhere which made Ireland suffer the greatest impact. Thus, even though the Scottish highlands were badly hit in 1846, only second in virulence to Ireland, they had emerged unscathed the previous year.[11] Therefore a combination of circumstances special to Ireland resulted – a succession

Mokyr, *Why Ireland Starved*, p. 9. See also S. Nicholas and R.H. Steckel, 'Heights and living standards of English workers during the early years of industrialization, 1770–1815', *Journal of Economic History*, 51, 4 (1991), 947, and in more detail in Nicholas and Steckel, 'Tall but poor: nutrition, health and living standards in pre-famine Ireland', *Working Paper Series on Historical Factors in Long Run Growth*, 39 (National Bureau of Economic Research, 1992).

10 An oft repeated idea, for which see Lee, *The Modernisation of Irish Society*, p. 1.
11 P.M.A. Bourke, 'Emergence of potato blight, 1843–6', *Nature*, 203 (22 August 1964), 805–8.

of failed crops, coupled with a level of consumption dependence on a form of monoculture not replicated elsewhere to the same degree. This was related generally to insufficient economic development and industrialisation, which is not the same as saying there was no industrialisation.[12] While potato dependency was not confined to Ireland it was not so widely prevalent in other European countries. Unfortunately the Irish had developed this dependency to a fine art. As often as not, in the two centuries or so before the mid-nineteenth, the introduction of potatoes in Europe came on the back of the failure of other more traditional grain-based staple foodstuffs.[13] To this extent it had developed as a fall-back crop, or as an emergency crop. In Ireland, in contrast, it had already become the mainstay crop, and there was in consequence no such insurance to fall back upon. Added to all of this was the actual severity of the crop failures, 'it was in no way a "normal" harvest failure, but was roughly twice as severe as the worst that might have been expected on the basis of recorded nineteenth-century experience in Ireland and other countries of western Europe'.[14]

After the Famine the potato declined in importance with the declining population, and this decline had important knock-on effects through the reduction in intensive methods of manual cultivation, involving the general decline of spade husbandry and also a decline in the labour-intensive application of manure. Taken together these had repercussions for the acreage under all crops and the yield from those acres, both of which declined down to the 1870s.[15]

Irish agricultural output did recover however, but an important concern is whether changes within agriculture were in motion before the Famine struck. If they were it would suggest that whilst Ireland was not necessarily at the forefront of European agricultural development, she was also by no means in the rear. If we take physical productivity for example, whilst wheat yields in England were higher than in Ireland at the mid-century, the reverse was the case for barley and oats, and even later in the century, when both economies had experienced a major shift towards livestock production, grain yields were broadly similar. At the same pre-Famine mid-point physical productivity in Ireland and Belgium

[12] Which is not to play down the impact of mortality in other European countries, but rather to play up the devastation in Ireland. For a few comparisons see Ó Gráda, *The Great Irish Famine*, p. 60; Mokyr, *Why Ireland Starved*, p. 276.

[13] See B.H. Slicher Van Bath, *The Agrarian History of Western Europe A.D. 500–1850* (London, 1963), pp. 266–71.

[14] P. Solar, 'The Great Famine was no ordinary subsistence crisis', in M. Crawford (ed.), *Famine the Irish Experience 900–1900: Subsistence Crises and Famines in Ireland* (Edinburgh, 1989), p. 118.

[15] P.M.A. Bourke, 'The average yields of food crops in Ireland on the eve of the Great Famine', *Journal of the Department of Agriculture*, 66 (1969), 26–39.

was also quite similar, even slightly higher in Ireland from the point of view of yields per acre of the main crops, and also higher for some animal products, and this situation had not significantly changed over the preceding decades.[16] When compared with other European countries in the second half of the nineteenth century, France and Italy for example, Irish yields were significantly larger, and, from nine northern European countries in the 1880s, Irish oat yields were second to the Netherlands, her wheat yields were third behind Denmark and Britain, but her potato yields were fourth behind Britain, the Netherlands and Germany.[17] By the eve of the First World War, Irish wheat yields were surpassed only by Denmark, and equalled by Belgium, whereas her barley yields were exceeded only by both the Netherlands and Belgium.[18] This is not altogether a satisfactory comparison since wheat cultivation in some European countries was not as important as rye, but then by the end of the century wheat production in Ireland was also unimportant. Average crop yields can be misleading indicators of land productivity however, because they say nothing about the dispersion about the mean or the skewness of the distribution. Thus in mid-century it looks as though both land and labour productivity were in fact greater in Belgium than in Ireland.[19] Before the Famine the high yields, which give credence to a high land productivity, were only achieved by a high labour input.

The real downturn in Irish land productivity could be seen in that once stalwart of the agrarian economy, the potato. With the loss of intensive methods – spade husbandry, weeding, and intensive manuring – the average potato yields in Ireland in 1909–13 came a poor seventh in the European league. The importance of this may be seen less in terms of human nutrition but more in terms of the input to livestock, particularly pigs. The three countries which perhaps bear the closest comparison with Ireland in terms of broad economic profile, Denmark, the Netherlands and Belgium (perhaps Belgium unfairly so since it was a relatively early industrialiser), all had significantly higher potato yields. As far as the data are available, two of those countries, Denmark and the Netherlands, were far advanced in terms of milk yields, a commodity which we shall

---

[16] P. Solar and M. Goossens, 'Agricultural productivity in Belgium and Ireland in the early nineteenth century', in B.M.S. Campbell and M. Overton (eds.), *Land, Labour and Livestock: Historical Studies in European Agricultural Productivity* (Manchester, 1991), pp. 367–8.

[17] Ó Gráda, *Ireland Before and After the Famine*, p. 53; P.K. O'Brien, D. Heath, and C. Keyder, 'Agricultural efficiency in Britain and France, 1815–1914', *Journal of European Economic History*, 6 (1977), 365. The 1880s evidence comes from Solar, 'The Great Famine', p. 119.

[18] For a comparative table see P. Lamartine Yates, *Food, Land and Manpower in Western Europe* (London, 1960), p. 197.

[19] Solar and Goossens, 'Agricultural productivity', p. 376.

see in later chapters was central to nineteenth-century Irish agricultural output until the 1870s or 1880s. Therefore, whilst the recovery from the Famine was swift, that recovery must be seen in the context of her European rivals. In addition, whilst that recovery took a directional turn out of arable into pastoral activity, some would say almost in anticipation of the essentially arable depression which certainly threw British agriculture sideways in the last quarter of the nineteenth century, again her European rivals clearly made a similar, and in some cases a better adjustment.

These European comparisons are not necessarily as illuminating as may at first sight seem to be the case. Certainly the structure of agriculture in Ireland could not be replicated very easily in Britain or on the Continent. The structural division of landlord, tenant and labourer which was in place in England and steadily reinforced as the nineteenth century proceeded was not the same as the landlord, tenant, labourer relationship which existed in Ireland before the Famine. The direct relationship which existed between landlords and tenants in England, no doubt at times with intermediary agents to conduct the actual business, did not have a strong counterpart in Ireland. In that country the intermediate position was held by middlemen, who in law were the tenants of the landlords, but they sub-contracted to under-tenants. Besides, the dominant absentee aspect of Irish landlords, often absentees from the country not just from the neighbourhood, divorced them from their lands and therefore from a very active interest in the running of those lands in a way which was quite unlike the influence of English landlords over their own possessions. In addition, there was no peasant proprietorship to speak of, if by peasant we mean owner-farmer. Whilst the tenants retained a form of ownership of the system of agriculture they were involved in, there was no great Irish ownership of the actual land. Even the ownership of the system was determined largely outside the hands of the tenants due to their obligation to meet ever rising rents. In Ireland it was the production of cash crops and livestock products that paid that rent, but it was the labour-intensive potatoes which were the means of sustenance and subsistence. In general, George Grantham has suggested that the dominance of labour intensive crops marks a phase in the economic development of European agriculture, 'when the demand and supply of cash balances influenced allocational decisions as well as the price level'. He put this down to a capital constraint, partly related to the need to front load the capital requirement in a society where agriculture could not command credit. He did suggest, however, that it was prevalent in isolated districts. Thus in eighteenth-century Champagne they grew hemp as the cash crop, but rye as the subsistence crop.[20] In Ireland, where it was more widespread, cereals acted as the cash crop leaving the production of potatoes, a crop which could be grown in great abundance especially with heavy labour input, spade husbandry and constant weeding to keep up yields, as the most common source of sustenance for most Irishmen and their families. This was the case for the small tenants but also for the underclass of Irishmen. The cottier system, whereby farmers provided small plots of land of under 5 acres and for which the cottiers paid high rents and often also provided labour services, was widespread. Finally the 'landless' labourers often rented yet smaller plots under a system known as conacre on short leases. It was the plots of land farmed by the cottiers and labourers which were the main sources of potato production. This system, which was so vastly different from that which prevailed in England, also had little in common with the proprietorial arrangements which existed on the Continent. While much of European agriculture was also small scale, nevertheless it was not operated in the same way. In Ireland, as on much of the Continent, it left the average plot of cultivation so small as to make the generation of an investible surplus very difficult.[21]

The Irish system was not a microcosm of a wider European practice, it was generically pretty well a wholly different practice. The Famine dealt it a savage blow. This is not to suggest that it was an ideal arrangement. Before the Famine the extremes of income distribution between the cottiers, or below them the labourers, and the landlords widened, and therefore the Irish land system perhaps bred within itself its own destruction, except that the Famine occurred and dealt the actual death blow. Thus the external agent represented by the Famine made it necessary to rethink the future of the prevailing system, and in ways which were not replicated elsewhere in Europe. The potato blight which was the source of the Famine was not special to Irish potatoes, but its repercussions were of a different order of magnitude and asked different questions of the agricultural system than were asked in Scotland and Belgium for example. The Irish agricultural system, as described, may sound like a stark representation of social relationships in the country-side, as if it affected everyone in this way. This was not the case, but it did affect most people. Alongside it a commercial sector of large graziers

---

[20] G. Grantham, 'Capital and agrarian structure in early nineteenth-century France', *Research in Economic History*, Supplement 5 (1989), 139.

[21] There may be a look-alike situation in France, but there it arose from peasant ownership and therefore a large element of personal independence. See O'Brien *et al.*, 'Agriculture', esp. 373–9.

and large mixed farmers existed with the small tenants, cottiers and labourers.[22]

In terms of the effect on population alone Ireland was a special case. In 1841 the population was more than three times the size of that in Scotland and over one-half the combined population of England and Wales. By 1921 the Scottish population was greater, and the Irish population was barely one-ninth the size of England and Wales combined.[23] In every inter-censal between 1841 and 1911 the Irish population fell whereas in the rest of Western Europe the opposite was the case (see table 1.1). It is thought that the events surrounding the failure of the potato crop on their own killed between 1 to 1.5 million people, added to which in the six years from 1847 to 1852 perhaps as many more Irish people emigrated, principally to North America, but also to England.[24] The slow-down in population in Ireland before the Famine was partly due to mounting emigration, but what the Famine did was to give this trend a mighty boost.[25] Whilst it slowed down thereafter, it did not stop. It continued at high rates – 90,000 per annum in the 1860s, 60,000 in the 1870s, 80,000 in the 1880s, 45,000 in the 1890s, and only 38,000 per annum thereafter down to the Great War – a level of emigration not matched by any other European countries until it was equalled and surpassed by Italy from the 1870s, and then increasingly by other countries towards the end of the century. Russia, Spain and German-speaking central Europe all eventually engaged in heavy out-migration, but all of these countries had a larger base of population to lose. Thus the fact of emigration was not special to Ireland, but its dynamic was.[26] This loss of labour, directly through the Famine but also through subsequent migration, in association with growing international adjustments in the terms of trade between grain and livestock products dictated a faster adaptation towards livestock farming in Ireland than perhaps was the case in much of Western Europe, and this in turn may have become a self-reinforcing process – a decline in labour availability leading to less labour-intensive livestock farming, but a movement in the

---

[22] See also Mokyr, *Why Ireland Starved*, pp. 19–21 on the likely distribution of pre-Famine farms from small to large, and the dominance in terms of numbers of the cottier and labouring class.

[23] See Kennedy et al., *The Economic Development*, p. 4.

[24] *Ibid.*, p. 4.

[25] On pre-Famine emigration issues see Mokyr and Ó Gráda, 'Emigration and poverty', *passim*.

[26] Mitchell, *European Historical Statistics*, pp. 145–53. See also Kennedy et al., *The Economic Development*, pp. 4–5. Set in a long-term and wider European context see S. Pollard, *Peaceful Conquest: The Industrialization of Europe 1760–1970* (Oxford, 1981), p. 152.

terms of trade towards livestock farming requiring fewer units of labour. Such labour issues are discussed at some length below in chapter 6.

Whilst this study of Ireland must be understood in a wider European context, in its subsequent exploration it is the special case of post-Famine Ireland which needs to be dissected. From the period just before the Famine until the Great War Ireland experienced one of the poorest real product growth records of any country in Europe, a rate of less than 1 per cent per annum (1841–1913) compared with well over 2 per cent in the rest of the UK and in most of north-west Europe. The closest parallel economy in terms of economic profile was Denmark, but that country experienced the highest annual growth performance of all at 2.7 per cent per annum (1830–1913). Yet to distinguish Ireland from the rest, to contextualise the special circumstances of her origins and the peculiar shock of the Famine, it will be necessary to explore in some depth the post-Famine recovery, and subsequent events in her agricultural structure, output and performance which culminated in a *per capita* real annual product growth of 1.6 per cent per annum placing her alongside Switzerland (1.6 per cent) and ahead of all her European neighbours except the remarkable economic development of Denmark (1.7 per cent).[27] Of course this apparently redeeming fact of a high per capita growth rate is partly illusory: it embraces a low base at the time of the Famine; it is derived from a poorish absolute growth combined with a dramatic fall in population (Ireland was the only country to lose population); and it must be recognised that the largest group who contributed to that fall in population were at the low income extreme of the population distribution; and finally this per capita situation had not been achieved by strong economic growth. The saving grace, if such it can be called, is that the economy, or more particularly the farming community, responded positively to adversity, and pragmatically pursued a strategy of survival based on a rational appreciation of its adjusted endowment of resources in a changing world market situation. In short, the farming community grew up in economic terms, it specialised in the face of adjusted terms of trade, and it eventually succeeded in solving much of its dependency on Britain and shaking off its otherwise self-reinforcing poverty trap.

Self-evidently some of the issues raised above were not unique to Ireland. Equally self-evidently some of them were, the Famine most obviously, but in reality more particularly the special circumstances of land tenure which as the century wore on brought conflicts between tenants and their landlords to a head. As Joseph Lee has put it, and in a

[27] Kennedy *et. al.*, *The Economic Development*, pp. 18, and pp. 19–21 for the origin of these figures, with analysis and argument as to their meaning.

way which distinguishes Ireland from its mainland European neighbours, 'Post-Famine Ireland had a land question. It had no peasant question.'[28] This land question took many forms. The ones which we discuss in this study relate mainly to the spoils which were derived from agriculture, spoils which were shared between the parties to the bargain but not necessarily equally. The knock-on effect was the emergence of a political confrontation which questioned the relationship between Ireland and Irishman and their political and economic partner Great Britain. Increasingly Irishmen began to question the nature of control over their futures. This study explores the agricultural build-up to that quest for independence. It looks in some detail at the changing structure of agriculture, at first forced by emergency circumstances, but then carried out through rational adjustments. The changing structure was an end in itself, but it can also be employed as the means to defining another end, and that is the estimation of agricultural output. This in turn leads quite naturally to an appreciation of Irish agricultural performance. At the end of the day, so the cliché goes, we go to work to earn a crust to buy some comforts. The final part of the story therefore will be concerned with the distribution of those earnings between the parties who had a call on the land - principally the landlords and the tenants. This is done not as an aid to understand the distribution of the comforts, though by implication this is possible, but rather to investigate some of the reasons why those distributions have been questioned, and why a country of tenant farmers in 1850 was transformed into one of owner-farmers by 1914.

The structure of the book

The analysis of the changing structure, output and performance of the Irish agricultural economy is achieved in three parts. The first is a discussion of the land itself. This is dealt with in two ways: the physical act of farming either under crops or raising livestock, or a combination of the two; and secondly the actual occupation of the land in landholding units. In Ireland it is often difficult to talk more narrowly of ownership and occupancy and therefore the more all-embracing term 'landholding' is adopted. Ultimately the business end of the industry was determined by occupiers and not the relatively passive absentee owners.

The second part of the book firstly measures the output of agriculture in the period, and secondly assesses the performance of the agricultural sector. The measurement of output is achieved in two ways, by monetary value and by physical product. The analysis of performance is achieved

---

[28] Lee, *Ireland*, p. 72.

mainly through actual or inferred productivity comparisons. This is mostly in terms of partial productivity measures involving known or relatively easily estimated land and labour factors, though the labour factor, in Ireland particularly, raises difficult problems of definition. A crude total factor productivity measure is also made but in truth the capital factor component of that measure is based on few hard facts. The measures of productivity lead on finally to an extended discussion of labour productivity in particular. In this a relatively untried measure of productivity is presented, based not on conventional statistics of the labour force, but rather on estimates of the number of labour units – standard man days – required to cultivate the known crop and livestock outputs which are revealed in the annual returns. The book is rounded off with part three, a series of appendixes of statistics and explanations of some of the procedures followed.

Whatever is achieved in this book there is no doubting the large voids in the story which remain. Discussion of the cooperative movement – the Irish Agricultural Organisation Society – which grew up at the end of the nineteenth century, is neglected. This may seem a large subject to ignore, touching as it did upon the economic development of the soil, the way it cut across the best interests of many traders, the way it helped to create a new kind of economic and social order in the countryside, and its importance in the political history of the late nineteenth and early twentieth century.[29] These areas of contact, however, had less obvious or testable influences on matters which concern us in this study than they did on the distribution of agricultural produce to and from the market. This study is driven by questions of production – land use, landholding, output and productivity – rather than marketing. Production and marketing are inseparable perhaps, but there is a more expert literature to fill this particular void.[30]

Perhaps a more substantial area of neglect is the social composition of agriculture and agriculturalists. The fullest possible study would inevitably embrace wider questions of local, national and Anglo-Irish politics. Apart from a superficial inquiry into the changing *broad* landownership

[29] See E. Barker, *Ireland in the Last Fifty Years (1866–1916)* (London, 1916), pp. 122–39, especially p. 137.

[30] L. Kennedy, 'Agricultural co-operation and Irish rural society, 1880–1914', D.Phil. Thesis, University of York, 1978. See also Kennedy, 'Retail markets in rural Ireland at the end of the nineteenth century', *Irish Economic and Social History*, 5 (1978), 46–61; Kennedy, 'Traders in the Irish rural economy, 1880–1914', *Economic History Review*, 32 (1979), 201–10; Kennedy, 'Farmers, traders, and agricultural politics in pre-Independence Ireland', in S. Clark and J.S. Donnelly (eds.), *Irish Peasants: Violence and Political Unrest 1780–1914* (Manchester, 1983), pp. 339–73.

and occupancy structure early in the volume, and a study of labour requirements later, these issues remain unexplored.

Fragments of these relationships are touched upon in the discussion of output and productivity, but in the concluding chapter there is an attempt to bring together some of the strands of argument and facts regarding post-Famine agriculture, particularly in relation to the all-important legacy of British colonialism, to give a partial economic context to the events which finally led to Irish independence. This is borne not of a desire to become involved with the polemics of imperialism, colonialism and the economic advantages which may or may not have accrued to the British from the sweat of the Irish, or from a desire to study the rise of Irish nationalism which led eventually to Irish independence. Ó Gráda's observation that 'Too often ... research on nineteenth-century Irish agricultural history ... has been dominated by just one part of the story, the politics of land tenure', indicates that this part of the story has been well rehearsed or even done to death.[31] There is a voluminous literature on the land issue, landlordism, agrarian violence and other related problems. The epilogue by S. Clark and J.S. Donnelly in their edited volume on *Irish Peasants* says it all. There are still many unexplored issues, or, as they put it, there is still an 'Unreaped Harvest' to confront.[32] Yet if there is something new to say then it should be said. So the conclusion is an attempt to relate the output of agriculture to the distribution of income, and to re-examine that income in terms of the proportion which went to the occupiers and that which went to the landlords as rent. It discusses well-known but conflicting views which seek to explain the rise of political consciousness, as expressed in the land reform which arose from the so-called Land War of the 1880s, in terms of the maldistribution of income prior to the 1880s. One school of thought has seen that maldistribution work in favour of the landlords, but another has seen it work in favour of the occupiers or tenants. On the basis of new output estimates a reiteration of an historic view which saw agricultural incomes favour the landlords over the tenants is proposed, and it is suggested that the relative deterioration in peasant incomes prior to land reform is worthy of further investigation.

[31] Ó Gráda, *Ireland Before and After the Famine*, p. 46.
[32] Clark and Donnelly (eds.), *Irish Peasants*, pp. 419–33, and for the period of the present book see especially pp. 425–31.

2   Agricultural change

Introduction

Cormac Ó Gráda's suggestion that by the turn of the century 'the humble farmyard hen and duck were adding more to agricultural output than wheat, oats and potatoes combined, crops which in the early 1840s accounted for more than half of output' summarises much of the story of post-Famine Irish agricultural change.[1] The arable sector most decidedly shrank, and there was an expansion of the animal sector. The increase in birds is not intended as the measure of this, because more importantly there was the rise of cattle and milk, and other livestock or livestock products. In land-use terms the country became greener. In addition, the inclusion of hay (meadow and clover) as a crop, and hence as a bolster to the tillage acreage, does not hide the fact that ultimately most hay output was intended for animal use. A doubling of the hay acreage from 1850 to 1910 from 1.2 million acres to over 2.4 million acres is therefore a ready expression of the overall land-use changes. The hay and pasture acreage combined increased from 9.995 million acres in 1851, to 12.08 million acres in 1881, and 12.42 million acres by the Great War. The arable acreage declined from 4.6 million acres in 1851, to 3.2 million acres in 1881, and 2.3 million acres by 1911. This chapter therefore looks more closely at land-use change by looking at the decline in arable farming and the countervailing rise of the livestock industry. National distributions in particular will be investigated, though there will also be a consideration of regional change. The agricultural depression of 1859–64 will be identified as a possible watershed in Irish agricultural history.

National land-use change

Table 2.1 summarises land use in Ireland for each population census year from 1841 to 1911. The 1841 population census gave some basic data on

[1]   C. Ó Gráda, 'Irish agricultural output before and after the Famine', *Journal of European Economic History*, 13 (1984), 154.

agriculture, and these are included here, though this is done with some hesitation. P.M.A. Bourke objected to reworking the figures for 1841 in this or a similar way, even though the nineteenth and early-twentieth-century enumerators regularly quoted the 1841 data without qualification.[2] They were incorrect to do so since the acreages were computed from the maps of the Ordnance Survey which were completed over the longer period 1829–41. Furthermore, although there was an apparent increase of 10 per cent in cultivated land from c. 1841 to 1851, the distribution of that change at the county level was so uneven as to be suspicious. Six counties, Dublin, Louth, Meath, Westmeath, Kilkenny and Roscommon, apparently experienced a fall in the extent of cultivated land. These were by no means *all* remote counties or those which were at the sharp end of the mid-decade Famine crisis. In ten counties, apparently, there was an expansion of the cultivated acreage which exceeded the national average. In Donegal, Kerry and Mayo this increase was 76, 49 and 30 per cent respectively. Such increases are hard to conceive given the trauma of the Famine.[3]

Thus, the 1841 census may not have been conducted in a systematic fashion, and in addition there seem to have been changes in the classification and definition of marginal grazing land. Joel Mokyr has pointed out that it was in the counties where this type of land was most abundant that the differences were greatest.[4] Bourke's revised estimate of the change in cultivated land from 1841 to 1851 added just 100,000 acres to the cultivated area over the decade. This does not compare with the 'official' increase of 1.3 million acres. Bourke's estimate of the 1841 acreage of crops, meadow and grass was 14.7 million acres.[5]

The total cultivated area rose during the first twenty years or so after the Famine, but thereafter it declined. Obviously the severe decline in population from 6.552 million in 1851 to 4.39 million by 1911 helps to explain the fall in the cultivated area.[6] The first interesting observation to make however is that the associated rise in acres per head of total

---

[2]   For example, *Agricultural Statistics of Ireland ... 1911, B[ritish] P[arliamentary] P[apers]* [Cd. 6377], vol. cvi, 1912–13, p. 3. P.M.A. Bourke on the 1841 census in, 'The agricultural statistics of the 1841 Census of Ireland. A critical review', *Economic History Review*, 18 (1965), 383–4.

[3]   Bourke, 'The agricultural statistics', 386–7.

[4]   J. Mokyr, 'Reply to Peter Solar', *Irish Economic and Social History*, 11 (1984), 119, but see Solar's criticism of Mokyr's use of the 1841 data in 'Why Ireland starved: A critical review of the econometric results', *Irish Economic and Social History*, 11 (1984), 113. Correctly speaking, Mokyr used the 1841 data, after adjustments, in *Why Ireland Starved: A Quantitative and Analytical History of the Irish Economy, 1800–1850* (London, 1983 1st edn, revised 1985), Solar criticised him, and Mokyr defended his use.

[5]   Bourke, 'The agricultural statistics', 391. Note, to the figures in table 2.1 should be added 0.195 million acres of fallow.

[6]   Population figures here and elsewhere taken from J.P. Huttman, 'Institutional factors in

Table 2.1: *Ireland, the division of the land, 1841–1911*

	Crops	Hay	Grass	Woods & Plantations	Grazed & Barren Mountain	Bog & Marsh	Waste
In Millions of acres							
1841	(	13.464	)	0.375	(	6.49	)
1851	4.613	1.246	8.749	0.305	(	5.416	)
1861	4.344	1.546	9.534	0.317	(	4.588	)
1871	3.792	1.829	10.071	0.325	(	4.311	)
1881	3.194	2.001	10.075	0.329	2.118	1.72	0.892
1891	2.759	2.06	10.299	0.312	2.211	1.744	0.949
1901	2.452	2.179	10.577	0.31	2.223	1.574	1.018
1911	2.349	2.512	9.847	0.3	3.084	1.198	1.061
Percentage Distribution							
1841		66.2		1.8		31.9	
1851	22.7	6.1	43.0	1.5		26.6	
1861	21.4	7.6	46.9	1.6		22.6	
1871	18.7	9.0	49.5	1.6		21.2	
1881	15.7	9.8	49.6	1.6	10.4	8.5	4.4
1891	13.6	10.1	50.6	1.5	10.9	8.6	4.7
1901	12.1	10.7	52.0	1.5	10.9	7.7	5.0
1911	11.5	12.3	48.4	1.5	15.1	5.9	5.2

Notes: Subject to rounding errors. The 1911 figures are misleading when compared with other years because from 1906 there was a change in definitions. In 1911 there were 2.584 million acres of grazed mountain land much of which would formerly have been counted as grass.
Source: Agricultural Statistics of Ireland ... 1911, British Parliamentary Papers [Cd.6377], vol. cvi, 1912–13, pp. 2–3.

population was a country-wide phenomenon. As will be emphasised below, structural change was a national phenomenon. We know that pastoral farming was more land extensive than arable cultivation, and a much smaller land area would have been adequate as the decades proceeded, purely for the purposes of self-sufficiency, but the population of animals actually grew in numbers and this was because a growing proportion of them were exported. Undoubtedly therefore, the British market for Irish livestock and livestock products was partly driving the land-use changes. If Ireland was part of the wider UK economy at the time of the Famine, there was no immediate and obvious reason to think of it otherwise afterwards.

the development of Irish agriculture, 1850–1915', Ph.D. Thesis, University of London, 1970, p. 414.

A cursory glance at the annual statistics actually reveals an increase, not a decrease, in the arable acreage from *c*. 1847 into the 1850s, in spite of the large fall in population associated with the Famine (see figure 2.1). This can be viewed as an attempt to return to what was considered a pre-Famine norm before the full impact of that Famine was realised. For example, exports of wheat, barley and oats to Britain were at a peak in the 1830s, and wheat at least continued to be an important export up to and including 1845.[7] In the immediate pre-Famine period wheat may have contributed 11.5 per cent to Irish agricultural output by value, and barley and oats 4 and 19 per cent respectively.[8] On the eve of the Famine, wheat and flour exports constituted about 40 per cent of the net output of these products.[9] In total, perhaps 27 per cent of agricultural output was exported on the eve of the Famine.[10] Therefore the export trade was vitally important, though it did not overwhelm the home use of the grain crops. The moves to return to the pre-Famine situation in some haste arose because the food deficiency at the time of the Famine through the loss of the potato crop could not have been solved simply by prohibiting the export of grain. The potato loss had been too great. Perhaps 80 per cent of oats production was already retained in Ireland at the time of the Famine.[11] The sheer bulk of food intake derived from potatoes could not have been easily replaced by retaining the remainder of the grain. Once the full and permanent impact of the Famine was established adjustments were bound to follow.

The acreage under wheat declined dramatically from 0.5 million acres in 1851 to 150,000 acres by 1881, and finally to 45,000 acres by 1911. The largest losses occurred in the 1860s, and at its lowest level in 1904 there were just under 31,000 acres of wheat.

The only significant extension of the arable occurred in conjunction with the rise of the animal populations; root and green crops were an important animal fodder. The turnip acreage however declined from nearly 400,000 acres to about 270,000. The turnip was formerly a neglected crop, but it was grown in quantity in 1847 as an emergency

7   M. Daly, *The Famine in Ireland* (Dublin, 1986), p. 22, based on J.M. Goldstrom, 'Irish agriculture and the Great Famine' in J.M. Goldstrom and L.A. Clarkson (eds.), *Irish Population, Economy and Society: Essays in Honour of the late K.H. Connell* (Oxford, 1981), p. 160. For a table of wheat, barley and oats exports to Britain see pp. 160–1.
8   C. Ó Gráda, *Ireland Before and After the Famine: Explorations in Economic History, 1800–1925* (Manchester, 1988), p. 48.
9   P. Solar, 'Agricultural productivity and economic development in Ireland and Scotland in the early nineteenth century', in T.M. Devine and D. Dickson (eds.), *Ireland and Scotland 1600–1850. Parallels and Contrasts in Economic and Social Development* (Edinburgh, 1983), p. 87 n. 32. See also J.S. Donnelly, *The Land and the People of Nineteenth-Century Cork: The Rural Economy and the Land Question* (London, 1975), p. 82.
10 Ó Gráda, *Ireland Before and After the Famine*, p. 51.
11 P.M.A. Bourke, 'The Irish grain trade, 1839–48', *Irish Historical Studies*, 20 (1976), 165.

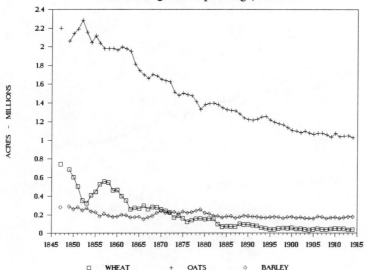

Root and green crop acreage, 1847–1914

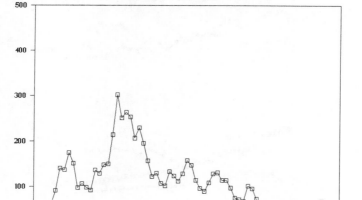

Tillage acreage, 1847–1914

2.1   Land use in Ireland, 1847–1914

Corn acreage, 1847–1914

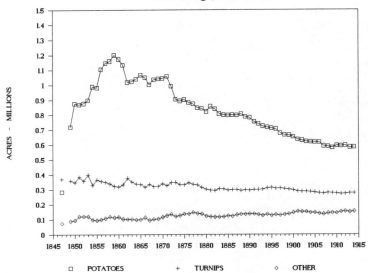

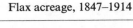

Flax acreage, 1847–1914

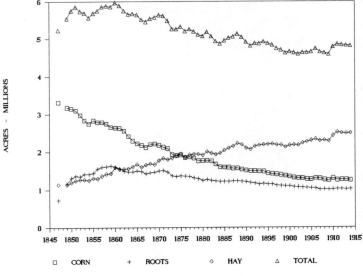

2.1   Cont.

human food. Subsequently it remained in cultivation at high levels as a fodder for livestock.[12]

The recovery in the potato acreage to 869,000 acres in 1851 continued up to 1861 to 1.134 million acres, but then it declined evenly to 590,000 acres by 1911. The potato acreage had fallen to just 284,000 acres in 1847. This large rise in acreage in the post-Famine decade was a successful bid to return to some kind of pre-Famine agricultural norm before the long-term changes took effect. Bourke's restructuring of the pre-Famine base was derived from the official returns which the administrative authorities produced as they reacted to the Famine crisis. This suggested a potato acreage of 2.378 million acres in 1844, 2.516 million in 1845 and 1.999 million in 1846. Such an extensive carpet of potatoes was never seen again in Ireland. The early 1840s represented just about the peak of potato cultivation with a probable actual peak in the early 1830s. There is some wariness, however, over the size of the pre-Famine potato acreage.[13] On yet another re-estimation of the data Mokyr has settled for a pre-Famine potato acreage of 2.187 million acres.[14] Whatever the truth of the situation, the consensus appears to be in favour of a pre-Famine potato crop of over 2 million acres, and certainly this was not approached in subsequent years.

Naturally the long-term downward trend in the potato acreage partly reflected the declining population, though the ratio of population to potatoes was by no means constant over time. Indeed, the potato acreage declined faster than the human population. For example, in 1861 there were 195.5 acres of potatoes per head of population, but by 1881 this had fallen to 165 acres per head and thence to 160, 143 and 135 acres per head in 1891, 1901 and 1911. That the potato acreage remained as high as it did was because there was an extension of animal husbandry and potatoes were an important animal feedstuff, especially for pigs. This therefore suggests a significant break from pre-Famine consumption

[12] See P.M.A. Bourke, 'The extent of the potato crop in Ireland at the time of the Famine', *Journal of the Statistical and Social Inquiry Society of Ireland*, 20; 3 (1959–60), 13.

[13] *Ibid.*, 11. R.D Crotty, *Irish Agricultural Production: Its Volume and Structure* (Cork, 1966), p. 315, suggested an alternative of only 1.4 million acres. Bourke stressed the likely confusion that the authorities made over the definition of the acre. Did they use statute, Irish or Cunningham acres? Crotty in particular has questioned Bourke's assumption that the acres were reported in Irish acres. Since the ratio of the statute acre to the Irish acres was 1:1.62, Crotty's assumption that statute acres were in use inevitably meant that his estimates were lower. On this general issue of standard and non-standard units of measurement see P.M.A. Bourke, 'Notes on some agricultural units of measurement in use in pre-Famine Ireland', *Irish Historical Studies*, 14 (1965), 236–45.

[14] J. Mokyr, 'Irish history with the potato', *Irish Economic and Social History*, 8 (1981), 20.

trends, a point which will be reviewed indirectly in subsequent chapters concerning output and performance.

This general decline in arable farming after the Famine was partly due to the decline in population, and this meant a decline in aggregate consumption, but there was also a response to external economic stimuli, notably the rise in demand for meat and dairy products generally in Western Europe, especially after the 1870s and particularly in England. The increase in North American grain reaching Western Europe at lower and lower prices from the late 1870s in an atmosphere of free trade may have led to the despairing decline of corn growing in England, but in Ireland the economy had already adjusted output to cash products other than wheat *before* the 1870s. This must be an important conclusion in the restructuring of Irish agriculture and its emergence, perhaps ahead of a number of European rivals, as a commercial economy, and a decisive break from its mid-century appearance as an impoverished agricultural economy. There is some supporting evidence for this from agricultural trends in Western Europe. It would make little sense to compare directly the area under cereal crops in the various European countries since they differ so greatly in size, but table 2.2 compares the trend of land devoted to cereals from 1851 to 1911, based on 5-year moving averages and centred on the years indicated.[15] Quite clearly the flight from cereal production was most pronounced in Ireland. In fact in both Denmark and Germany there was actually an increase in the land devoted to cereals, and in France and Holland the cereal acreage held up very well. As in Ireland, the move away from cereals was great in both Belgium and Great Britain. What was special about Ireland was the trend away from cereals early in the chronology. The table does not measure like with like since there were international differences in the distributions of the different cereals. Whilst Irish producers left wheat, in some European countries wheat was not nearly as significant as some other cereals. Rye was insignificant in Ireland, but it dominated in Germany and Holland, and was important in most other countries apart from Britain. In Ireland and the UK in general oats became dominant. The decline in wheat production set Ireland apart, especially before about 1880 and the flood of North American imports which then hit western Europe.

There is a second way to make this European comparison. Demonstrably the Irish population declined severely from 1851 onwards with two important consequences: a decrease in the amount of aggregate consumption, assuming little adjustment in the composition of that consumption; and a possible crisis for labour supply in what was a

---

[15] B.R. Mitchell, *European Historical Statistics 1750–1975* (2nd revised edn, London, 1981), pp. 211–36.

Table 2.2 *Trends in European cereal acreages, 1851-1911*

	Belgium	Denmark	France	Germany	IRELAND	N'lands	GB
Trend in cereal area based on 5-year averages – 1879–83 = 100							
1851	112.2		106.1	85.4	172.2	88.2	
1861		87.2	106.2	92.2	147.0	94.3	
1871		93.4	96.5	94.1	120.3	99.4	109.1
1881	100.0	100.0	100.0	100.0	100.0	100.0	100.0
1891	86.6	104.0	101.2	101.2	87.0	95.6	90.0
1901	80.4	103.2	97.1	103.1	76.6	93.4	81.8
1911	79.4	103.7	93.2	104.9	73.6	91.1	80.4
Trend per capita							
1851	136.7		111.8	115.8	136.2	115.4	
1861		107.1	107.1	117.5	131.8	113.7	
1871		103.4	100.8	103.9	115.3	111.3	123.9
1881	100.0	100.0	100.0	100.0	100.0	100.0	100.0
1891	78.8	94.5	99.9	92.6	94.5	85.1	80.5
1901	66.5	83.2	95.8	82.8	88.1	73.5	65.7
1911	59.1	74.3	90.9	73.4	85.7	62.3	58.3

Note: Belgium 1851 in fact the single observation for 1856, in 1881 the single observation for 1880 and in 1891 the single observation for 1892.
Source: B.R. Mitchell, *European Historical Statistics 1750-1975* (Cambridge, 1981 edn), pp. 211-36.

traditionally labour-intensive arable sector. In all other European countries the population actually rose. In the lower half of table 2.2 therefore, the cereal trends are reworked on a per capita basis. Whilst the flight from cereals is still evident in Ireland in the first three decades after 1851, thereafter it was less impressive compared with the same trend in Belgium, Holland and Great Britain, and still less pronounced than in Denmark and Germany, though more so than in France. The important point is still well made however: there was a restructuring of agricultural production away from cereals, *and* in advance of the late nineteenth-century depression of cereal prices, putting Ireland in the vanguard of European change.

If this is evidence of rational behaviour by farmers the reason for it can be detected fairly easily and crudely by an inspection of price relativities. Figure 2.2a shows the trend of wheat prices relative to cattle prices. If this can be viewed as a 'terms of trade' index, then that index moved in favour of various types of cattle products as the century proceeded. The two main blemishes in the trends occurred firstly during the grain supply problems associated with the Crimean War of the mid-1850s, when

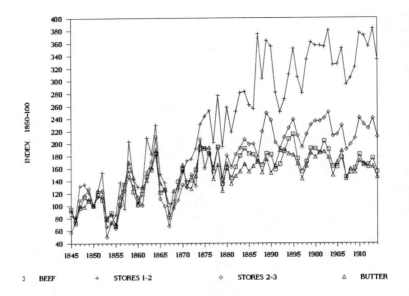

2.2a   Wheat and cattle price relatives

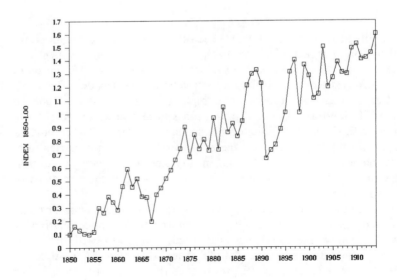

2.2b   Livestock and crops price relatives

Western European grain prices rose, and secondly during the depression of 1859–64 when for different reasons grain prices rose relative to animal prices. During these years a series of poor seasons with low crop yields, associated with bad weather, produced both a rise in grain prices and also a fodder famine, and this had the knock-on effect of forcing the premature sale or slaughter of livestock, glutting the market and lowering its price.[16]

More generally, figure 2.2b is a price index comparing crops with animals.[17] The general trend moved in favour of animal production, though there were periodic adjustments back to crops, notably during the Crimean War, but also during the American Civil War which was coincident for much of the time with the agricultural depression of 1859–64, and finally, suddenly and sharply at the end of the 1880s, during a time of industrial depression in Britain with its associated short-term decline in real incomes.

Farmers are environmentalists. If they switch crops they are mindful of the suitability of the soil for those crops. In Ireland the desertion of wheat and the other grains in favour of other land uses tended to increase the long-term yields per acre of those grain crops, in spite of the decline in labour intensive methods (allowing for weather and other short-term problems). Relatively poor wheat/barley/oats lands were converted early and therefore average wheat/barley/oats yields increased.[18] For some crops this 'had the effect of neutralizing the decline in acreage' and this was 'particularly noticeable with oats, potatoes and turnips'.[19] This is demonstrated for selected crops in figure 2.3. The output of oats and potatoes in particular was maintained, roughly speaking, from the 1870s.[20]

[16] See also T. Barrington, 'A review of Irish agricultural prices', *Journal of the Statistical and Social Inquiry Society of Ireland*, 15 (1927), 249–80.

[17] This has been constructed by adopting the procedures outlined in chapter 4 and appendixes 4–6 for estimating annual output and constructing a composite price index. The separate product prices have been weighted by their percentage contribution to gross output. See also Mokyr, *Why Ireland Starved*, 1985 edn, pp. 147–9.

[18] This was demonstrated in the *Agricultural Output of Northern Ireland, 1925, BPP*, [Cmd. 87], 1928, p. 11, in a retrospective look over the previous 50 years. 'It is well known that when the area devoted to any crop is reduced it is the least suitable soils which go out of cultivation first. The land on which these crops continued to be grown was presumably the best land available, consequently a tendency for an upward movement in yields might be expected.' See also T. Barrington, 'The yields of Irish tillage crops since the year 1847', *Journal of the Department of Agriculture*, 21 (1921), 212–13, 289–305.

[19] *Agricultural Output of Northern Ireland, 1925*, p. 11. See also L. Kennedy, 'The rural economy, 1820–1914', in L. Kennedy and P. Ollerenshaw (eds.), *An Economic History of Ulster, 1820–1939* (Manchester, 1985), p. 23.

[20] See also P.M.A. Bourke, 'The average yields of food crops in Ireland on the eve of the Great Famine', *Journal of the Department of Agriculture*, 66 (1969), especially 27–30. See also Barrington 'The yields of Irish crops'; B. Solow, *The Land Question and the Irish Economy, 1870–1903* (Harvard, 1971), pp. 112–13; D.L. Armstrong, *An Economic History of Agriculture in Northern Ireland 1850–1900* (Plunkett Foundation, Oxford, 1989), pp. 163–4.

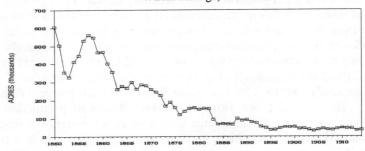

Wheat acreage, 1850–1914

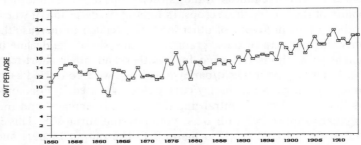

Wheat yields, 1850–1914

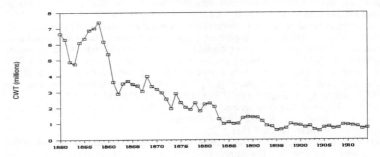

Wheat output, 1850–1914

2.3   Wheat, oats and potatoes – acreages, yields and outputs, 1850–1914

Oats acreage, 1850–1914

Oats yields, 1850–1914

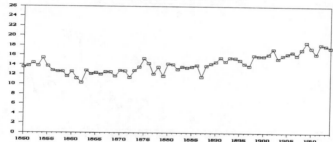

Oats output, 1850–1914

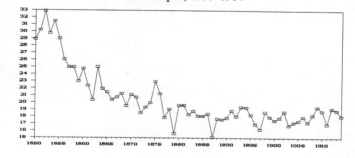

Potatoes acreage, 1850–1914

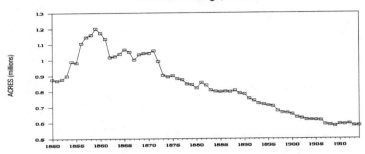

Potatoes yields, 1850–1914

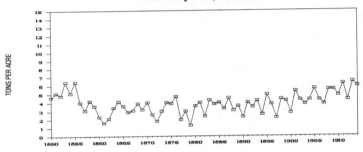

Potatoes output, 1850–1914

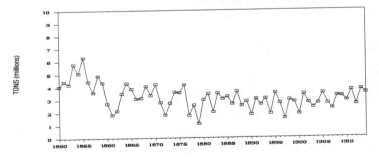

Pre-Famine yields were almost certainly higher, placing Irish crop yields in the premier division of European agriculture. Spade labour, with a large manual labour force, and prodigious applications of manure characterised much pre-Famine cultivation. While the system was generous in its use of labour, the manuring and the constant weeding and hoeing ensured low weed proliferation and high yields. The combination of spade labour, manure and large potato acreages, points to a crude though effective circle of pre-Famine land and labour productivity. This could not be maintained in the post-Famine period, and inevitably less intensive agricultural practices took over.[21] The locked-up and well-maintained fertility of the soil, therefore, was progressively dissipated in the post-Famine decade. As far as potatoes were concerned there was the ever-present problem from the mid-1840s of potato blight, and this was not eradicated until anti-fungal spraying was used on a large scale in the early twentieth century.[22] The Registrar General became positively obsessed about the neglect of weeding in his annual reports which prefaced the Agricultural Returns. He was reflecting on the comments of the constabulary in their inquiry and estimation of local average yields. By the 1870s at least this problem was so common, and prevailed apparently without any attempt to do anything about it, that the typesetters of the annual returns surely had no need to reset their pages from year to year. Choose almost any year in the 1870s and the following unchanging remarks were made:

There can be no doubt that if due attention were bestowed on the destruction of weeds, the lands of Ireland would afford a largely increased yield; but unfortunately *luxuriant crops of weeds*, which are to be seen in almost every part of the country during Summer and Autumn, not only rob the farmer himself, but often inflict a vast amount of injury on his neighbour.[23]

The multiplication of yield times acreage produces the physical production of crops, and what was seen from about 1870 was the lowest level of potato output achieved, probably in that century, or for quite some time for more than a century. Bourke's estimates of potato acreages and yields at the time of the Famine, when yields were enhanced by spade

---

[21] See Daly, *The Famine*, pp. 23, 55; P. Solar, 'Harvest fluctuations in pre-Famine Ireland: evidence from Belfast and Waterford newspapers', *Agricultural History Review*, 37: 2 (1989), 157–65.

[22] P. Solar, 'The Great Famine was no ordinary subsistence crisis', E.M. Crawford (ed.), *Famine the Irish Experience 900–1900: Subsistence Crises and Famines in Ireland* (Edinburgh, 1989), chapter 3, pp. 112–33, especially 114, 118–22. T.P. O'Neill, 'The food crisis of the 1890s', in Crawford, *Famine*, pp. 176–97, gives much detail of the food crisis of the 1890s, what he refers to as the last of the Irish famines.

[23] For example, see *Agricultural Statistics of Ireland ... 1878, BPP* [C. 2347], vol. lxxv, 1878/9, p. 11. See the same words in the returns for 1876, 1879, 1880, 1881.

cultivation and heavy manuring, suggests there might have been 14.8 million tons of potatoes from the 1844 crop, 10 million tons in 1845 but a slump to 2.999 million tons in 1846.[24] Whilst it recovered to 6.3 million tons in 1855, the average in each of the six separate decades from the 1850s ranged from 4.7 to as low as 2.6 million tons.[25] Whatever else were the consequences of the Famine, a once very important and integral part of the agricultural economy was lost. The heavy labour input into potato production, with weeding and manuring, had a positive effect on keeping other crop yields high. That they remained high after the Famine had everything to do with reduced acreages. In a wider context the decline in potato production in Ireland was a complete contrast to the trend in Western Europe as a whole. While the potato area recovered in size down to the late 1850s, so it did in all other countries. Thereafter it declined dramatically in Ireland, whereas in all other countries it advanced more or less evenly, though in the least pronounced fashion in Great Britain. The one exception to this rule was Belgium. In per capita terms, the decline in potato growing was equally severe in Germany, Ireland and the Netherlands, less so in Denmark, but most severe in Belgium and Great Britain, and in France there was actually an increase. The one feature which might distinguish Ireland from the rest is that the unremitting decline in potato production had begun by 1860, while in other countries, in both absolute terms and acres per capita it was still on the rise until about 1880. The coincidence of this sharp break in trend with the depression which hit Ireland in the late 1850s and early 1860s is a compelling reason for the change.

The agricultural depression of 1859–1864

The post-Famine decades were not years entirely of recovery and smooth adjustment. The first crisis to hit the agricultural economy was the agricultural depression of 1859–64. It has some claim to be regarded as a watershed in Irish agriculture because it heralds the final re-alignment of the economy towards pastoralism.[26] It began with a severe drought in the critical months of July to September in 1859 when rainfall was 30–40 per cent below average. This was followed by inundations of rain in each of the three years 1860–62, 'almost certainly the wettest three consecutive years of the entire nineteenth century in Ireland, with the significant exception of 1846–8'. In 1861 it was especially bad because of the

---

[24] Bourke, 'The extent of the potato crop', 11–14.
[25] Derived from appendixes 1 and 2.
[26] J.S. Donnelly, 'The Irish agricultural depression of 1859–64', *Irish Economic and Social History*, 3 (1976), 33–54.

Table 2.3   *Trends in European potato production, 1851–1911*

	Belgium	Denmark	France	Germany	IRELAND	N'lands	GB
Trend in Potato Area based on 5-year averages – 1879–83 = 100							
1851	75.4		67.8	57.4	101.9	62.6	
1861		68.6	81.2	73.6	133.3	75.0	
1871		95.1	86.8	88.0	121.0	89.0	104.5
1881	100.0	100.0	100.0	100.0	100.0	100.0	100.0
1891	93.0	115.0	112.6	105.4	90.9	105.7	97.8
1901	72.0	119.5	113.6	115.7	77.0	109.9	102.5
1911	79.0	135.0	117.3	119.6	70.7	117.5	104.9
Trend per capita							
1851	91.8		70.8	77.7	80.5	82.2	
1861		84.0	81.3	93.6	119.0	90.9	
1871		104.9	89.9	96.9	115.7	99.7	119.1
1881	100.0	100.0	100.0	100.0	100.0	100.0	100.0
1891	84.6	104.3	110.5	96.5	100.0	94.0	87.9
1901	59.4	96.0	110.5	92.8	89.3	86.4	82.3
1911	58.7	96.4	112.0	83.3	83.4	80.5	76.4

Notes and Source: See Table 2.2

concentration of rain from July to September.[27] Then in 1863 there was an unusually dry Spring, and the drought continued into the Summer. When it broke in September the grain harvest had hardly begun.[28] Finally, although the Spring of 1864 brought luxuriant pastures, in the Summer of that year there was an intense drought – the grasslands shrivelled and food for livestock was in severe short supply.[29]

Thus for six continuous seasons either the grassland suffered with drought, or the arable and fodder sector suffered with at times drought, and more particularly heavy inundations of rain. Crop yields turned down dramatically and the existing trend of declining corn acreages was hastened. This was especially so for wheat, but the acreages under green and root crops, and hay, either stabilised or actually increased during the period, accelerating the move which was already in motion out of corn-based tillage and into livestock. The effect of the depression on cash crop prices was not to cause them to rise – the price for wheat and to an extent other grain crops was influenced more by the larger British or European market than by conditions in Ireland itself – but instead they actually

[27] *Ibid.*, 35.
[28] Though the 1863 grain yields were slightly up on the pre-depression average of 1856–8. *Ibid.*, 37.
[29] *Ibid.*, 35–6.

fell.[30] After an astronomical price rise in 1859, the price of hay fell almost as dramatically in 1860 and settled down into the early 1860s at much the same price it had enjoyed during the four years before the depression, though with a higher acreage. This stabilisation of hay prices was an advantage to the livestock economy, as was the dramatic fall in potato prices in 1862–5. Potato prices had been at high levels in the previous two years precisely because of the rain damage which promoted the blight. Potato yields were reduced to 61, 42 and 55 per cent of the average yield of 1856–8 in each of the three years 1860–62.[31] The pig population could not be sustained without potatoes and their numbers fell from 1.1 million of less than 1 year of age in 1858, to 858,000 in each of 1863 and 1864 but with an as yet incalculable mortality loss before maturity.[32] Other aspects of the economy were touched by the depression. The prolonged wet had an effect on turf production: it could not dry out and contributed to a fuel famine.[33]

The only good thing to emerge during this period, and quite unrelated to the depression, was the effect of the cotton famine spilling over from the United States Civil War which encouraged the increase in Irish flax production.[34] If for this reason the depression hit Ulster less severely than elsewhere, it was nevertheless still a national depression with a clear downturn in deposits at banks and the imposition of credit restrictions. At a time when they needed it most, farmers found credit was most expensive.[35]

Regional land-use change

The broadest change which has been identified so far was the switch from tillage to pasture. At times hay was referred to as a tillage crop since it was often part of a rotational system of crop husbandry. If hay is counted along with grass as part of the stock of pasture, then the ratio of pasture to arable rose from just over 2 in 1851 to over 5.5 in 1911. The arable acreage had declined from 4.6 million acres to 2.2 million acres over the same period, and the pasture acreage had increased from 9.995 million acres to 12.359 million. At the next regional level, the four provinces, there were some subtle differences. Figure 2.4 looks at

---

[30] See Crotty, *Irish Agricultural Production*, p. 69 on supply price determination.
[31] Donnelly, 'Irish agricultural depression', 37. See also Barrington, 'A review of prices', 251.
[32] Donnelly, 'Irish agricultural depression', 38–9.
[33] *Ibid.*, 47.
[34] See also Kennedy, 'The rural economy', pp. 30–1.
[35] P. Ollerenshaw, *Banking in Nineteenth-Century Ireland: The Belfast Banks, 1825–1914* (Manchester, 1987), pp. 102–5.

Provincial corn acreage 1851–1911

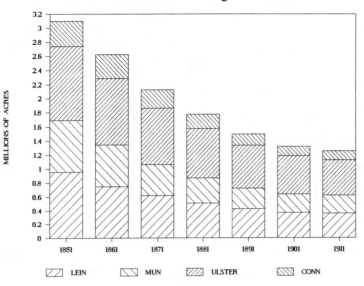

Provincial green crop acreage 1851–1911

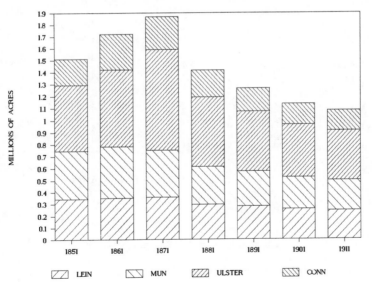

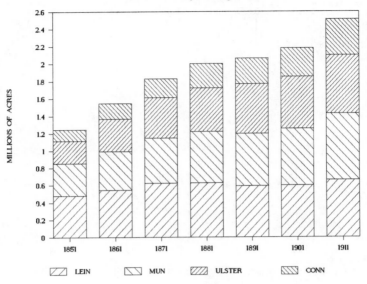

Provincial hay acreage, 1851–1911

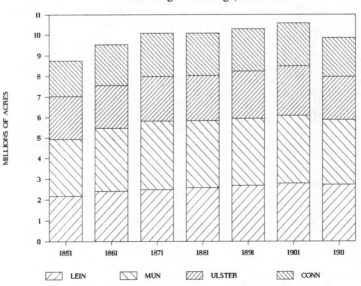

Provincial grass acreage, 1851–1911

2.4　Provincial land-use change, 1851–1911

simplified tillage and pasture distributions for the four provinces. Ulster here is defined as it was in the period before Partition as nine counties.

Connacht and Munster were the least tilled provinces in 1851 (counting hay as tillage), and they remained the least tilled throughout the period. Taking corn, green and root crops together they were still the least arable provinces. The patterns thereafter are not nearly so well defined. For example, the least or most tilled did not necessarily also mean the most or least grassed. This broad rule did apply to Ulster but not to Connacht which was the least tilled and also one of the least grassed provinces, and consequently most given over to bog, waste and water. In fact throughout the period 35 per cent or more of Connacht was bog, waste or water. Leinster by comparison was one of the most tilled and most grassed and had the smallest proportion throughout the period under bog or waste (only 15–16 per cent for the whole period against a national average of 21–26 per cent). Nevertheless, the trends in land use detected so far at the national level were roughly replicated at the provincial level.

Broadly speaking, corn declined from leading crop in all provinces to become second most important crop (and even third in Connacht), and everywhere hay became the most prominent cultivated crop, though marginally so in Ulster. In Leinster and Munster this reversal of roles was complete by 1871 or thereabouts, but in Ulster it was not achieved until the turn of the century, and in Connacht not until *c.* 1881. Meanwhile everywhere there was a steady rise in grass cultivation up to *c.* 1901. In Leinster and Munster particularly, and Connacht marginally, one half or more of all land was under permanent grass. This general move towards pastoralism was slower to proceed in Ulster than elsewhere. This was attributable to a number of factors related to physical endowments (soils and climate), and socio-economic factors such as the prevalence of small farms with their emphasis on farm-family labour, and the local product emphasis on flax. More than other regions of the country, mixed farming came to characterise Ulster in the second half of the nineteenth century.[36]

Adjustments to land use principally came about by changing existing land uses, rather than by bringing in extra cultivation from woods, bogs and wastes. The exception to this rule was the initial encroachment into marginal lands in the first decade after the Famine when over half a million acres came out of bog and waste into cultivation. To some extent this reverted in use by the end of the century. In these simple decadal terms the high point of crop and grass cultivation was 1871, though in

[36] Kennedy, 'The rural economy', 18–20.

fact for crops, including hay, it was 1860. The national increase in cultivated land from 1851 to 1871 was 7 per cent: it was as high as 17 per cent in Connacht and 10 per cent in Munster, but only 3 per cent in each of Leinster and Ulster. Thus it was the west and the south of the country which experienced the greatest fluctuations in the movement of the margin. T. Jones Hughes has demonstrated this by constructing distributional maps of wheat and barley cultivation in 1851. It was precisely the broad west and the south of the country which had the most arable with which to contribute to that margin of change. North of a line from Dublin to Galway (excluding Mayo and parts of the east coast), wheat and barley rarely exceeded 10 per cent of the cropped area whereas to the south of this line, almost everywhere, those crops occupied over 10 per cent, reaching 30 per cent in parts of the most southerly and south-easterly counties.[37]

So Connacht was the least tilled province, but it was also one of the least grassed, and so the pasture to tillage ratio for this province, of all the provinces, became the highest, rising from 3.2 in 1851 to 8.0 in 1901 and 7.7 in 1911.[38] In Munster, agricultural change behaved in much the same way, and in Leinster agricultural change just about replicated the national average. Yet this is an odd average, and therefore to balance the changes Ulster remained more determinedly a tillage province. That having been said, the rate of change from tillage to pasture was heaviest in the already dominant pasture areas of Connacht and Munster. An important conclusion therefore is that regional specialisation did not so much come about afresh; these post-Famine changes may have reflected Famine or pre-Famine developments which were already underway. The

[37] T.J. Hughes, 'Society and settlement in nineteenth-century Ireland', *Irish Geography*, 5: 2 (1965), 88.

[38] It depends, as in much of what is said, on what is the best base to use. The use of 1851 reveals that in Leinster there was a 175 per cent increase in the pasture/tillage ratio, the largest change of any province.

The Provincial Pasture/Tillage Ratio 1851–1911

Date	Leinster	Munster	Ulster	Connacht	Ireland
1851	2.0	2.8	1.5	3.2	2.2
1861	2.7	3.4	1.6	3.4	2.6
1871	3.2	4.6	1.8	4.3	3.1
1881	3.9	5.6	2.1	5.5	3.8
1891	4.6	6.6	2.6	6.7	4.5
1901	5.4	7.4	3.0	8.0	5.2
1911	5.6	7.5	3.0	7.7	5.3
%change 1861–1911	108.6	120.6	87.5	126.5	104.0

Sources: Derived from *Annual Agricultural Statistics*, various years.

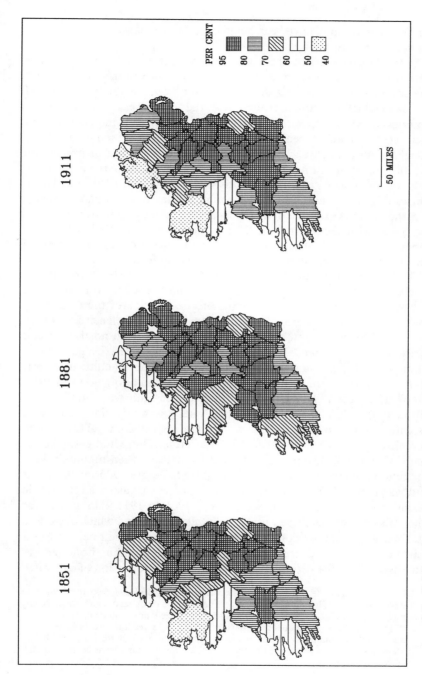

PER CENT

95
80
70
60
50
40

1911

1881

1851

50 MILES

2.5   Cultivated area: percent of county area, 1851–1911

degree of change may have been heightened by the mid-century crisis, but not necessarily caused by it.

At the county level in 1851 there was a clear east to west split. The eastern side of the country had more land in cultivation than the west. The counties of Connacht and Munster stand out for the proportion of their land which was not cultivated, with the singular exception of county Limerick in Munster, but in Ulster there was also a significant area of uncultivated land concentrated in the north-west in the counties of Donegal, Fermanagh and Londonderry. The distribution of population and the transportation network partly helped to form these patterns, with greater concentrations of cultivated land around the population centres of Belfast and Dublin. Environmental factors also had a bearing with the obvious example of Wicklow and the relatively barren expanses associated with the Wicklow hills, just to the south of Dublin. To this extent, in some counties, the ability to change land use by bringing uncultivated land into cultivation was severely limited.

The high point for cultivation was *c.* 1871 in all but eleven counties of which five had a high point within a decade one way or the other of 1871. Not too much can be claimed for the precise location of these counties. They vaguely have a broad distribution: either they encircle Dublin (Dublin, Kildare, Carlow and Wexford); or they stretch north-west and north-north-west from Dublin (Westmeath, Longford, Sligo, Mayo, Monaghan, Tyrone and Donegal). The low point of cultivation was 1851, except in eleven counties, seven of which had a low point in 1911. Six of these seven were Ulster counties (the seventh was Dublin), and although the difference between high and low cultivation was often insignificant, in the cases of Antrim, Fermanagh and Donegal the drop in the cultivated area from 1901 to 1911 was large. This could easily be the result of changed definitions employed from 1906 when mountain land was differentiated as either barren or grazed. Before 1906 much of this land was probably returned as pasture.[39] There was also a large drop in the cultivated area of Londonderry from 1901 to 1911. Along with the other Ulster counties there were large tracts of highland which acted as a tidal margin of cultivation. The post-Famine recovery in cultivation is well represented in Clare, Galway, Kerry, Roscommon, Londonderry and Sligo, where there were large increases in the proportion of cultivated

---

[39] This was certainly the inferred conclusion from the *Agricultural Output of Northern Ireland, 1925*, p. 2. There it was reported that after 1906 much land which was formerly comfortably classified as pasture was reclassified as grazed mountain land. This, coupled with improvements in the quality of pasture in the 10–15 years before 1925, led to an increase in the amount of land designated as rough pasture, which for the period before 1906 would have been called pasture. Such problems make comparisons before and after 1906 very difficult.

land from 1851 to 1861. The problem to face is not having a reliable pre-Famine datum, and so it cannot be said whether there was a recovery to former levels of cultivation, or not yet a full recovery, or finally whether new lands were brought into cultivation.

Other than these descriptive changes there was not a great deal of movement in the broad percentages of county areas which were culti-vated over time. Perhaps two features of Irish history and inheritance help to explain the regional changes: there was an inheritance of environ-ment in terms of land and climate which brought about a dispropor-tionate distribution of unusable land, either because it was too wet or because it was topographically unsuitable; and there was the dispropor-tionate impact of, and therefore recovery from, the Famine.

A broad east to west split in the degree of cultivation can be identified, from Meath in 1851 as the most intensively cultivated county (91.1 per cent of its area under agriculturally defined uses), to Mayo as the least intensively cultivated (48.1 per cent). The degree of cultivation across counties, relative to one another, was remarkably fixed through time.[40]

The broad conclusion is plain to see, whilst there were great national changes in the agricultural face of the country, overall the relative geography of agriculture hardly varied at all, especially after the im-mediate post-Famine recovery period. Yet if the relative position of the counties with respect to *total cultivation* did not change a great deal over time, this belies the important land-use changes which did take place. There was a dramatic flight from arable cultivation, especially corn (see figure 2.6, the declining proportion of cultivated land devoted to corn, root and green and potato crops, and figure 2.7, the overall change in tillage cultivation). Post-Famine land-use changes were only regionally specific by relatively minor degrees; the country was more or less moving as one.[41]

The move out of arable and into hay and grass was truly a national movement in which only one county refused to 'move'. Waterford remained with under 10 per cent of its cultivated area under hay throughout the period. The pattern in 1851 was a concentration of hay production, relatively speaking, in Dublin and the circle of counties around the city. A simple relationship springs to mind, simply the need for bedding and feeding materials for the urban horse population in a

[40] A matrix of thirty-two counties across seven population census years measuring the proportion of land under cultivation produces Spearman rank correlation coefficients of between 0.92 and 0.99.

[41] In terms of the percentage of county area under arable cultivation the Spearman rank correlation coefficients for seven population census years is always over 0.76, and as high as 0.99. The rank order of the arable as a percentage of cultivated area produces much the same pattern with rank correlation coefficients of between 0.71 and 0.99.

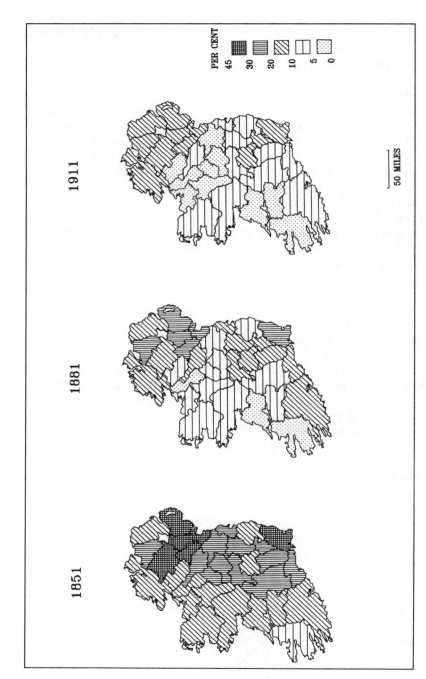

2.6a   Corn: percent of cultivated area, 1851–1911

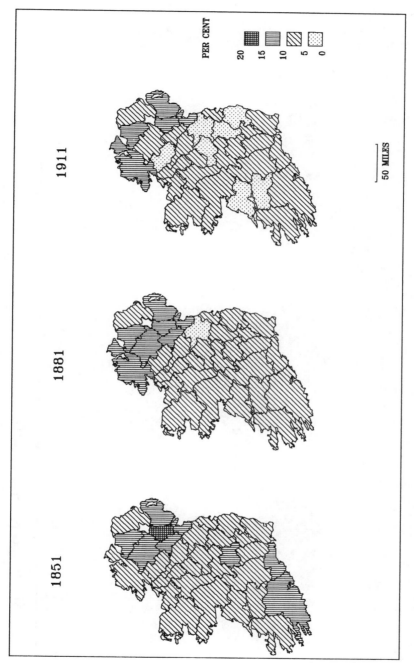

1851　　　　　　1881　　　　　　1911

PER CENT

20
15
10
5
0

50 MILES

2.6b　Root and green crops: percent of cultivated area, 1851–1911 (including potatoes)

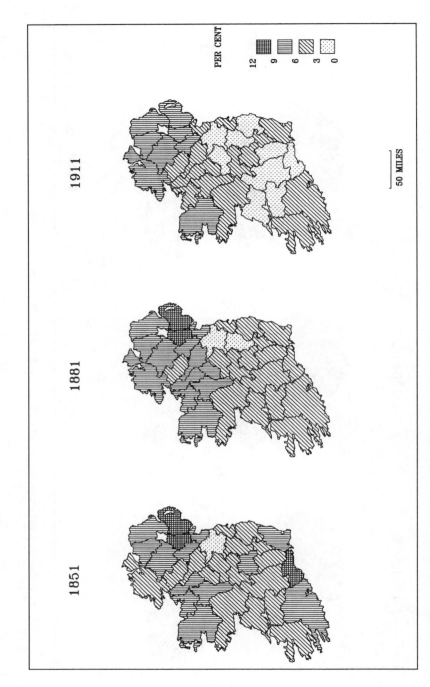

PER CENT

12
9
6
3
0

1851　1881　1911

50 MILES

2.6c　Potatoes: percent of cultivated area, 1851–1911

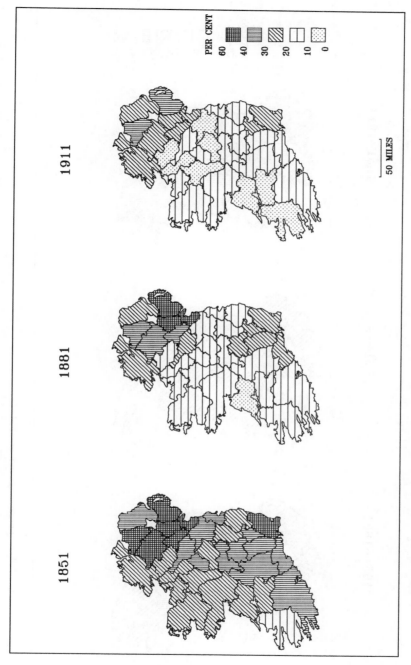

1851   1881   1911

PER CENT

60
40
30
20
10
0

50 MILES

2.7a Tillage: percent of cultivated area, 1851–1911 (includes corn, root and green crops and flax)

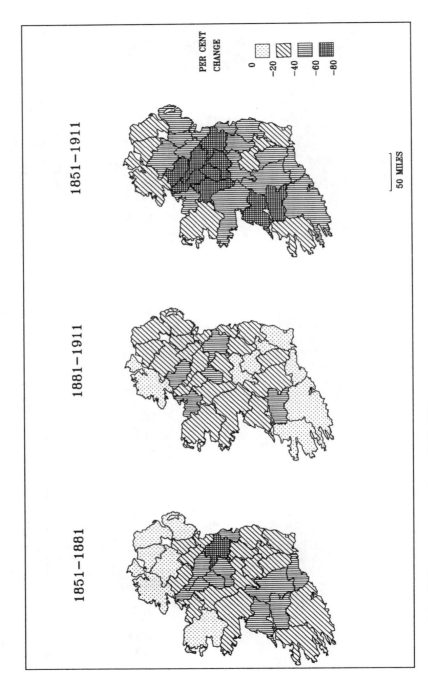

PER CENT
CHANGE

0
−20
−40
−60
−80

1851−1911

1881−1911

1851−1881

50 MILES

2.7b Tillage: percentage change, 1851–1911

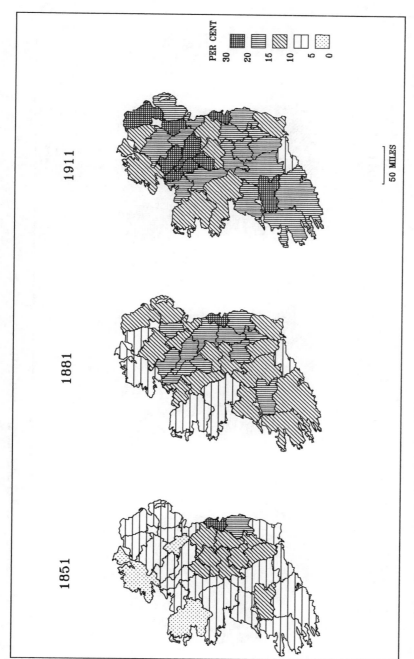

2.8a   Hay: percent of cultivated area, 1851–1911

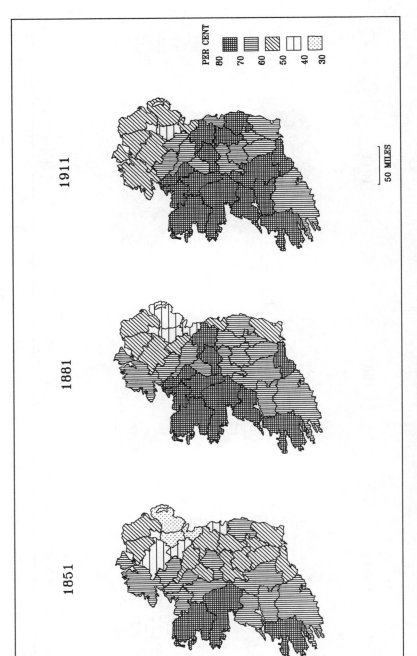

2.8b   Pasture: percent of cultivated area, 1851–1911

large city dominated by horse transport. There may also have been a relationship with the growing livestock export trade through Dublin.

By 1881 there was the more widespread, more intensive cultivation of hay well beyond Dublin, but it was still not yet extensive in the north-west counties. From Galway coastwise to Londonderry, excluding Sligo, less than 10 per cent of the cultivated area was devoted to hay. Still in all, in 1881, the area surrounding Dublin was the most intensively cultivated with hay, with a pronounced corridor of hay cultivation stretching north-westwards to Sligo. In all of these counties between 15 and 20 per cent of all cultivation was hay. By 1911 the national change from arable to livestock was all but complete and to an extent this was reflected in hay cultivation. Indeed, by this date the dominance around Dublin was waning and the move to hay had become a national move, though the west-north-west coastal counties still emerged as relatively non-hay cultivators, but then these counties score very low on all measures of cultivation.

The regional differences in the change in hay cultivation were greater than for arable; some of the least hay counties in 1851 had become some of the most hay counties by 1911.[42] Thus the apparently uncomplicated substitution of hay for arable which occurred nationally hides regional differences. Factors such as labour supply and proximity to the urban areas influenced whether hay or grass replaced arable. Hay was relatively labour intensive, with a derived demand which was partly related to non-agricultural usage (urban horses), but pasture was generally meant to refer to permanent grass, non-mown and directly fed-off by livestock.

Livestock trends

Unlike the equivalent discussion of land use, historians are fortunate to have a pre-Famine datum regarding livestock which they can employ in discussions of long-term distribution and change. This was incorporated into the 1841 population census. The bad news is that there are doubts over the accuracy of the livestock component of that census, especially with regard to cattle numbers. There is also a doubt as to whether the 1841 census is a good guide to the pre-Famine pig population. The pig was bound closely with the fortunes of the potato harvest, and since there was a scarcity of potatoes in the spring of 1841 prior to the June census, a scarcity which was related to relative potato crop failures in 1838 and 1839 and a below normal year in 1840, the size of the pig enumeration in 1841 is considered to underestimate the normal pre-

---

[42] The Spearman rank correlation coefficient of the most to the least hay counties comparing 1851 with 1911 is as low as 0.4.

Table 2.4   *Irish animal numbers, 1841–1914* (to nearest 000)

	Cattle	Sheep	Pigs	Horses	Goats	Poultry	Mules & Asses
1841	2,250[a]	2,106	1,413	576	[b]	8,459	93
1847	2,591	2,186	622	558	164	5,691	126
1849	2,771	1,777	795	526	183	6,328	140
1850	2,918	1,876	928	527	201	6,945	145
1851	2,967	2,122	1,085	522	235	7,471	159
1852	3,095	2,614	1,073	525	278	8,176	165
1853	3,383	3,143	1,145	540	296	8,661	170
1854	3,498	3,722	1,343	546	311	8,630	169
1855	3,564	3,602	1,178	556	284	8,367	172
Ave 1850s	3,412	3,131	1,161	563	257	8,646	171
Ave 1860s	3,557	3,959	1,164	564	185	10,373	190
Ave 1870s	4,039	4,212	1,310	542	254	12,536	203
Ave 1880s	4,084	3,399	1,275	551	271	13,910	223
Ave 1890s	4,430	4,320	1,328	606	315	16,590	254
Ave 1900–14	4,739	3,949	1,261	600	270	22,013	273

Notes: [a] According to the 1841 Census this should be 1,863, 000. Bourke's re-estimation of cattle numbers suggests a total of between 2,220,000 and 2,330,000.
[b] Goats were not included in 1841.
Sources: Derived from *Census of Ireland for the year 1841*, *British Parliamentary Papers*, vol. 24, 1843, pp. 454–7; P.M.A. Bourke, 'The agricultural statistics of the 1841 Census of Ireland. A critical review', *Economic History Review*, 2nd series, 18 (1965), 381–2; *Annual Agricultural Statistics*, various years.

Famine pig population. In his attempt to accommodate these problems Bourke re-estimated the pre-Famine animal population in broad animal groups, and his estimates are included in table 2.4.[43]

If the 1841 figures and Bourke's re-estimates are reliable then the cattle, sheep and horse/mule/ass populations were not seriously touched, or touched at all, by the events surrounding the Famine, but the pig and poultry populations were severely depleted, and the pigs greatly so if it is believed that in 1841 their numbers were low compared with the temporal trend. These were the two animals with reproductive cycles and 'litter' sizes which were best able to withstand exogenous shocks because they were best able to effect a rapid recovery.[44] Pigs were fed almost exclusively on the potato, and the poultry were fed on potato refuse,

---

[43] Bourke, 'The agricultural statistics', 381–2.
[44] For local studies see M.E. Turner, 'Livestock in the agrarian economy of counties Down and Antrim from 1803 to the Famine', *Irish Economic and Social History*, 11 (1984), 19–43; Donnelly, *The Land and the People of Cork*, 76–9.

amongst other things. For both animals the relationship with the potato crop was critical.[45]

In the main, however, the analysis begins not in 1841 or in the 1840s at all, but in 1850 at a time when the *immediate* effects of the Famine had passed, and beyond the time when abnormal circumstances, other than the obvious ones associated with the Famine, had also passed. These abnormalities included the pilfering of growing crops, and more importantly the stealing or killing of cattle, sheep and poultry which in Cork reached the level of organised gangs of cattle stealers.[46]

As Ireland became greener, so this was reflected in a rise in animal numbers, especially of beef cattle, but less so in milk. Milch cows constituted about 70 per cent of all cattle over 2 years of age in 1855, reducing to less than 60 per cent by the end of the century. Measured another way, they represented about 44 per cent of *all* cattle in the 1850s reducing to 33 per cent by 1883 and thereafter remaining at about the same level. There was a rise in cattle numbers of all ages from 1847 onwards. The rise was greatest for non-milch cattle, and indeed milch cows actually fell in numbers. There was a general downturn in milch cattle numbers to 1883 from a previous high point in 1859. They recovered from the mid-1880s, but this recovery was a long slow business and they never again attained that high point of 1859 before the Great War. This represented the relatively declining fortunes of the milk and butter trades giving way to the rising tide of the fat and store cattle trades round about 1880.[47] All other cattle however, whilst exhibiting a common downturn in 1859–64 during the depression, recovered and grew with minor hiccups throughout the period.

The seriousness of the depression of 1859–64 can be seen in the recovery period. On the export side the worst years were at the end of the depression in 1864 and 1865. The stock of cattle less than 1 year of age, and then those of 1–2 years of age were built up in numbers. They matured as beasts greater than 2 years of age in 1867. This was the year when the least number of cattle greater than 2 years of age disappeared from the enumeration, disappeared in the sense that they were slaughtered, sold or exported. This gives rise to what O'Donovan referred to as a paradox. There was a decrease in the number of milch cows, and

[45] P.M.A. Bourke, 'The use of the potato crop in pre-Famine Ireland', *Journal of the Statistical and Social Inquiry Society of Ireland*, 21 (1967–8), 84, 86.

[46] Donnelly, *The Land and the People of Cork*, pp. 87–8.

[47] For more detail on butter and the trade in butter see, Peter Solar, 'The Irish butter trade in the nineteenth century: new estimates and their implications', *Studia Hibernica*, 25 (1989–90), 134–161. For a local study see Donnelly, *The Land and the People of Cork*, pp. 135–58 and J.S. Donnelly, 'Cork Market: its role in the nineteenth-century Irish butter trade', *Studia Hibernica*, 11 (1971), 130–63.

Trend of cattle numbers, 1849–1914

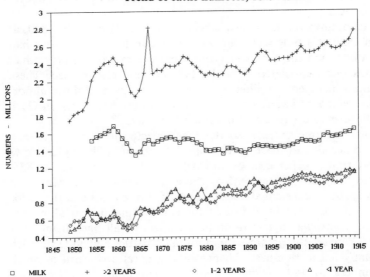

| □ MILK | + >2 YEARS | ◇ 1–2 YEARS | △ ◁ YEAR |

Trend of sheep numbers, 1849–1914

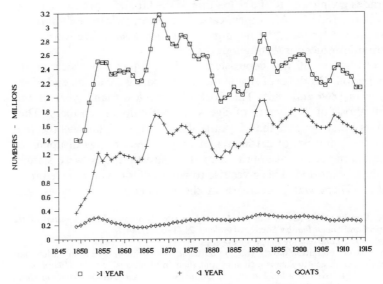

| □ >1 YEAR | + ◁ YEAR | ◇ GOATS |

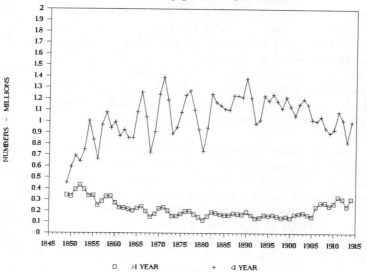

Trend of pig numbers, 1849–1914

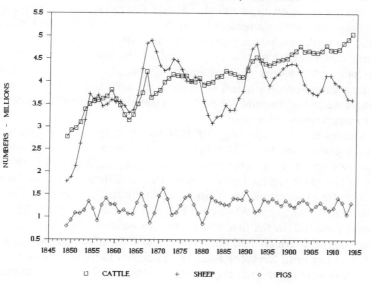

Trend of animal numbers, 1849–1914

2.9   Livestock trends, 1849–1914

therefore a likely decrease in the number of calves born, *but* an increase in the number of cattle of all ages. This is partly explained by a decrease in the trade for veal: the calves lived longer and therefore had a greater chance of surviving long enough to be included in the annual enumeration, and hence of becoming animals of 1–2 years of age and meeting the demand for fat and store cattle.[48] This explanation is not altogether verified by the trade in store cattle to Britain (appendix 1, part C), the numbers of which fluctuated haphazardly, though calves as a proportion of all cattle exports fell from about 10 per cent in 1870 to under 5 per cent by 1914.

There were two major choices facing the farmer with his newly born calves: did he rear them as calves; or did he dispose of them and concentrate on the milk/butter trade; or indeed did he do a combination of the two? From 1854 to 1861 there was a downturn in the ratio of calves to milch cows, a downturn which preceded the agricultural depression of 1859–64. In 1854 there were 45 cattle of less than 1 year of age to every 100 milch cows. By 1861 there were only 34. Thereafter this ratio rose dramatically until in about 1865 there were 74 and their numbers remained on or about 70 thereafter. The store cattle trade came into its own as the second half of the century proceeded, and with the decrease in the veal trade those cattle born between censuses had a greater chance of being enumerated.[49] It was Raymond Crotty who said starkly that they were slaughtered, but he said it in the context of an emphasis on milk and butter production early on, giving way to the fat and store cattle trades as the second half of the century proceeded.[50] Calves had small value, but in terms of feed they were high cost consumers. Therefore it was not so much a decline in veal consumption as an increase in the demand for mature cattle which caused the ratio of calves to milch cows to increase.

The annual enumeration suggests that the change in cattle numbers was fairly even over time. The other dimension is to look at those changes over space. Our descriptions of changing land use capture the essence of the main regional emphasis of animal distributions, but there is one specific observation on regional agricultural development which might be made. For every 100 acres of hay and pasture, that is per unit of their main feedstuff, in the first half of the chronology at least, it was the northern counties which had the greatest density of cattle (see figure 2.10). It was not until the second half of the chronology that the counties of the south-west came into their own as substantial cattle producers.

[48] J. O'Donovan, *The Economic History of Live Stock in Ireland* (Cork, 1940), p. 207.
[49] See also Pringle, 'A review of Irish agriculture', 32–3.
[50] Crotty, *Irish Agricultural Production*, p. 85.

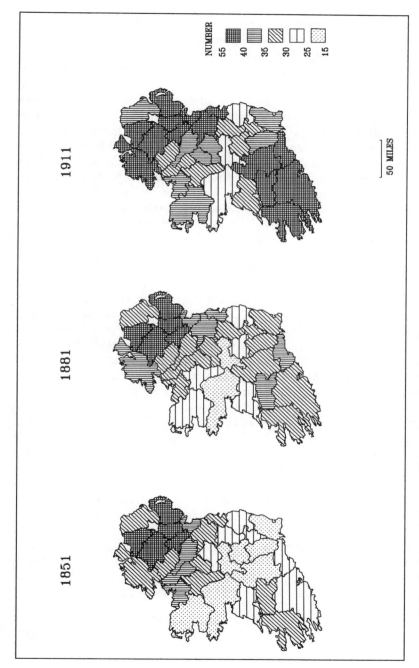

2.10   Cattle: per 100 acres of hay and pasture, 1851–1911

This reflects the more mixed, or perhaps more highly developed, farming systems in the north in the mid-century, but it also suggests the scope for greater development which existed in the rest of the country.

The trend in sheep numbers was more variable. The gestation period of a ewe in lamb is less than for a cow and therefore any crisis in numbers was relatively shortlived. In addition, the incidence of multiple births is more common with sheep than with cattle. Stocks are both more easily reduced, but also they more easily recover. This is clear to see during the 1859–64 depression, or rather during the subsequent sharp recovery. From 1868 until the Great War the trend in numbers for sheep greater than 1 year of age was generally downwards, whereas for sheep less than 1 year of age the trend was flatter, or even slightly on the increase. There were pronounced medium-term swings in numbers, the long downward trend of 1868–82 and sharp recovery to 1892 are evident. The sharp rise from 1864–8 was probably a ripple effect from the United States Civil War and the associated cotton famine which gave a fillip to both the flax (linen) and wool industries.

The Civil War began in 1861 but it was preceded by the disruptions to cotton supplies emanating from the Indian Mutiny of 1857–8. Sheep may have been retained longer than otherwise might have been the case. Wool prices increased by over 100 per cent from the mid-1850s to the highest wool price year of 1864, or in general terms by 40–60 per cent for the duration of the United States Civil War compared with the mid-1850s. The low point in sheep numbers in 1880 or thereabouts coincided with a near continuous downturn in wool prices from 1871 to 1881, but while sheep numbers recovered from the early 1880s wool prices continued to decline into the twentieth century. Once a market trend for sheep had been identified by breeders, only a fairly small, and, more importantly, fixed, breeding stock of older sheep was subsequently required, and this was reflected in a decline or stability in older sheep numbers.

Potatoes were the main source of food for pigs, and therefore it was widely believed that their numbers fluctuated with the fortunes of the potato harvest. O'Donovan claimed that variations in the size of the pig population in the nineteenth century could be traced through variations in potato yields, and Hans Staehle made similar claims.[51] Thus, in '1881 there was an excellent potato harvest in Ireland, and the number of pigs kept in the year 1881–2 was as a result much greater than the number of

---

[51] O'Donovan, *Live Stock in Ireland*, p. 266; H. Staehle, 'Statistical notes on the economic history of Irish agriculture, 1847–1913', *Journal of the Statistical and Social Inquiry Society of Ireland*, 18 (1950–51), 444–71. See also Solar, 'The Great Famine', pp. 123–4; Armstrong, *An Economic History of Agriculture*, pp. 169–70.

Table 2.5   *Pigs and potatoes, 1877–1883*

	No of pigs millions	Average potato yield tons per acre	Potato acreage 000s acres	Potato output 000s tons
1877	1.469	2.0	0.873	1.746
1878	1.269	3.0	0.847	2.541
1879	1.072	1.3	0.843	1.096
1880	0.850	3.6	0.821	2.956
1881	1.096	4.0	0.855	3.42
1882	1.43	2.4	0.838	2.011
1883	1.348	4.3	0.806	3.466

Sources: Derived from *Annual Agricultural Statistics*, various years.

pigs in 1880–1 and the preceding years'.[52] Annual pig numbers, average potato yields, potato acres and hence potato output are reproduced in table 2.5.

The relationship between pig numbers in one year and potato yields in the previous year is more complicated than at first meets the eye. Surely it was not the yield on its own which was important, but also the acreage. The product of the two gave the availability of food. It also depended on the slaughter and disposal of animals in the face of a glut or shortage of food available to them. Although pigs have large litters and are able to recover quickly from crises of mortality, in some particularly bad years it may have depended on the need to slaughter some of the breeding stock as well. From casual observation there was a lower potato output in 1877 than in 1878, and sure enough by 1878 the pig population had fallen, but the good potato year of 1878 was followed by a reduction, not an increase in the pig population by 1879. In other ways the rate of change of food availability was not so obviously related to the rate of change of the pig population. When variations in the pig population and variations in the output of potatoes are tested together – whether expressed as yields or as total tonnage of output – the correlation between them is very weak.[53]

The pig population did depend on potatoes, but the relationship of the input with the output is not the simple one that earlier writers imagined.

[52]   O'Donovan, *Live Stock in Ireland*, p. 266 based on an inspection of the data in Saorstát Éireann, *Agricultural Statistics 1847–1926: Reports and Tables*, Department of Industry and Commerce (Dublin, 1930), p. 16.

[53]   This occurs whether testing yields in year *t* with pigs in year *t* or yields in year *t* with pigs in *t + 1*, or *t + 2*. The strongest correlation coefficient obtained is −0.28 with no time lags, but this has the wrong (negative) sign suggesting that an increase in potato output was accompanied by a decrease in pigs.

A better specified model would also have to include the provision of milk for pigs, but for this there would have to be annual milk yields available. There is some fragmentary evidence. In Cork, for example, in 1884 a reduction in milk yields as a result of a long summer drought was followed by a sharp downturn in pig numbers as farmers were obliged to curtail pig breeding.[54] The relationship between potato output and young pigs would also have to be considered. After all, even in extremis it would be unwise to slaughter the breeding stock. Even so the correlation coefficient between variations in potato output and young pigs (those less than 1 year of age) is modest and also negative.[55]

Pig numbers show certain trends over time. Pigs have larger litters than sheep and cattle and therefore there can be large yearly fluctuations in their numbers. A fodder crisis may have led to a large slaughter, but recovery was rapid. Once again problems associated with the depression of 1859–64 are evident, but the depression did not override this ability of the animal to recover quickly in numbers. Therefore, what was seen in the period 1850–70 was a long-term rise in pig numbers. Graphically it is represented by a gentle arc. This trend moved against the generally downward movement of pigs greater than 1 year of age. This reflected the smaller breeding stock which it was necessary to keep and yet still maintain buoyant high numbers of young pigs. The perverse observation is that once the number of older pigs recovered from about the year 1900, so the number of younger pigs declined. This is partly illusory. Before 1906 the distinction was made between pigs greater than or less than 1 year of age, but from 1906 the enumeration changed and the distinction was made between those greater than or less than 6 months of age. This change reflected a reduction in the age at which pigs were slaughtered compared with the previous fifty years or so.[56]

There is also a problem over the accuracy of the enumeration of the pigs. In every year from 1861 to 1865 and 1871 to 1908, exported pigs constituted over 30 per cent of all pigs enumerated, with a peak of 56 per

[54] Donnelly, *The Land and the People of Cork*, p. 293.

[55] The closest statistical result which fits the expectations of the literature is a correlation coefficient of -0.46 between variations in potato output and the annual 'disappearances' of pigs. The coefficient now has the correct negative sign suggesting that an increase in available potatoes produces a decrease in the annual disposal of pigs. That is, it suggests that sufficient food was available to keep the pigs and not glut the market.

[56] *Agricultural Output of Northern Ireland, 1925*, p. 20. There were adjustments in the enumeration of poultry at this time. In 1907 the poultry numbers suddenly increased because in that year and thereafter separate details were collected regarding the numbers of young birds. J. Bourke says that this sudden rise in poultry numbers in 1906–7 was indeed due to a change in the method of collecting the statistics. Surprisingly, an allowance for this reveals 'no unexpected acceleration in the number of poultry', J. Bourke, 'Women and poultry in Ireland, 1891–1914', *Irish Historical Studies*, 25 (1987), 307.

cent in 1900. It is uncertain, however, if all the exported pigs (and young pigs in general) had been enumerated in the first place. The enumerator of the *Agricultural Statistics* for 1891 suggested that, 'Some of the younger animals included in the Statistics of Exports may of necessity escape enumeration in June each year when the returns of livestock are collected'.[57]

Horse numbers remained fairly stable over time; the 1859–64 depression is evident, but thereafter there was a recovery from the mid to late 1860s to the Great War, with a pronounced downturn to the late 1890s. Looking at the 'disappearance' of horses over time and allowing for a mortality of about 5 per cent, reveals that of those over 2 years of age in the years 1850–2 about 12,000 were evidently disposed of, but that in 1853 this number rose to 30,000, and in 1854 amounted to 24,000. A connection with the requirements of the Crimean War is suggestive here. Thereafter the disposals varied quite considerably, but if short-term movements are ignored they rose from 20–30,000 in the 1850s to 60–70,000 or even 80,000 in the early twentieth century. At times the breeding of animals went into decline although the raising of animals increased and therefore horses were imported.[58] For example, in the period 1890–1 some 4,137 horses were imported into Ireland.[59]

Looking at the net disappearance of horses of all ages, and allowing for a 5 per cent mortality of the older horses, reveals that disappearances rose to 30,000 by 1860, slumped dramatically by the late 1860s, and fell unevenly to a peak of about 50,000 in the late 1890s, and fell again thereafter to about 30,000 by about 1910. In the first ten to fifteen years of the present century the country exported about 30,000 horses to Britain annually. Table 2.6 summarises the net horse trade between Ireland and Britain. The net exports rose until about 1900, but then slumped dramatically into the new century.

Within the horse population, horses used for agricultural purposes dominated. They constituted four-fifths of all mature horses (those over 2 years of age). Horses for traffic and in manufacturing rose in proportion from 5 to 10 per cent of all mature horses while those used for recreational purposes stayed at about 6 or 7 per cent throughout. As a proportion of all horses, those employed in agriculture declined from 72 to 62 per cent over the 50 years from 1861 to 1911, while horses of 1–2 years of age increased from about 10 to 16 per cent over the same period, and horses less than 1 year of age fluctuated in proportion from 8 to 13 to 9 per cent from 1861 to 1891 to 1911. In terms of work performed by

[57] *Agricultural Statistics of Ireland ... 1891* [C. 6777], vol. lxxxviii, 1892, p. 23.
[58] O'Donovan, *Live Stock in Ireland*, pp. 279–80.
[59] *Agricultural Statistics ... 1891*, p. 23.

Table 2.6   *Net horse trade between Ireland and Britain*

	Annual average horse exports	Annual average horse imports	Net exports
1878–80	24,723	2,408	22,315
1881–5	28,453	2,336	26,617
1886–90	30,444	2,880	27,564
1891–5	32,883	4,041	28,842
1896–1900	38,955	6,189	32,766
1901–5	27,362	8,021	19,341

Sources: Agricultural Statistics of Ireland. . . . 1907, British Parliamentary Papers [Cd. 4352], vol. cxxi, 1908, pp. 272-5. See also K. Chivers, 'The supply of horses in Great Britain in the nineteenth century', in F.M.L. Thompson (ed.), *Horses in European Economic History* (British Agricultural History Society, Reading, 1983), p. 43; *Census of Ireland for the Year 1841, BPP*, vol. 24 for 1843, pp. 454–7.

horses there was a rise to 1871, then a decline in the number of tilled acres per horse. This may suggest paradoxically both a rise and a decline in productivity. The more acres there were to till the more efficient the horses became. This is an economies of scale argument, but with the transition in agriculture in the third quarter of the century the opposite applied, with fewer acres to till there was probably a needless hoarding of horses. Horses are a cost even when they do not work because they have to be housed and fed, while human labour, mostly, is paid only when it works.

Animal exports to Britain

It was the British market which was partly, if not mainly driving agricultural change in Ireland, and this is most easily demonstrated by looking at the broad trends of animal exports to Britain. By 1908, 58 per cent of the net value of livestock production came from exports. In the 1850s between 35 and 40 per cent of the cattle which 'disappeared' each year from the annual enumeration were exported to Britain. This increased to 50 per cent in the mid-1860s and was up to 70 per cent by the end of the century. Over the broad period 1875 to 1891 this export trade was in the ratio 40:53:7, fat cattle to store cattle to calves. From 1850–75 between 30 and 50 per cent of the sheep were exported, and over 30 per cent of the pigs were exported as live pigs and an untold proportion in the form of bacon.[60]

[60] R. Perren, *The Meat Trade in Britain 1840–1914* (London, 1978), p. 96; *Agricultural Statistics of Ireland . . . 1891*, p. 23; *Agricultural Statistics of Ireland . . . 1901, BPP* [Cd. 1170], vol. cxvi, 1902, p.xvii; *Agricultural Statistics of Ireland . . . 1911, BPP* [Cd. 6377], vol. cvi, 1912–13, p. xxi; *Agricultural Statistics of Ireland . . . 1916, BPP* [Cmd. 112], vol.

Table 2.7 *Livestock exports from Ireland to Britain 1854–1914*

	Cattle	Sheep	Pigs
Annual Averages (thousands)			
1854–6	242	483	241
1861–4	394	459	355
1865–9	407	584	350
1870–4	558	635	421
1875–9	656	710	493
1880–4	670	569	435
1885–9	687	633	464
1890–4	681	935	530
1895–9	759	780	626
1900–4	803	865	605
1905–9	807	690	413
1910–14	790	677	269
As a proportion of the annual enumeration			
1854–6	6.8	13.1	21.1
1861–4	12.1	13.4	32.4
1865–9	10.7	13.1	26.9
1870–4	13.8	14.6	31.9
1875–9	16.2	17.4	36.7
1880–4	16.7	17.3	36.5
1885–9	16.5	18.0	34.4
1890–4	15.4	20.9	40.4
1895–9	17.1	18.7	46.9
1900–4	16.7	21.4	47.4
1905–9	17.2	18.0	34.0
1910–14	16.3	18.2	21.2

Sources: Annual Agricultural Statistics, for the years 1891, 1901, 1911 and 1916. See also R. Perren, *The Meat Trade in Britain 1840–1914* (London, 1978), p. 96.

Table 2.7 summarises the main livestock exports from Ireland to Britain. The longer-term history of the livestock trade, however, is in some doubt. The cattle trade appeared to take off in the decade preceding the Famine, with an annual average export of 47,000 head of cattle in 1821–5 rising to 98,000 in 1835 and over 200,000 annually in 1846–9, but the precise numbers are doubted in the light of Peter Solar's investigation

li, 1919, p. xiv; *Report from the Committee … Transit of Animals by Sea and Land, BPP* [C. 116], vol. lxi, 1870, appendix XXV, p. 110; *Minutes of Evidence … Upon the Inland Transit of Cattle, BPP* [C. 8929], vol. xxxiv, 1898, appendix III; *Agricultural Statistics of GB, 1907. Vol. XLII, Part III Prices and Supplies of Corn, Live Stock and Other Agricultural Produce, BPP* [Cd 4264], vol. cxxi, 1908, where pp. 272–5 is 'Trade in Livestock with Ireland'. See also Armstrong, *An Economic History of Agriculture*, pp. 133–4, 182–6.

of the agricultural trade statistics contained in the Irish Railway Commissioners' Report of 1835.[61] For other agricultural products the Commissioners' figures were clearly overestimated. In another study, for
example, for the butter trade specifically, Solar is emphatic about the
misrepresentation in the Railway Commissioners' Report. He suggests
that instead of expanding spectacularly from the 1820s, the butter trade
stagnated or even declined.[62]

Another influence on the trade which has also been reinvestigated is
the role of transport developments in initiating change. It was once
thought that the development of an efficient cross Irish Sea steam
navigation before the Famine was complemented by railway development in Ireland after the Famine, and had an impact on agricultural
change.[63] This is now doubted. For example, Liam Kennedy subscribes
to the view that post-Famine trends had roots 'firmly established' in the
early nineteenth century.[64] In this formulation the rise in the animal
trades and the *relative* decline in the grain trade (although that trade
seemed to peak in the 1830s and was still high up to 1845), to a
considerable extent should have pre-figured the less ambiguous trends of
the post-Famine years towards the livestock economy. The extent of the
pre-Famine change out of tillage into livestock has now become a
recurring subject of debate: was or was not the livestock sector on a
rising tide of importance, and was it or was it not related to transport
developments, especially steam? Peter Solar casts oceans of doubt over
the idea of a rising tide, at least in the thirty years before the Famine, 'a
shift from tillage to pasture in the pre-famine period receives absolutely
no support from the evidence of butter exports', but he does see a
possible relationship subsequently with transport developments in the
second half of the nineteenth century, with the pattern of export
concentration of the butter trade through the ports of Belfast, Cork,
Dublin and Waterford.[65]

---

[61] O'Donovan, *Live Stock in Ireland*, pp. 212–13; P. Solar, 'The agricultural trade statistics
in the Irish Railway Commissioners' Report', *Irish Economic and Social History*, 6
(1979), 24–40, especially 30–2.

[62] Solar, 'The Irish butter trade', esp. 135–6, 155–6.

[63] O'Donovan, *Live Stock in Ireland*, pp. 213–14. See also Donnelly, *The Land and the
People of Cork*, pp. 137–8.

[64] L. Kennedy, 'Regional specialization, railway development, and Irish agriculture in the
nineteenth century', in Goldstrom and Clarkson (eds.), *Irish Population, Economy, and
Society*, pp. 173–93, especially 187, 191.

[65] Solar, 'The Irish butter trade', esp. 134, 140–48, quote from 151. See also P. Solar,
Growth and Distribution in Irish Agriculture before the Famine, Ph.D. Dissertation,
University of Stanford, 1987, especially pp. 94–120. For a statement about moderate
pre-Famine adjustments out of tillage into animal production see Daly, *The Famine*, p.
23. For the more extreme view which sees an emphatic shift out of tillage and into
livestock from 1820 see Crotty, *Irish Agricultural Production*, chapter 2. For a middle of

While in general there seems to have been a rise in live cattle exports from 1815 to 1845, the step from a schedule of exports to a general statement about the volume and realignment of agricultural production in the country at large is a questionable move to make, and it may unduly focus attention on the Famine as a watershed in Irish agricultural history.[66] Whatever post-Famine developments there were may have been accelerated by the repercussions of that event, but they may have had an origin in earlier decades.[67] In addition, there are related complicating factors in play when it comes to assessing the size of production. The carcass weight of animals improved in the middle decades of the century, and therefore a trend which relies on numbers alone belies the true increase in Irish meat hitting the British market.[68] Furthermore, foot and mouth disease broke out in 1869. It was particularly virulent in 1871 and 1872, and reappeared in 1884, but it is uncertain if it had much affect either on the cattle trade or on total cattle numbers. In addition, although a policy of slaughtering diseased animals was eventually introduced in 1884, it would be difficult to say what effect the disease had on carcass weights in the meantime.[69]

Conclusion

Thus there occurred a radical change out of tillage production. This was a change which preceded the counterpart move in Britain and on the Continent, and this may account for some of the productivity differences which are revealed in later chapters. The substitution for tillage was a

---

the road view with respect to Ulster see Kennedy, 'The rural economy', pp. 17 and 55 note 51. Ó Gráda finds little scope in his estimates of agricultural output to allow for the switch to pasture so emphasised by Crotty. His estimates suggest that tillage output – after deducting for intermediate use as fodder – constituted over 60 per cent of total output, Ó Gráda, *Ireland Before and After the Famine*, p. 50. His general scepticism is affirmed in C. Ó Gráda, 'Poverty, population, and agriculture, 1801–45' in W.E. Vaughan (ed.), *A New History of Ireland, V, Ireland Under the Union, I 1801–70* (Oxford, 1989), pp. 122, 130, and the trade statistics outlined on pp. 132 and 136.

[66] Goldstrom, 'Irish agriculture', especially pp. 162–3.

[67] *Ibid.*, pp. 170–1.

[68] O'Donovan, *Live Stock in Ireland*, p. 215.

[69] *Ibid.*, p. 257, and in relation to this chronology of the disease compare with appendix 1 and table 2.2. In Britain there was much debate over contagious diseases just prior to legislation in the 1860s which then required foreign livestock which entered the British market to arrive dead rather than alive. Ireland for these purposes was counted as part of Britain, therefore Irish animals could enter Britain on the hoof. This led to much criticism and suspicion over the health of Irish animals whenever there were outbreaks of animal diseases. See in this context, *Report of Mr Chambers and Professor Ferguson on Pleuro-Pneumonia among Cattles recently imported from Ireland into Norfolk, BPP* vol. lx, 1875, pp. 621–30. See also Armstrong, *An Economic History of Agriculture*, pp. 178–81.

move into livestock, and in particular into the cattle trade in its widest senses, that is the export of store and fat cattle, and also milk products. Once British agriculture had adjusted its production and output from about 1870 to accommodate the changing international market for agricultural products, so an emphasis on cattle production in Ireland superseded the earlier relative emphasis on milk production.

It seems obvious to seek and see a connection between agricultural change and the crisis generated by the Famine, but this is not as clear as at first sight might seem to be the case. To some extent there had already occurred a swing towards livestock production before the Famine. In what is to some extent still a statistical dark age some historians have seen this pre-Famine change as emphatic, others as merely symptomatic of slower adaptation to the wants of the British market. The debate over pre-Famine changes brings the question of the Famine as a watershed in Irish economic history into sharper focus. In the post-Famine decade the emphasis of production was still on tillage. If anything, a more important watershed, or catalyst for economic change, was the depression of the first half of the 1860s. From then onwards the country-wide move to greater 'greenness' accelerated. By 1914 the country was emphatically based on livestock production. The national changes identified in this chapter, in turn, were pretty well replicated at the regional (provincial and county) level. The country at large moved largely in the same direction, though not everywhere at the same rate. The market for Irish produce continued to be, or became even more, dependent on Britain. This dependence is in sharp contrast to the moves described in a later chapter towards greater Irish ownership of Irish soil.

The picture which emerges as far as the switch to livestock production is concerned is one of rational behaviour and rational responses to economic circumstances and stimuli. Adjustments in the prices terms of trade saw a rapid and complete move towards a livestock economy – a feature which will be emphasised later in the discussion of output. In isolation this looked like an emphatic move, and so it was in isolation, but like the rest of the Irish economy it was constrained by final product markets and also labour markets. The final product market in Britain was of course expanding, but it was a free market situation, and once the problem of livestock product perishability was capable of a solution, if not wholly solved, any protection which Irish suppliers received was necessarily eroded over time.

Yet perhaps of greater importance was the diminished size of the home Irish market as its population fell, further increasing the dependency on the British market. The declining population had an influence on labour supply which held in check the degree to which agricultural adjustments

Table 2.8 *European livestock trends, 1851–1911*

	Belgium	Denmark	France	Germany	IRELAND	N'lands	GB
5-year averages of livestock unit equivalents – 1879–83 = 100							
1851			103.4		78.1	83.5	
1861	89.9	78.1	101.8		87.7	90.8	
1871	92.9	87.8		105.2	102.3	97.3	100.5
1881	100.0	100.0	100.0	100.0	100.0	100.0	100.0
1891	112.9	116.2	104.6	116.1	110.5	107.4	108.9
1901		129.3	110.9	131.3	115.8	121.0	109.1
1911	147.4	170.8	110.3	154.3	117.9	154.2	111.5
Livestock unit equivalents per capita							
1851			108.0		61.7	109.6	
1861	102.8	95.6	101.9		78.3	110.1	
1871		96.9		115.9	97.8	109.0	114.6
1881	100.0	100.0	100.0	100.0	100.0	100.0	100.0
1891	102.7	105.3	102.6	106.3	121.5	95.5	97.9
1901		103.9	107.9	105.4	134.5	95.1	87.6
1911	109.6	122.0	105.3	107.5	139.0	105.6	81.2
Livestock unit equivalents per economically active male and female member of the agricultural, forestry and fishing workforce (Rebased on 1911)							
1851					33.5		
1861	45.9	32.3	108.3		46.9	80.6	
1871	46.9	32.3			62.2		79.7
1881	51.3	31.4		77.7	67.1		85.0
1891	67.6	38.1	95.6	89.7	80.9	82.4	100.7
1901		73.1	104.5		93.9	84.8	106.4
1911	100.0	100.0	100.0	100.0	100.0	100.0	100.0

Note: Based on 5-year averages as close to the years indicated. This works for most countries though the data for Belgium are the least satisfactory. See also footnote 70 and the forward reference to chapter 6 where the principle of livestock unit equivalents is explained, and footnote 71 for a note on the economically active population.
Source: B.R. Mitchell, *European Historical Statistics, 1750–1975* (Cambridge, 1981 edn), pp. 309–26.

could take place. The situation in Europe was quite different. Table 2.8 is a summary of the trend in livestock numbers across a range of European countries based on 5-year averages from 1851–1911. It summarises cattle, pig and sheep trends in terms of livestock unit equivalents.[70] By 1911 the trend towards animals in Ireland was not particularly strong. It was on a

[70] Details of animal numbers taken from Mitchell, *European Historical Statistics*, pp. 309–26. Because there was an inadequate breakdown of the animals into types and ages it was not possible to construct the livestock unit equivalents with any sophistication, certainly not as much as is used in chapter 6 below.

par with the equivalent trend in Britain and France, but eventually languished behind a group of northern European countries – Belgium, Holland, Denmark and Germany. However, this history should be seen in two parts. The strongest growth rates (especially when annual data are used for those countries where they are available) occurred in Ireland early on, especially before 1871, and in many respects it was all but complete by the turn of the century. The final achievement for Ireland may not have been so great, but it was certainly front loaded and represented both a rapid recovery and a simultaneous change of direction in the twenty years or so after the Famine.

Ireland was the only country to suffer a net loss of population. Therefore the middle part of table 2.8 shows the simple per capita trend in livestock changes. In these terms, no other country could touch Ireland for growth. She started from a low base, and she reached the greatest height. All but Denmark could barely keep pace with their growing populations, and then Denmark could only do so towards the end of the period, and the British situation particularly exposed her need for external supplies to maintain consumption standards. There was in this history a certain mutual dependency. The market which Britain tapped was international, but in the light of Irish exports to Britain, the trends in table 2.8 begin to question the weight of direction of that dependency across the Irish Sea. Indeed, in these crude per capita terms alone, a combined UK total exposes a serious downturn in the UK livestock industry, especially after 1891.

Finally, these livestock trends can be related to the agricultural labour force, thus placing the Irish experience in another context. As the bottom part of table 2.8 shows, the growth in livestock per capita was stronger in Ireland than in all the other countries until 1891–1901, from which broad period the evidently rapid rise in Denmark (from 1891), Holland (from 1901) and Belgium (from either 1891 or 1901) can be seen.[71]

---

[71] *Ibid.*, pp. 161–71. For the economically active agricultural population for Ireland I have used Huttman, 'Institutional factors', p. 414, because the figures in Mitchell are missing for 1891 and 1901. The equivalent index estimates of output per capita of the economically active population from using Mitchell for the years which are available are 35.2, 49.2, 64.3, 66.7, n.a., n.a., and 100. The dates are all based on the UK census with the other European figures fitted in as close as possible to these dates.

3   The occupation of the land

Introduction

The changes, developments and trends described and analysed in chapter 2 were not achieved with the same landholding and tenurial structure that existed prior to the Famine. That eventually exacted a particularly heavy toll at the lower end of the socio-economic ladder, from which there resulted an important redistribution of the myriad of tiny agricultural plots. To call them holdings would be to elevate them to a status of possession which flatters to deceive. Yet however small they were, and however tenuous a stake they gave their occupiers in the agrarian structure, it was vital to their survival. Its loss, and their loss through the circumstances and consequences of the Famine, whether by death or emigration, left behind a slimmed down agricultural economy, with, at first, land to spare. If catastrophies like the Famine have one saving grace it is that they enter and leave the system very quickly, and though they render enormous damage while in residence the subsequent reconstruction is of necessity swift. The lasting effect is that a new structure emerges, and it is one which has to adjust quickly to the radical realignment of factors of production – in the Irish case the human capital of labour, and the physical capital in land itself. This chapter is concerned with the second of those two adjustments, and labour issues are discussed later.

The annual agricultural census collected information on landholding distribution in great detail. It gave categories of landholders from those with less than 1 acre to those with over 500 acres, with seven subdivisions in between. For the years 1847–50 inclusive, only five categories were given; less than 1 acre, 1–5 acres, 5–15 acres, 15–30 acres and over 30 acres. This was also the categorisation given in the 1841 population census. The present study is less concerned with the immediate post-Famine years and more with the long-term trends. To this extent, necessarily, the early data will be treated less than completely, only sufficiently to give some background to the remainder of the post-

Famine period. The main problem which has to be confronted is the definition of landholding in terms of ownership and occupancy. The main issues to explore relate to the long-run trends in landholding distribution between the different landholding groups, and the differences in land use between the landholding groups. It was within those different landholding groups that the realignment of the agricultural future of the country took place, but it was not an adjustment which affected all size groups equally.

The Famine and pre-Famine statistical base

The problems associated with the 1841 census in the interpretation of animals numbers were outlined in chapter 2, but there are also problems with the census when it comes to analysing landholding distributions. The census enumerators may have employed local definitions regarding the size of the acre, and in addition they defined a holding to include arable and pasture, but not necessarily any waste.[1] Any understanding of the effects of the Famine on the occupation of the land must confront these problems. The probable use of Irish acres in 1841, together with the omission of waste had the combined effect of underestimating the recorded farm size compared with the actual farm size, and of misrepresenting the real numbers of holdings in the various size categories. P.M.A. Bourke has suggested that the overall effect was to reduce by about one-half the apparent size of farms. Irish farms, therefore, were not as small as at first sight they seemed. Taking these problems into account he re-estimated the number of holdings in 1841, as in table 3.1.[2]

Therefore he solved the problem, effectively, by doubling the size categories and redefining all holdings as in fact twice as large as originally reported. He was prompted to do this partly because the 1841 figures stood uneasily in comparison with the post-Famine assessments. Comparing the annual statistics from 1847 with the original 1841 figures he noted a fall in the number of landholdings of 1–5 acres from 310,000 to 140,000 (a decline of 55 per cent). This was during full Famine conditions, but it was followed by a further decline in numbers to only 88,000 by 1851 (a decline of 37 per cent) under still very severely distressed conditions. He disbelieved this trend arguing that the initial fall could not have been so great.

In his study of county Cork J.S. Donnelly found that holdings of less than 1 acre fell in number from 8,052 in 1844 to 6,720 in 1847, a fall of

[1]   P.M.A. Bourke, 'The agricultural statistics of the 1841 Census of Ireland. A critical review', *Economic History Review*, 18 (1965), 378.
[2]   *Ibid.*, 379.

Table 3.1 *Distribution of agricultural holdings in 1841*

1841 size categories (acres)	1841 holdings (nos.)	1841 new size categories (acres)	1841 new holdings (nos.)
1– 5	310,375	2–10	310,436
5–15	252,778	10–30	252,799
15–30	79,338	30–60	79,338
> 30	48,623	> 60	48,623
	Total 691,114		Total 691,196

Source: P.M.A. Bourke, 'The agricultural statistics of the 1841 Census of Ireland. A critical review', *Economic History Review*, 2nd series, 18 (1965), 377 and 379

17 per cent, and fell further to 4,342 in 1849 before recovering to 5,727 in 1851, a fall of 15 per cent from 1847 to 1851. This particular local history seems more sensible in view of Bourke's apprehensions about the national trend. In addition, the number of holdings of 1–5 acres in Cork declined from 7,468 to 4,605 to 2,943 and 2,855 over the same years, or by 36 per cent from 1844 to 1847 and 41 per cent from 1847 to 1851. The number of holdings of 5–15 acres declined from 13,000 to 10,000, 6,000, 7,000 and 6,000 over the same years, and holdings of over 15 acres declined in number from 29,000 to 26,000. For each of the four groups in Cork there was a total decline from 1844 to 1851 of 28.9, 61.6, 53.5 and 11 per cent respectively, and for all holdings a decline of 29.7 per cent. Donnelly used the evidence from the Devon Commission to establish his 1844 bench mark.[3] Yet extreme changes might be expected in Cork; it was one of the five worst affected counties in terms of mortality loss in the period 1846–50.[4]

At this stage there is no need for a rehearsal of the full arguments surrounding the insecurity of the 1840s base.[5] This insecurity also attaches to the immediate Famine years (the 1847 annual statistics for example). For all holdings over 1 acre there was a rise from 690,000 in 1841 to 730,000 in 1847 and then a fall to 570,000 in 1851. In those early

[3] J.S. Donnelly, *The Land and the People of Nineteenth-Century Cork. The Rural Economy and the Land Question* (London, 1975), p. 119.

[4] *Ibid.*, p. 121.

[5] For a substantial criticism of Bourke's doubts over the accuracy of the 1841 statistics see the discussion by T.P. O'Neill, K.B. Nowlan and R. Dudley Edwards appended to P.M.A. Bourke, 'The extent of the potato crop in Ireland at the time of the Famine', *Journal of the Statistical and Social Inquiry Society of Ireland*, 20:3 (1959–60), 26–35, where 30–35 is an appendix on the 'Uncertainties in the statistics of farm size in Ireland, 1841–51'; see also P.M.A. Bourke, 'Notes on some agricultural units of measurement in use in pre-Famine Ireland', *Irish Historical Studies*, 14 (1965), 236–45.

post-Famine years there was a great deal of adjustment in the land-holdings of the smallest occupiers or holders, producing an exceptional rate of turnover. Though this is significant for assessing the immediate shock of the Famine, in a long-term study both the 1841 and 1847 datums are insecure.

There is an intermediate datum which has been used by historians. In 1845 the Land Commission Office used returns provided by the Poor Law Commissioners to assess farm size distribution before the Famine. The method of collection was not the same as subsequently employed in the annual agricultural statistics since in 1845 the authorities were concerned with the number of *persons* holding land of a specified size, and this produced problems of definition regarding joint tenants, amongst other things. Nevertheless, the distribution of holdings from this survey suggest that 54 per cent of holdings were under 10 acres, 20 per cent 10–20 acres, 15 per cent 20–50 acres, 7.5 per cent over 50 acres and a residue of about 3.5 per cent were unclassified.[6] These are unhelpful divisions with respect to the subsequent collection of statistics in the annual returns, but Bourke has attempted to reclassify the data. The revised figures for 1845–51 are given in table 3.2, along with Joel Mokyr's estimate of pre-Famine farm size distributions.[7]

According to Mokyr, 14.8 per cent of Irish farms were less than 1 acre in extent and 19.7 per cent were between 1–5 acres. If the 1841 census data can be believed there should have been 45 per cent of Irish farms of 1–5 acres. Mokyr's revision of the numbers of the smaller holdings is quite considerably downwards. He estimated that 40.2 per cent of all holdings were of 1–10 acres in size, and Bourke's re-estimate of the 1841 base suggested that 45 per cent of all holdings were of 2–10 acres in size. The *order* of magnitude arrived at by these two historians is comfortably

---

[6]   Bourke, 'The agricultural statistics', 380; see also Bourke, 'The extent of the potato crop ... Uncertainties in the statistics', 20–26.

[7]   J. Mokyr, *Why Ireland Starved: A Quantitative and Analytical History of the Irish Economy, 1800–1850* (London, 1985 edn), pp. 18–19. This is based on *Appendix to the Minutes of Evidence taken before Her Majesty's Commissioners of Inquiry into the State of the Law and Practice in respect to the Occupation of Land in Ireland, B[ritish] P[arliamentary] P[apers]* [672], vol. xxii, 1845, 280–3 and 288–9, known more familiarly as *The Devon Commission*, for which the *Report*, *Evidence* (in 3 parts), and *Index* can be found in [605, 606, 616, 657, 673], vols. xix-xxii, 1845. See also P. Solar, 'Agricultural productivity and economic development in Ireland and Scotland in the early nineteenth century', in T.M. Devine and D. Dickson (eds.), *Ireland and Scotland 1600–1850. Parallels and Contrasts in Economic and Social Development* (Edinburgh, 1983), p. 77. See also D. Fitzpatrick, 'The disappearance of the Irish agricultural labourer, 1841–1912', *Irish Economic and Social History*, 7 (1980), 72, who remarked that, 'Despite many ingenious attempts by others to amend or render plausible statistics of holdings for the "forties", *we have despaired* [my emphasis] of constructing comparable figures for 1841.'

Table 3.2   *Distribution of holdings and farm sizes, c. 1845–1851*

Size categories (in acres)	1845	1847	1851	% change 1845–7	% change 1847–51
(a) Distribution of holdings 1845–51					
< 1	135,314	73,016	37,728	−46.0	−48.3
> 1– 5	181,950	139,041	88,083	−23.6	−36.6
> 5–15	311,133	269,534	191,854	−13.3	−28.8
> 15	276,618	321,434	290,401	+16.2	− 9.7
Total	905,015	803,025	608,066	−11.3	−24.3

(b) Farm sizes in pre-famine Ireland – after Mokyr

	%
Farms < 1 acre	14.8
Farms 1– 5 acres	19.7
Farms 5–10 acres	20.5
Farms < 10 acres	55.0
Farms 10–20 acres	20.2
Farms < 20 acres	75.2
No. of farms	915,513
Mean farm size (acres)	14.69

Sources: P.M.A. Bourke, 'The agricultural statistics of the 1841 Census of Ireland. A critical review', *Economic History Review*, 2nd series, 18 (1965), 380; J. Mokyr, *Why Ireland Starved: A Quantitative and Analytical History of the Irish Economy, 1800–1850* (London, 1985 edn), p. 19.

close. At these remarkably low acreages however, the precariousness of peasant survival is all too apparent, and their margin for error was exceedingly slim.

When all is said and done, however, the main conclusion will be inescapable: there was a severe adjustment in landholding distribution during the course of the Famine, and during its immediate aftermath, before a fairly stable pattern took a grip; and even if it was not absolutely stable, a casual glance at appendix 3 shows that the changes from about 1851 were gradual rather than cataclysmic.

The focus in this book is on long-term change, structure and adjustment, and therefore there is little analysis of the immediate trauma of the Famine itself. Yet if Bourke's revisions of statistics are correct then the savage reduction in the numbers of the smallest farms or holdings can be easily explained. It came about by death and emigration, and also because of the financial hardship which faced the survivors. Those who ate potatoes but grew cash crops to earn their rents were forced to eat those grains and default on their rents. Small-holder farming – which never was viable without supplementary employment – became totally unviable for many. In particular, the crisis left the small man short of the barest capital if only for seed. In other words there were resulting desertions, as well as evictions.[8] Mary Daly's conclusion is inescapable, 'The decline in the number of small farms is therefore not surprising, even ignoring the active role of the landlord in evicting farmers'; and generally the gap between the rich and the poor widened.[9]

The Famine questioned the stability of the pre-Famine tenurial relationships. In Cork the opportunity was taken to oust 'bankrupt middlemen' and weed out 'struggling or broken tenants, and [of] enlarging the farms of those who remained'. This county was particularly badly hit in terms of mortality and therefore an inordinate amount of slack was created in the land market. More generally there was a residual indebtedness amongst landlords which was worsened by cash-flow problems arising from the Famine. Many insolvent landlords ended up in the Encumbered Estates Court under the Encumbered Estates Act of 1849, and were allowed to dispose of their property. Their source of income was rents, but their capital stock was the land itself, and, like eighteenth and nineteenth-century British landlords who used their lands to raise mortgages to carry out their social and other functions, they became vulnerable whenever a depression struck the land. The entails on British estates could be relaxed or removed, if necessary through recourse to parliament, and so in Ireland, with a disproportionately exaggerated depression in the mid-1840s, there was offered a method of extrication from debt by the removal of the equivalent of entail, or in its own language, of encumbrance. Peasant owner-farming on the Continent had its own fair share of worries and threats to its survival, but not the particular one which faced Ireland in the 1840s, and not on the same scale.

The landlords who availed themselves of the act, as often as not, were Anglo-Irish in origin and they were absentees. To all intents and purposes they were English and a proportion of the rents which came from their estates had usually left the country. Very often the new incoming landowners were Irish rather than English. Disrupting though this was, at least it set in motion the move towards Irish land

[8]   M. Daly, *The Famine in Ireland* (Dublin, 1986), pp. 65–6. See also Donnelly, *The Land and People of Cork*, pp. 100–3. For a comprehensive analysis and county-by-county breakdown of the history of evictions from the 1840s to the 1880s see W.E. Vaughan, *Landlords and Tenants in Mid-Victorian Ireland* (Oxford, 1994), chapter 2 and appendixes 1–4.

[9]   Daly, *The Famine*, pp. 65–6, also 94–7; J.S. Donnelly, 'Production, prices, and exports, 1846–51' in W.E. Vaughan (ed.), *A New History of Ireland, Ireland Under the Union I. 1801–1870*, V, (Oxford, 1989), p. 292.

which became Irish owned, a move which accelerated later in the century.[10]

Table 3.2, along with contemporary qualitative evidence, seems to point to a decrease in the smallest holdings of less than 1 acre, a decrease of the order of 46 per cent from 1845 to 1847, and a further 48 per cent by 1851. Across the Famine years there were decreases of 24 and 13 per cent in the next two size groups, but then an increase of 16 per cent in the number of holdings of over 15 acres. There were wide regional differences in these changes. In Wexford the number of holdings of less than 1 acre actually increased from 1844 to 1847, by as much as 67 per cent, while the total number of holdings in each of the size groups 1–5, 5–15 and over 15 acres all declined by 27, 15 and 13 per cent respectively. It has been suggested that the small farmers in this county were forced to give up their holdings. They became day labourers, exchanging their farms for small plots or vegetable gardens of less than 1 acre.[11] Local relief schemes were successful enough in creating employment and in allowing labourers to survive on the land, albeit at much reduced farming levels. In contrast, in Cork, where starvation and eviction were at higher levels, there was a decrease in the number of holdings under 1 acre.[12]

The fund of land which was made available by the crisis has also been attributed to a policy of deliberate intervention by the government. Theo Hoppen describes this policy as 'social engineering.' His uncompromising view is that intervention by the government in creating a network of soup kitchens, and by the Poor Law Amendment Act of 1847 which authorised outdoor relief in breach of a central principle of the Irish poor law which allowed for indoor relief only, was the fulfilment of a British political ambition. He calls it the 'sovereign remedy for Irish rural discontent, namely the reduction of cottiers and smallholders to the status of wage labourers'. This policy was assisted by the Gregory clause in the Irish poor law which '[denied] relief to all those occupying more than a quarter of an acre of land [and] represented a decisive step towards the achievement of such a policy'. Using more neutral sounding language, but nevertheless conveying a similar picture of change, though not necessarily inspired by such stark motives, there is Collison Black's

[10] Donnelly, *The Land and People of Cork*, p. 100, also 114–20. More generally see J.S. Donnelly, 'Landlords and tenants' in Vaughan (ed.), *Ireland Under the Union*, pp. 332–49. From the large scale to the local see the study of a Tipperary parish in W.J. Smyth, 'Landholding changes, kinship networks and class transformation in rural Ireland: A case study from county Tipperary', *Irish Geography*, 16 (1983), 16–35, especially 23–4.

[11] M. Gwinnell, 'The Famine years in county Wexford', *Journal of the Wexford Historical Society*, 9 (1983), 47 citing the *Devon Commission* as evidence.

[12] *Ibid.*, 47; and the Cork evidence based on Donnelly, *Land and People of Cork*, p. 119.

observation that, 'the inability to pay rents during the Famine years gave a powerful stimulus to clearances, later reinforced by the "quarter-acre clause", and once people began to emigrate in large numbers, the possibility of consolidation was greater than ever before'.[13]

Post-Famine longitudinal changes

The very smallest holdings, those under 1 acre, were the only ones to rise in numbers over the sixty years or so up to the Great War. Other sub-groups fluctuated within fairly narrow limits. Holdings of less than 1 acre constituted 6–7 per cent of all holdings in all but two years of the 1850s. There were between 35–38,000 holdings of this size. By 1864 they constituted about 8 per cent of all holdings (49,000) and remained at that level until 1890. The depression of 1859–64 accelerated the general upward trend in numbers. In 1856 there were 37,000 holdings under 1 acre, but by 1867 their number had risen to 51,000. There were over 50,000 such holdings throughout the 1870s reducing to 48,000 by the late 1880s. From 1890 onwards the proportion of holdings of this size grew from about 8 per cent to 18 per cent by the Great War, rising in number from 50,000 in 1890 to 70,000 by 1900, and 100,000 by the Great War. In the decade before 1914 the trend accelerated with the move to greater owner-occupancy. This began with the Purchase Acts of 1891 and 1896 and was mostly enhanced by Wyndham's Act of 1903. In 1870 only 3 per cent of Irish holdings were owner-occupied, but by 1908 the proportion was 46 per cent and rising rapidly.[14]

For holdings of 1–5 acres there was a countervailing decline in numbers from 80,000 or so in the 1850s and early 1860s to 70,000 by the mid 1870s, declining further to 60,000 or so from the late 1870s through to 1911. By 1912/13, as the number of the smallest holdings had dramatically increased, so the number in this second group equally dramatically had decreased. A similar pattern of change took place in the trend of holdings of 5–15 acres, and a less pronounced one for those of 15–30 acres. In the middle range of size groups – 30–50 acres, 50–100 acres and 100–200 acres – the general tide of movement was in the other direction, and the numbers of these holdings increased.

[13] K.T.Hoppen, *Ireland Since 1800: Conflict and Conformity* (London, 1989), p. 54. See also Hoppen, *Elections, Politics, and Society in Ireland 1832–1885* (Oxford, 1984), p. 96; R.D. Collison Black, *Economic Thought and the Irish Question 1817–1870* (Cambridge, 1960), p. 35. For a local study see the actions of the Earl of Lucan in county Mayo in, D.E. Jordan Jnr, *Land and politics in the West of Ireland: County Mayo, 1846–82*, Ph.D. Dissertation, University of California at Davis, 1982, pp. 92–3.

[14] B.L. Solow, *The Land Question and the Irish Economy, 1870–1903* (Harvard, 1971), p. 193. See also chapter 7, table 7.1, below.

Whilst these changes were significant to the individuals concerned, the fact is that quite large changes in the numbers of the smaller holdings led to the redistribution of only a small area of land. Conversely, if one holding of over 500 acres had been redistributed into 500 holdings of 1 acre each, the modest change in personnel at the top end of the ownership spectrum would have been exaggerated at the lower end in terms of landownership redistribution.[15] The broad changes are summarised in Figure 3.1. These changes were small, and more or less evenly distributed over time.

The definition of landholding is not as clear as may at first sight seem to be the case.[16] Most of the time a landholding defined the autonomous integrity of, literally, a holding on the ground, but the national figures were constructed by the census enumerators who first investigated at the county level and then aggregated to find the national figure. At a county boundary, for example, what otherwise would have been single 'holdings' may have been sub-divided by that boundary and therefore have been counted as more than one holding. Thus there were instances of double-counting of what otherwise would have been single holdings. Those holdings might then have been allocated to the wrong size group. It was also possible that some landholders had more than one holding, wherever they were located. In order to obtain a more rounded, if not more accurate, picture of land distribution, the enumerating authorities in 1861 decided to report the number of occupiers in various size groups (as well as holders), regardless of within which county their individual holdings were located, or of how many holdings they occupied. In order to aid an understanding of the spatial distribution of these occupiers they were nevertheless also assigned to specific counties. Thus in 1861 there were 41,561 holdings of less than 1 acre (including some double counting) though 39,210 occupiers of such holdings (now without double counting). At the other end of the scale there were 1,591 holdings of over 500 acres, but 2,437 occupiers of holdings of over 500 acres.

In all size groups below 200 acres the number of holdings exceeded the number of occupiers. This arose because of double-counting. However, for holdings greater than 200 acres the number of occupiers was greater than the number of holdings. There is a ready explanation of this in, for example, the annual return for 1916.

---

[15] See also Smyth, 'Landholding from county Tipperary', 22.

[16] Note that the figures for holdings and occupiers before and after 1909 are not directly comparable as the enumeration went through some subtle changes. See *Agricultural Statistics of Ireland ... 1916*, BPP [Cmd. 112], vol. li, 1919, p. xvi, and *Agricultural Statistics of Ireland ... 1914*, BPP [Cd. 8266], vol. xxxvi, 1916, p. xvii.

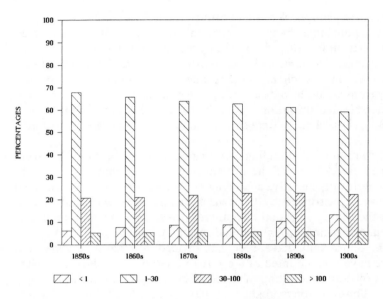

3.1a   Landholding distribution, 1850s–1900s (in size groups in acres)

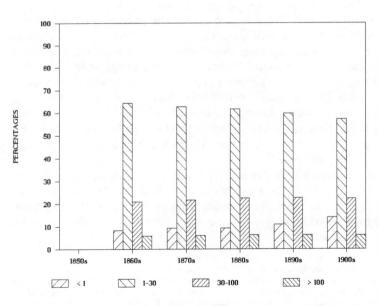

3.1b   Land occupancy distribution, 1860s–1900s (in size groups in acres)

the number of 'occupiers' of 200 acres and upwards is greater than the number of 'holdings' of 200 acres and upwards. The reason for this difference is that where two or more separate holdings are in the hands of one occupier, the holdings are included separately in their respective classes according to size while the occupier appears in the class indicated by the total amount of land in his occupation. For example if one person is the occupier of two separate holdings of 90 and 150 acres, these holdings will appear in the table of holdings under the 50 to 100 acre class and the 100–200 acres class respectively: whereas the occupier will be included in the 200 to 500 acre class as being in occupation of 90 + 150 or 240 acres.[17]

The trends described so far are not seriously affected when comparing the number of occupiers with the number of holdings over time.

The census enumerators were careful to draw a distinction between landholdings and their occupiers, and they were also mindful of the small sizes of many holdings, the inclusion of which might distort their analysis. Very often they discounted holdings of under 1 acre. At this size it was difficult to distinguish between gardens and agricultural premises (however small), and equally important, it was difficult to distinguish between agricultural holdings, and gardens, town gardens and allotments. It was felt that holdings of under 1 acre contributed very little to the agricultural output of the country, and, though they were large in number, it was felt that they could be safely excluded from many discussions concerning Irish agriculture.[18]

The main trends in the changing pattern of the distribution of holdings were undramatic in the short term. The broad trends seem clear enough; holdings of 1–5 acres and 5–15 acres decreased in importance, and there was a countervailing increase in the importance of holdings larger than this. There seems to have been a gentle, but nevertheless identifiable regressive redistribution of land. This was a pattern more or less repeated for occupiers.

Yet a good deal of history is lost in this kind of analysis. A summary approach such as this hides the rate of property turnover; these aggregate numbers have no discernible faces. An important element which is missing, and which Liam Kennedy amongst others has identified, is the increasing access to the land gained by traders and merchants towards the end of the nineteenth century from both the country and the towns. It has been suggested that *if* there was an alliance of interests between traders and farmers during difficult years, such as during the depression of 1859–64 and during the Land War period of the late 1870s and 1880s, when it was mutually beneficial for

[17] *Agricultural Statistics ... 1916*, p. 14n.
[18] *Agricultural Statistics of Ireland ... 1911*, BPP [Cd. 6377], vol. cvi, 1912–13, pp. xxiv, 12; *Agricultural Statistics ... 1916*, p. xvi.

the farmers and the traders who serviced them to press for rent reductions for the farmers, then by the end of the century many of those traders had become rivals for land; the shopkeeper-grazier emerged as an important force in the countryside.[19] Ill-feeling developed between the farmers and the traders which became worse when the farmers embraced the cooperative system which burgeoned in the late nineteenth and early twentieth centuries.[20] There is some confusion in the directional flows here because there was also a move by farmers and their families to move to off-farm occupations, such as trading, especially towards the end of the century. They were responding to changes in farming, changes which had moved away from relatively labour-intensive forms.[21]

There were other important changes. As the country became 'greener' and the economy was driven more by the production of dry cattle and other livestock, there began to emerge the grazier-cum-rancher. They farmed large acres, in excess of 300 acres, and normally their land was held in scattered multiple holdings; to a large degree they were absentees; and they acquired the small-holdings which were formerly occupied by those tenants who had lived precariously close to subsistence. The acquisition of such land by the graziers and the elimination of a farmer-tenant class led inevitably to an adjustment in the socio-economic relationships in this traditional rural society. In the most practical of terms it reduced employment opportunities. Conflict, therefore, was never far away. It emerged and it grew between the peasants and ranchers in the grazing regions of Connacht, north Munster and north Leinster. The result was a ranch war in which 'cattle driving' became rampant between 1906 and 1909.[22] This was the illegal removal of cattle from the graziers' lands, whereby those cattle were left to wander. The characteristic of ranching was low output per acre of land, and even though the ranchers' costs were also small, there was in this low land productivity system the seeds of the conflict. 'Narrow profit margins impelled territorial expansion', and land, ironically, became a scarce commodity in some districts. The opportunities to do this, such

---

[19] L. Kennedy, 'Farmers, traders, and agricultural politics in pre-Independence Ireland', in S. Clark and J.S. Donnelly (eds.), *Irish Peasants. Violence and Political Unrest 1780–1914* (Manchester, 1983), pp. 346–7; D.S. Jones, 'The cleavage between graziers and peasants in the land struggle, 1890–1910', in Clark and Donnelly, *Irish Peasants*, pp. 394, 401, 412–13. On the fact that this cleavage predated the crisis of the late 1870s in county Mayo see Jordan, 'Land and politics: County Mayo, 1846–82', p. 156.

[20] Kennedy, 'Farmers, traders, and agricultural politics', pp. 347–9, 356.

[21] L. Kennedy, 'Traders in the Irish rural economy, 1880–1914', *Economic History Review*, 32 (1979), 201–10, especially 209.

[22] In general see Jones, 'The cleavage between graziers and peasants'; also S. Warwick-Haller, *William O'Brien and the Irish Land War*, (Dublin, 1990), pp. 159–60, 174.

as during those times of heavy eviction from 1849 to 1856, and to a much lesser extent thereafter, were limited in time and localised in space.[23]

A further complication in the long-term analysis arises from the system of 11-month lettings, whereby the landlords let their untenanted pastures for 11 months.[24] This had two effects: first the graziers tended to outbid other would-be holders of that land; but second, those graziers could not claim formal tenancy or interest in the land. The landlord remained occupier in law and his 'tenant' – not really the desired term – was a temporary occupant. In terms of the annual collection of statistics there is uncertainty whether this system was fully registered, but in 1906 as much as 2.6 million acres was devoted to it. This was nearly 13 per cent of the country, and more importantly, nearly 18 per cent of the cultivated area (crops and grass). In general there seems to have been a move in the late nineteenth century for landlords to increase the amount of 'untenanted' land in their possession, and therefore of extending the 11-month system. For example, in the Kells barony of County Meath in 1901, anything between 30 and 50 per cent of the commercial pasture and ranch land was untenanted in this fashion. With the terms of trade moving in favour of livestock production this feature inevitably grew and the graziers' appetites for more pasture increased. In a countervailing fashion, and increasingly, the peasantry (the term used in the literature) was deprived. This resulted in peasant overcrowding in the west, particularly in the congested counties, and in time the fresh supply of untenanted land was acquired only by drastic measures such as eviction.

Land and livestock distributions and the average size of holdings

This then is the aggregate statistical analysis overlain by adjustments in the traditional tenurial habits of the Irish countryside. Economic considerations may always have determined landlord/tenant and grazier-rancher/peasant relationships, but in the second half of the nineteenth century they came on a broad front as the agricultural economy adjusted to its emphasis on pastoral activities. Perhaps a more revealing and useful way to investigate landholding changes and adjustments is to look at the average size of holding within each size group.

---

[23] Jones, 'The cleavage between graziers and peasants', pp. 392, 394. For a clear chronology of the history of evictions see Vaughan, *Landlords and Tenant in Mid-Victorian Ireland*, appendixes 1–4.

[24] Much of this following section on 11-month lettings derived from Jones, 'The cleavage between graziers and peasants', especially pp. 396–404.

This is not possible from the way the census enumerators presented the details for publication, though they themselves, in very summary terms, were conscious of the problem. Thus in 1861 the 41,561 holdings of less than 1 acre occupied a total land area of 25,470 acres, but the 1,591 holdings of over 500 acres occupied a total land area of 1,988,374 acres.[25] Thus, the average size of the smallest holdings was 0.6 acres, and the largest was 1,249.8 acres. In addition, the specific aggregated land use under the different holdings size groups was also indicated. For example, of the 25,470 acres in holdings of less than 1 acre as much as 21,856 acres was under crops, or 86 per cent, but of the 1.988 million acres in holdings of over 500 acres only 75,976 acres was under crops, or less than 4 per cent. By 1871 the number of the smallest holdings had increased to 48,448 but the number of acres they occupied had declined to 25,334 acres, while the number of the largest holdings had decreased slightly to 1,568 but the acreage they occupied had increased to over 2 million acres. Average farm or holding size in the two size groups therefore had become 0.6 acres (the same as earlier because of rounding) and 1,320.2 acres respectively. This particular example, therefore, exposes one important feature of the Irish land economy, and that was the regressive nature of the changes. The smallest became more numerous, but smaller, and the largest became more exclusive but bigger.

This practice of reporting the distribution of land by both the numbers of landholders and the total acreage within size groups ceased in 1874, and then in 1902 a review was undertaken of average size of holdings within groups.[26] Table 3.3 summarises the long-term trend in the average size of holdings. There was remarkable long-term continuity in each group, and this continuity was evident at the national aggregate level, as shown in table 3.4.

Table 3.4 highlights one problem of definition. What exactly constitutes a holding, and with what kind of land uses, and how should land which by common observation was in a form of 'cold storage' or underutilised be treated in the analysis? For example, should barren ground, common, waste and so on be included? In terms of the agricultural statistics such land was allocated to holdings. It would be more meaningful to define holdings as enterprises which convert usable land into produce. In this case bog, barren rock and other unusable lands should be eliminated from the calculation or somehow weighted to diminish their importance. Land could then be classified simply into two uses, crops or grass, though that would then

[25] *Agricultural Statistics of Ireland ... 1861*, *BPP* [3156], vol. lxix, 1863, p. xv.
[26] *Agricultural Statistics of Ireland ... 1902*, *BPP* [Cd. 1614], vol. lxxxii, 1903, p. xxvi.

Table 3.3 *Long-term trends in landholding distribution and average size of holding, 1853, 1861, 1871, 1902*

Size groups	Holdings as a % of all holdings	Average size of holding	Holdings as a % of all holdings	Average size of holding
Acres	1853	1853	1861	1861
< 1	6.1	0.7	6.8	0.6
1–5	13.6	3.5	14.0	3.5
5–15	30.5	10.2	30.1	10.4
15–30	23.7	22.2	23.1	22.3
30–50	12.0	40.5	11.9	40.3
50–100	8.8	74.5	8.9	73.3
100–200	3.5	154.1	3.5	149.8
200–500	1.4	357.7	1.4	340.4
> 500	0.3	1,344.5	0.3	1,244.7
Total	100.0	34.5	100.0	33.1
	1871	1871	1902	1902
< 1	8.2	0.5	12.6	0.5
1–5	12.6	3.6	10.6	3.0
5–15	28.9	10.5	26.2	10.0
15–30	23.4	22.3	22.7	22.5
30–50	12.3	40.4	12.6	40.0
50–100	9.3	73.5	9.7	75.0
100–200	3.6	150.2	3.9	150.0
200–500	1.4	342.0	1.4	350.0
> 500	0.3	1,320.2	0.3	1,295.9
Total	100.0	34.3	100.0	34.5

Sources: Derived from *Annual Agricultural Statistics*, various years.

ignore the turf from the peat bogs, which was demonstrably a valuable commodity.[27]

Table 3.4 shows a familiar tale; there was a steady decline in the average size of tillage cultivation into the twentieth century and an initial rise in grass up to a plateau level from the 1880s. In terms of land use per holding in the different size groups, however, there were some significant developments. Of every 100 acres cultivated on holdings of less than 1 acre in 1851, 33.4 per cent was under cereals, 63.2 per cent under root and green crops – with potatoes alone accounting for 50 per cent – and less than 3 per cent under meadow. At the opposite extreme, for every

[27] R. O'Connor and C. Guiomard, 'Agricultural output in the Irish Free State area before and after Independence', *Irish Economic and Social History*, 12 (1985), p. 93 put the value of turf at over £3 million in 1912–13, or 7 per cent of gross agricultural output.

Table 3.4   *Average size of holdings, 1851–1914 (in acres)*

	Total no. of holdings	Extent of land (a) (millions)	Extent of cult area (1) (b) (millions)	Extent of cult area (2) (c) (millions)	Average size of holding in terms of (a)	Average size of holding in terms of (b)	Average size of holding in terms of (b + c)
1851	608,066	20.329	5.859	8.749	33.43	9.64	24.02
1856	592,489	20.329	5.753	9.545	34.31	9.71	25.82
1861	610,045	20.329	5.891	9.534	33.32	9.66	25.29
1866	597,628	20.329	5.521	10.004	34.02	9.24	25.98
1871	592,590	20.329	5.621	10.071	34.31	9.49	26.48
1876	581,753	20.329	5.207	10.507	34.94	8.95	27.01
1881	577,739	20.329	5.195	10.075	35.19	8.99	26.43
1886	564,352	20.329	5.034	10.251	36.02	8.92	27.08
1891	572,640	20.333	4.818	10.299	35.51	8.41	26.40
1896	575,664	20.333	4.843	10.334	35.32	8.41	26.36
1901	590,175	20.351	4.631	10.577	34.48	7.85	25.77
1906	597,344	20.351	4.739	10.063	34.07	7.93	24.78
1911	607,960	20.371	4.861	9.847	33.51	8.00	24.19
1914	566,137	20.371	4.815	9.928	35.98	8.51	26.04

Notes: b Total crops including hay (meadow and clover). c Extent of grass. b + c Total cultivated area, crops + grass. Note that the separation of grazed from ungrazed mountain land after 1906 gives the impression that the grass acreage declined. In 1888 the average size of holding was 36.13 acres, the largest size achieved in the chronology.
Sources: Derived from the *Annual Agricultural Statistics*, various years.

100 acres on holdings of over 500 acres in extent, 35.2 per cent was under cereals, but only 18.4 per cent under root and green crops – with now less than 5 per cent under potatoes – and 45.6 per cent under meadow. One interesting feature was the stable proportion of land under cereal crops. Holdings of less than 1 acre were dominated by potato cultivation, but on holdings of 1–200 acres between 47–57 per cent of the land was under cereals, and as wide a fluctuation as 19–36 per cent was under root and green crops. This was pretty well perfectly graded from the largest to the smallest holdings, and there was a graduated trade-off with the increasing cultivation of potatoes on smaller holdings. Conversely, the extent of meadow increased from 9 per cent on holdings of 1–5 acres to 33 per cent on holdings of 100–200 acres.[28]

For some years this type of analysis can be taken a stage further. In 1861 on holdings of over 500 acres, of which there were 1,591 occupying

[28]   See, *The Census of Ireland for the year 1851. Part II. Returns of Agricultural Produce in 1851, BPP* [1589], vol. xciii, 1852–3, p. vii, for more detail and comment.

1.988 million acres, as much as 1.174 million acres was defined as bog and waste and a further 61,000 acres was in woods and plantations, leaving just 753,000 acres as crops, grass and fallow (in the ratio 1:9, crops to grass). An average sized holding for this group was composed of 473.6 acres of cultivated land of which only 48.7 acres was cropped, but 424.9 acres was grassed. The bog and waste, woods and plantations, no doubt, were important in the economy of these holdings, but such land comprised a disproportionate share of the total. It seems a coincidence of timing to suggest that this accumulation of unused land came about as a result of the Famine and was associated with abandonment as less pressure was brought to bear on it. In truth, before the Famine, the unimproved land was concentrated in the hands of the very largest properties already, and therefore these lands were not necessarily abandoned because they had not necessarily been reclaimed and 'occupied' in the first place.[29]

Table 3.5 is a useful summary of land use in terms of different landholding size categories.[30] On all holdings of under 100 acres over 80 per cent of the land was intensively used under a regime of crops or grass. This ranged from the comparatively intensive smaller arable holdings to the more extensive and larger grassland farms, with farms of somewhat equal intensity of crops and grass occurring around the 15 acre level. On holdings of 100–200 acres the amount of 'surplus' land – bog and waste – was equal to the extent of the arable. On holdings of 200–500 acres the bog and waste exceeded the arable but it was not yet the dominant form of land use, but on holdings of over 500 acres it became the dominant land use.

The average size of all holdings increased by about 1 acre over the period 1861–71, but this was almost entirely due to the increase in the average size of the very largest holdings. This was close enough to be a mirror image of the trend from 1853 to 1861. More striking still was the land-use change within holding size groups. Overall, the average size of the grassland area increased, mainly by what appears to have been a conversion of bog and waste into grass. Over the whole period 1853–71 the combined acreage of these marginal lands declined from 4.702 million to 4.26 million acres, and the grass increased from 9.381 million

[29] See J.H. Andrews, 'Limits of agricultural settlement in pre-Famine Ireland', in L.M. Cullen and F. Furet (eds.), *Ireland and France 17th to 20th Centuries: Towards a Comparative Study of Rural History* (Ann Arbor, Michigan and Paris, 1980), especially p. 49.

[30] See also R.O. Pringle, 'A review of Irish agriculture, chiefly with reference to the production of live stock', *Journal of the Royal Agricultural Society of England*, 2nd series, 8 (1872), 21 for a breakdown for 1869, and Solow, *The Land Question*, p. 108 for a breakdown for 1871.

Table 3.5 *Land use by holding size groups, 1853, 1861 and 1871* (in acres)

Size Groups	Total average size			Average extent of crops and fallow		
	1853	1861	1871	1853	1861	1871
< 1	0.7	0.6	0.5	0.6	0.5	0.5
1–5	3.5	3.5	3.6	2.3	2.3	2.3
5–15	10.2	10.4	10.5	5.1	5.1	5.0
15–30	22.2	22.3	22.3	9.4	9.4	9.1
30–50	40.5	40.3	40.4	14.9	14.9	14.3
50–100	74.5	73.4	73.5	22.3	22.3	21.3
100–200	154.1	150.0	150.2	34.1	33.3	32.2
200–500	357.7	341.2	342.0	47.6	45.4	43.8
> 500	1,344.5	1,249.9	1,324.5	55.1	48.7	42.3
Total	34.5	33.3	34.3	9.9	9.7	9.5

	Average extent of grass			Average extent of other land		
< 1	0.04	0.04	0.03	0.04	0.05	0.05
1–5	1.0	0.9	1.0	0.2	0.3	0.3
5–15	4.4	4.3	4.5	0.7	1.0	1.0
15–30	10.4	10.2	10.7	2.4	2.8	2.5
30–50	19.9	19.5	20.6	5.7	5.9	5.5
50–100	38.1	37.3	40.3	14.1	13.8	11.9
100–200	80.9	79.4	85.9	39.1	37.4	32.2
200–500	170.3	170.9	179.5	139.8	124.9	118.7
> 500	384.5	424.9	438.7	904.9	776.1	840.5
Total	16.0	15.6	17.0	8.6	8.0	7.7

Sources: Derived from *Annual Agricultural Statistics*, various years.

to 10.071 million acres. There was also a marginal decrease in cropland, from 5.8 to 5.624 million acres, including fallow, which itself decreased in area during 1853–61 from 103,000 to 41,000 acres, and which then halved in area to 21,000 acres by 1871. Within all size groups the average area of bog and waste declined, except on the very largest of holdings where its size fluctuated wildly. More and more land was brought into cultivation per holding by reclaiming the waste, principally for extending the grass, but there was also a general rationalisation of the number of holdings after 1861 from 610,000 to 593,000 by 1871. Those landholders and occupiers who remained on the land were using that land better, and the largest owners were mopping up the residual unimprovable waste.

The distribution of livestock by holding size groups can be treated in a similar way. By reducing all animals to a common unit – livestock unit equivalents – based on estimates of the different feed requirements of different animals, the distribution of animals over time can be shown in terms of stocking densities (as in table 3.6).[31]

There were more livestock units on the larger holdings than on the smaller, which is not at all surprising, and in terms of stock densities there was a nearly perfect gradation from high stocking levels on small holdings to low levels on large holdings. If cattle are isolated there seems to have been something akin to a filter at play; it was as though they were bred on holdings of less than 50 or 100 acres and passed on to larger holdings for fattening and final disposal. This 'market' in cattle cannot be precisely interpreted from the annual statistics, but it does seem significant that in 1854 on all holdings of over 200 acres the number of cattle over 2 years of age was greater than the combined total of cattle of 1–2 years and below 1 year, while on holdings of less than 200 acres the reverse situation prevailed. In the long run similar patterns can be observed. In 1861, for example, the number of cattle over 2 years of age on all holdings over 100 acres was greater than the combined total of cattle of less than 2 years. Conversely, on all holdings of less than 100 acres the younger animals predominated. Even allowing for time lags – for example as the animals grew older and therefore were enumerated in a different age group – there is still the appearance of a filter from the smaller to the larger holdings. However, given that 1861 was in the midst of the depression of 1859–64 with the problem that presented of retaining animals during the fodder famine, it would be foolish to draw too many hard conclusions from that particular year. By 1871 it was only on the largest holdings, those in the two categories over 200 acres, that the pattern survived.

In addition, there were important regional variations, suggestive of a second filter from the west to the east of Ireland. This was detectable at the national level in 1854 on or around holdings of 100–200 acres but in Connacht it occurred at around 50 acres. This pattern was repeated in 1861 when it was at its greatest extreme in Leinster. There, on holdings of 200–500 acres, there were 101,000 cattle over 2 years of age but only 25,000 cattle of between 1 and 2 years and a further 12,000 of less than 1 year, and on holdings of 100–200 acres the numbers were 91,000 34,000 and 23,000 cattle respectively.

In modern farming distributions, farm size groups are not always a useful basis for classification because they imply that all acres are treated equally. In actual fact land quality varies enormously, say, from the intensive arable to the extensive hill farm. The former might be small in

[31] The data for 1854 are employed whereas in table 3.5 those for 1853 are used. This is because milch cows were not separately enumerated until 1854.

Table 3.6 *Livestock densities by holding size groups* (in livestock unit equivalents)

Size groups (acres)	1854		1861		1871	
	Per holding	Per 100 acres	Per holding	Per 100 acres	Per holding	Per 100 acres
1–5	1.0	28.8	0.9	26.4	1.1	31.6
5–15	2.6	25.1	2.4	23.3	2.9	27.8
15–30	5.0	22.7	4.8	21.4	5.6	25.1
30–50	8.5	20.9	8.1	20.1	9.3	23.0
50–100	14.2	19.1	13.9	18.9	15.5	21.0
100–200	27.0	17.5	26.7	17.8	28.8	19.2
200–500	49.6	13.9	49.7	14.6	51.9	15.2
> 500	79.2	5.9	82.7	6.6	85.5	6.5
Total	6.7	18.1	6.4	18.0	7.5	20.1

Sources: Derived from *Annual Agricultural Statistics*, various years.

size but large in output, and conversely the latter might be large in size but small in output.[32] One solution to this problem is to reduce all crops and animals into a standard factor form, and one way to do this is to employ standard labour requirements. Such an exercise produces table 3.7.[33]

There is at least one interesting feature of this exercise. One man with his own labour and with up to 365 days available, *theoretically*, could have farmed a holding of around 15 to 30 acres. This cannot be a precise estimate because at certain times of the year some farming activities competed with one another for labour time. If 15 to 30 acres defines a 'peasant' or self-sufficient family holding then it also points to the precarious position of holders of less than say 15 acres. Assuming there was not a large leisure preference, doubtless these small men had to supplement their incomes by working for others. It looks as though a holding of 150 acres required three full-time units of labour, and a holding of 350 acres about five units of labour. This can be viewed in another way. The larger holdings with their larger animal populations and relatively smaller emphasis on tillage could operate with quite low labour densities per acre (last column of table 3.7). This is the closest this particular analysis comes to looking at Irish agriculture and all of its *major* elements together, that is land use and the relative importance of tillage and grass, farm sizes and labour.

[32] See J.T. Coppock *An Agricultural Atlas of Scotland* (Edinburgh, 1976), p. 42.
[33] The crop and livestock labour input weights employed are listed in chapter 6, table 6.4.

Table 3.7 *Standard man days labour requirements by holding size group*

Size groups (acres)	1854		1861		1871	
	Per holding	Per 100 acres	Per holding	Per 100 acres	Per holding	Per 100 acres
1–5	79.2	2,263.7	79.0	2,257.3	81.8	2,272.7
5–15	170.8	1,674.9	172.8	1,661.3	177.6	1,691.2
15–30	305.5	1,376.3	306.2	1,373.1	306.5	1,374.5
30–50	470.6	1,162.1	472.0	1,171.2	468.5	1,159.7
50–100	694.7	932.5	700.2	955.2	690.8	939.9
100–200	1,055.3	684.8	1,052.2	702.4	1,034.7	688.9
200–500	1,516.6	424.0	1,505.2	442.2	1,459.4	426.7
> 500	2,024.5	150.6	2,058.3	165.4	1,989.7	150.7
Total	338.4	919.6	338.1	947.1	346.8	929.8

Sources: Derived from *Annual Agricultural Statistics*, various years.

Once the trauma of the Famine had passed the long-term changes were undramatic (see table 3.8). There may not be much confidence in the 1841 base, but by interposing the pre-Famine estimates for 1844/5 from Bourke and Mokyr, it looks as though the Famine did produce a large redistribution of holdings, especially those of 1 to 5 acres in size. This was in the manner of movements along, in this case up, the agricultural ladder, a sorting process which swelled the ranks of those with holdings greater than 30 acres. An important question arises from this conclusion. Was land vacated by the main casualties of the Famine, and was it mopped up by holdings in the higher landholding categories, not necessarily to be put to productive use, but simply to hold? Then within about five years, now with a much reduced 'peasant' sector, a pattern emerged which was to change only gradually over the next several decades. The compilers of the *Agricultural Output of Northern Ireland, 1925* stressed that change was rapid and great down to about 1853. By that year, however, in northern Ireland at least, the effect of the Famine and the major reorganisation of agriculture which followed it was worked out.[34]

Post-Famine regional considerations

These national patterns were not necessarily replicated at the provincial level. Bearing in mind the doubts over the 1841 base, nevertheless the

---

[34] *The Agricultural Output of Northern Ireland, 1925* [Cmd. 87], 1928, p. 43.

Table 3.8   *Distribution of landholding by size groups, 1841–1911* (percentages)

Size groups	1841	Pre-Famine		1847	1849	1850	1851
		(*a*)	(*b*)				
(i) Short term							
1–5 acres	44.9	23.1	23.6	19.0	15.8	15.5	15.5
5–10 acres		24.1					
5–15 acres	36.6		40.4	36.9	34.5	34.2	33.6
10–20 acres		23.7					
> 20 acres		29.1					
15–30 acres	11.5		35.9)	22.5	24.2	24.5	24.8
> 30 acres	7.0		)	21.5	25.3	25.7	26.1
Number > 1							
Acre (in 000s)	691	780	770	730	619	593	570

	1861	1871	1881	1891	1901	1911
(ii) Long term						
1-5 acres	15.0	13.7	12.7	12.3	12.2	12.0
5-15 acres	32.4	31.5	31.1	30.3	29.9	29.6
15-30 acres	24.8	25.5	25.8	25.9	26.0	26.3
> 30 acres	27.8	29.3	30.4	31.5	31.9	32.1
Number > 1						
Acre (in 000s)	568	544	527	517	516	521

Notes: (*a*) After Mokyr, recalculating his percentages and eliminating holdings (farms) of less than 1 acre. (*b*) After Bourke. 1841 figures from Bourke.
Sources: Derived from *Annual Agricultural Statistics*, various years; P.M.A. Bourke, 'The agricultural statistics of the 1841 Census of Ireland. A critical review', *Economic History Review*, 2nd series, 18 (1965), 380; J. Mokyr, *Why Ireland Starved: A Quantitative and Analytical History of the Irish Economy, 1800–1850* (London, 1985 edn), p. 19.

following observations can be made. In all of Ireland between 1841 and 1911 the numbers of landholdings of 1 to 5 acres decreased from 310,000 to 62,000, a decline of 80 per cent, but of this decline a little over 220,000 purportedly occurred in the Famine decade. The post post-Famine changes were modest by comparison, and fairly evenly spaced over time. Subsequently the decades of largest change were the 1860s and 1870s with reductions of 12.5 and 10.3 per cent respectively. There may have been accelerated changes in the two depressions of 1859–64 and 1879–82. Within those depressions, 1861 itself is an unfortunate datum to use in that apart from the years 1847–51 inclusive, it was the year when there were more landholdings of 1–5 acres than in any other year throughout the chronology. In its turn, 1881, in the midst of the Land War associated with the second major agricultural depression, there was a dramatic increase in landholdings of this group. These observations accentuate the effects of depression on landholdings above, not below 5 acres, on those landholdings which nudged up against that 15 acre margin of peasant survival. They were the ones to struggle for self-sufficiency, and they were exposed during crises.

In regional terms there were some interesting variations. From 1841 to 1911 the decline in the group holding 1–5 acres was graded from the less severe in the south and east (Leinster and Munster) to the more severe in the north and west (Ulster and Connacht). If the Famine decade is ignored and a new datum started in 1851, then there was only a small long-term change in Munster (10.3 per cent), and the changes in the remainder of Ireland took place without any further strong regional bias. After 1861 all of the provinces experienced a reasonably even-paced decline in numbers, though in Munster from 1881 to 1911 there were three successive inter-decadal increases.

There was a decrease in the number of landholdings of 5–15 acres from 253,000 in 1841 to 192,000 in 1851 and 154,000 in 1911 (or a decrease of 19.5 per cent from 1851 to 1911). Leaving aside the Famine decade itself, the greatest decrease was in the 1860s, during the first post-Famine depression. These holdings were at or close to the margin of subsistence, and they were particularly vulnerable during crises. At the broad regional level there was actually an increase in numbers from 1841 to 1911 in the west in Connacht (though a modest decrease of 5.4 per cent from 1851 to 1911). Indeed, in the Famine decade, and modestly thereafter for two more decades, there was an increase in the numbers in this group in Connacht. This was the beginnings of the sorting process along the landholding ladder and should perhaps be viewed more as the subsummation of the very smallest landholders in the aftermath of the Famine. It is more realistic to characterise it as an enlargement of holdings in a certain size group rather than a takeover, a more passive less active process. This is not to deny an unmeasurable degree of active aggrandisement. This process or gradation continued into the next size groups, the holdings of 15–30 acres, and then again into the group with over 30 acres. While it operated over all of Ireland, in Connacht it was particularly pronounced by beginning lower down the landholding structure. From the 1870s, even in Connacht, there was a decrease in the number of these holdings (5–15 acres), though more helpfully it is true to say that their numbers stabilised from the mid-1880s. In the other three provinces the decline in the number of these holdings was much the same and the loss was evenly spread over time.

An analysis of the larger landownership groups shows both a clear national pattern emerging – but with a movement up the agricultural ladder to larger holdings – and also some interesting, though broad, regional patterns. For example, from 1841 in all provinces except Munster there was an increase in the number of holdings in the two groups of 15–30 acres and those over 30 acres, and this was most pronounced in Ulster and Connacht. This is the mirror image of the decline in the number of holdings of 1–5 acres. This conclusion arises by employing the doubtful information from 1841 as the datum. Employing 1851 as a new datum shows that while there was a continued increase in numbers in Connacht in the group of holdings of 15–30 acres, in the other provinces there was a decrease. In these provinces the trend was in the direction of larger holdings.

There was a pronounced movement towards larger holdings in both Leinster and Munster (that is towards holdings of over 200 acres). Whereas this movement also occurred in the other two provinces, hence completing the national patterns observed so far, it was also the case that in the two western/north-western provinces there was an emphasis on farming broadly on farms of 5–15 and 15–30 acres, that is on family farms. It would be incorrect to attribute this too heavily to a trend over any great period because quite clearly the changes wrought during the Famine decade were of paramount importance. For example, in both Connacht and Ulster there was an increase in the number of holdings over 1 acre in the 1850s.[35] In one study Cormac Ó Gráda sees this as an example of seasonal migration acting as a potent means of 'consolidating peasant property in the post-Famine period', and in another he says it indicates that 'pre-Famine conditions persisted in the west', which 'remained to a considerable extent, a potato economy, dependence on the root being almost as great as in the pre-Famine years'. That is, it was a pre-Famine pattern of temporary or partial emigration which was not necessarily replaced by full emigration in the short term, or by the desertion and consolidation of holdings into large units.[36] In fact the average size of holdings in Connacht, on the whole declined, in the period 1851–61, and did not attain the average size of 1851 again until 1891. Explanations related to seasonal migration must be partly generated by the economic conditions of the receiving economy, in this case the seasonal migration required in Britain, which in turn meant that a change in those conditions, such as the late-nineteenth-century depression had a countervailing impact on the donating economy.

[35] For selected parts of Mayo see C. Ó Gráda, 'Demographic adjustment and seasonal migration in nineteenth-century Ireland', in Cullen and Furet (eds.), *Ireland and France*, p. 190.

[36] *Ibid.*, pp. 190–1; C. Ó Gráda, 'Seasonal migration and post-famine adjustment in the West of Ireland', *Studia Hibernica*, 13 (1973), 49. Following Ó Gráda's line for Mayo see Jordan, 'Land and politics: County Mayo, 1846–82', pp. 112–15.

Table 3.9 *Average size of holdings > 1 acre by province* (in acres)

	Average size	Average size cultivated land only	Ratio tillage to grass	Average size	Average size cultivated land only	Ratio tillage to grass
A Leinster				B Munster		
1841	36.2	29.4	1: 0.2	37.0	23.6	1: 0.6
1851	39.2	32.7	1: 1.2	49.2	35.7	1: 1.8
1861	41.2	34.6	1: 1.15	50.1	38.2	1: 2.1
1871	43.2	36.4	1: 1.5	51.6	40.7	1: 2.4
1881	45.1	37.5	1: 1.8	52.9	40.4	1: 2.6
1891	45.8	37.7	1: 2.1	53.2	39.8	1: 2.7
1901	46.2	38.3	1: 2.3	52.2	39.3	1: 2.8
1911	46.0	39.9	1: 3.7	51.3	44.5	1: 5.8
(1911 excluding mountain land Grazed)		37.9	1: 3.1		38.3	1: 4.9
C Connacht				D Ulster		
1841	28.2	14.3	1: 1.0	23.1	14.4	1: 0.6
1851	36.2	21.0	1: 2.3	25.2	19.0	1: 1.1
1861	33.7	22.1	1: 2.4	25.2	19.4	1: 1.1
1871	34.7	23.2	1: 2.8	27.1	20.8	1: 1.1
1881	35.3	23.0	1: 2.9	28.2	21.1	1: 1.2
1891	36.3	23.2	1: 3.2	28.8	21.6	1: 1.4
1901	36.3	23.2	1: 3.3	29.2	21.9	1: 1.5
1911	35.3	28.0	1: 7.5	29.2	25.2	1: 2.7
(1911 excluding Mountain land Grazed)		21.7	1: 5.6		20.2	1: 2.0
E Ireland						
1841	30.0	19.5	1: 0.5			
1851	35.6	25.9	1: 1.4			
1861	35.7	27.1	1: 1.6			
1871	37.3	28.8	1: 1.8			
1881	38.5	28.9	1: 1.9			
1891	39.2	29.2	1: 2.1			
1901	39.3	29.4	1: 2.3			
1911	38.9	33.0	1: 4.3			
(1911 excluding mountain land Grazed)		28.1	1: 3.5			

Note: In general see the text over the doubtful accuracy of the 1841 data.
Sources: Derived from the *Annual Agricultural Statistics*, various years.

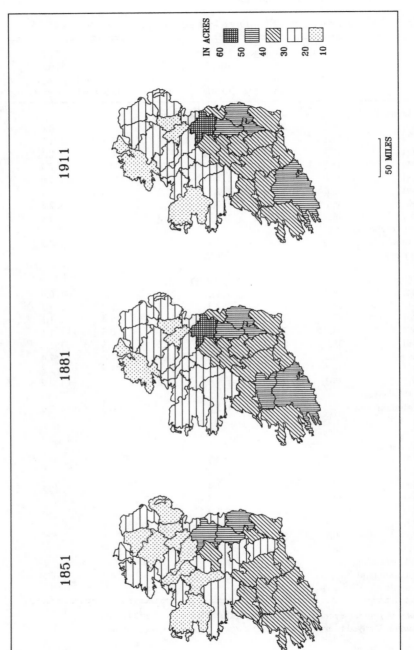

3.2 Average size of holdings, 1851–1911 (excluding holdings <1 acre)

It might seem obvious to seek an explanation from the changing crop to pasture ratio in order to explain the provincial differences in average holding sizes. In Ulster, with a *relative* abundance of small intensively farmed arable units, this seems to be the case in comparison with the larger extensive farms, whether arable or pasture, of Leinster and Munster. By the twentieth century, Connacht was both in small holdings and it was dominantly pastoral compared with other parts of Ireland. The relationship between size of holding and land use is by no means clear. However, in the case of Connacht the acreage under crops, including hay, was almost the same in 1911 (710 ,000) as it had been in 1851 (712,000). In Ulster, at least to begin with, the presence of much rural industry and the generally minimal impact of the Famine were factors which helped to slow down the trend towards larger average farm sizes.[37]

The trends identified so far are apparent when the average size of holdings greater than 1 acre are mapped (figure 3.2). Over time, there was an increase in average size, especially in the south and south-east of the country, in Munster and Leinster, with larger than average holdings always in the counties surrounding Dublin and in Cork. This pattern and the trend which underlined it was virtually complete by 1881. The counties of west Leinster, along the Munster border (Kings, Queens and Kilkenny), merge with the surrounding counties to form a large group amounting to about half the country in which the average size of holding by 1881 was greater than 30 acres. The exceptions were the counties of Longford and Louth. From 1851 to 1911 continuously they each had an average size of holdings of 20–30 acres.

In terms of the broad groups displayed on the maps, only three counties did not remain in their 1851 groups or in fact rise to a higher group. In Mayo the average size of holding increased from 1851 to 1881 but then decreased from 1881 to 1911; and in a different group, in Donegal and Limerick the average size increased from 1851 to 1881 and decreased from 1881–1911. In 1851 there was not a great deal of variety within each province except for Leinster. In Munster all counties were within the same average size group. By 1881 all of the counties of Connacht were much the same. The counties of Leinster, however, throughout the period, maintained some variety, from the generally smaller holdings of Louth and Longford in the outer ring of the province to the larger holdings, upwards of 50 acres, in an inner ring in those counties surrounding Dublin.

Holdings under 1 acre are excluded from figure 3.2. If they are

[37]   L. Kennedy, 'The rural economy, 1820–1914', in L. Kennedy and P. Ollerenshaw (eds.), *An Economic History of Ulster 1820–1939* (Manchester, 1985), p. 20.

included, however, then as well as having a predominance of large holdings in 1911 for example, the counties of Leinster and Munster also had a very large proportion of holdings of less than 1 acre. By 1911 the counties of Connacht and Ulster were dominated by holdings of between 1 and 30 acres. In 1911 over 70 per cent of all holdings in Connacht were under 30 acres in extent. Connacht is the region to which the Congested Districts legislation of the 1890s was directed. While the average size of holdings was not necessarily much smaller than elsewhere in Ireland, there was a larger than average, or congested concentration of smaller holdings which on their own offered an insecure economic living.[38]

Conclusion

At the national level, purely in terms of numbers, there was a rationalisation of holdings producing a modest decline in the numbers of all holdings greater than 1 acre from 1851 to 1911 of less than 10 per cent. This decline was greatest in the 1860s and 1870s corresponding to a large decline in the group holding 1–5 acres and more modest declines in the group holding 5–15 acres. No doubt a subtle relationship existed between these movements and the depressions of 1859–64 and 1879–82. It might be helpful to view these trends as a form of socio-economic process of filtering up and down the agricultural ladder. In addition, a large number of these smaller owners were displaced down the ladder into holdings less than 1 acre. Such conclusions highlight the greater redistributive effects at the lower landholding level when the personalities involved are counted rather than weighed. Over the period 1851–1911 the numbers of all landholders stayed remarkably stable. Numbers stood at 608,000 in 1851; though this choice of date hides the dramatic fall from 803,000 in 1847 to 585,000 in 1853 (a fall of 27 per cent). Total numbers recovered to 610,000 by 1861, followed by a steady fall of nearly 8 per cent to 563,000 in 1887 and an equal, and equally steady rise to 608,000 in 1911. In the first decade or so of the twentieth century the trends were influenced by the changes in tenure towards greater owner-farming, especially from 1902.

In 1851 the country divided roughly into two; the counties of Connacht and Ulster formed one group, and those of Leinster and Munster another. The dividing line ran roughly east north-east to west–south-west from Dundalk Bay to Galway Bay. Over 70 per cent of the holdings in all of the counties of Connacht and Ulster, and including Louth and Longford, were under 30 acres in extent. Although as many as 15 per cent of

---

[38]   E. Barker, *Ireland in the Last Fifty Years (1866–1916)* (London, 1916), pp. 116–20.

all holdings in Leinster were under 1 acre, generally speaking and in numerical terms, the counties of Munster and Leinster were dominated by holdings of over 30, or even of over 100 acres. The only counties of Ulster and Connacht with many of holdings of over 30 acres were Donegal and Antrim.

By 1881 the movement towards larger holdings was evident. The opposite movement, which swelled the ranks of the smallest holdings as a percentage of all holdings, was also under way, and naturally it was the groups in between with holdings of from 1–30 acres which provided the pool of supply for greatest outward movement. Holdings up to a threshold of about 30 acres could be served by the labour of the individual household. The more the average size fell below 30 acres the more it can be deduced that there was increasingly not enough work to occupy more than one person, and anything above 30 acres probably required the employment of additional labour. It might be instructive to relate these broad trends to changing settlement patterns.[39]

By 1911 the numbers of the very smallest holdings, those under 1 acre, were thick on the ground in all counties south and south-east of a line from Limerick-Belfast-Coleraine. From the yearly movement in holdings of this size this trend seems to date from about 1890. In 1880 there were about 51,000 holdings of this size, by 1890, but with an intervening decline, there were still 51,000, and by 1895 there were nearly 60,000 rising steadily to 87,000 by 1911. This national trend therefore was concentrated in Munster, Leinster and eastern Ulster. To this extent, in terms of the percentage of holdings, in some counties of the south there was a discernible decline in the emphasis on the larger holdings. There is a clear connection here with the emergence of owner-farming on a large scale, especially from the early twentieth century.

For holders of 30–100 acres, the greatest increase in numbers was in Connacht, and to a less degree in Ulster, but every county showed an increase in numbers in this group. This was modest in the south-east corner of the country, but there, in Waterford, Wexford and to a lesser extent in Tipperary and Kilkenny, and also at the other end of the

[39] Mary Cawley's study of Roscommon 'clachan' settlements was not, however, related to landholding sizes. This seemed an important lost opportunity. Instead she reports 'that there was no *a priori* relationship between population decline and the demise of clachan settlement', not even during the height of population change in the 1840s. Furthermore, in the second half of the nineteenth century not only did clachan settlements survive with no discernible, or at least *constant* relationship with economic change – whether that change was induced by evictions, migration or depression – but also there was the emergence of new settlement clusters. The pattern of settlement clusters in Roscommon, therefore, was characterised as much by continuity as it was by change. M.E. Cawley, 'Aspects of continuity and change in nineteenth-century rural settlement patterns: findings from county Roscommon', *Studia Hibernica*, 22:3 (1982–3), 106–27.

country in Ulster, there were significant increases in the very largest holdings of over 100 acres.

In truth there may have been as many histories of landholding change as there were townships, parishes or other communities.[40] In their own ways all of the provinces showed signs of turbulent change, but if anything it was in the middle of the country, in the clutch of counties straggling west and north-west of Dublin, in Meath, Westmeath, Kildare, Leitrim, Sligo, and to a lesser extent Longford, Cavan, Louth, Mayo and Donegal, which showed the greatest stability in the long-term structure of landholding distribution. Donnelly's study of nineteenth-century County Cork poses an interesting speculation on this idea of stability. To what extent were the changes that have been identified less important than they might have been if Ireland had been a closed economy? The rate of emigration usually increased in times of agricultural distress, or even during times of moderate potato shortage, and this periodically created a fund of land for subsequent redistribution. In Cork at least, there were times when the flow of remittance funds, from exiled family members, from America in particular, propped up otherwise evictable holders of land or otherwise potential emigrants. ' "American money", as distinct from "passage money", made it possible for others in the west [that is west of county Cork] to maintain an otherwise hopeless position on uneconomic holdings and to persist in an irrational, if understandable, resistance to emigration.' A margin for redistribution, change and consolidation was thereby held in check in this the premier county for emigration, though the extent to which this was a situation which existed more widely will have to await the larger research project it deserves.[41]

Yet the specific experience of Cork or the west of Ireland in general, does not confuse a clear enough conclusion. The post-Famine changes in agricultural structure were essentially complete by the 1870s, a conclusion which is, of course, perfectly compatible with the agricultural product changes implied in the adjustments in land use and the changed agricultural emphasis outlined in chapter 2. If turning the agricultural economy upside down can be seen as a revolution, then the Famine provoked a challenge in Ireland which the agriculturalists accepted and through which they survived. The proof of that challenge and the strategy for survival may be seen in the trend of agricultural output which follows.

[40] Certainly Smyth's data from the Tipperary parish of Clogheen-Burncourt does not present a pattern of change which is at all familiar from the national trend from 1851 to 1900, though it does conform more closely with the provincial trend for Munster as a whole. Smyth, 'Landholding changes', 22.

[41] Based on Donnelly, *The Land and People of Cork*, pp. 227–32. On depopulation in general see also K. H. O'Rourke, *Agricultural change and rural depopulation: Ireland, 1845–1876*, Ph.D. Dissertation (Harvard University, 1989).

4   The product of the land: output

Introduction[1]

If 'the statistics of agricultural production' for a country like nineteenth-century Ireland are equal to 'an index of national prosperity', then for this country there is a better opportunity than for most to construct that index.[2] Agricultural historians of post-Famine Ireland are well provided with annual agricultural statistics which in their diversity are richer than British sources. While the English and Welsh June Returns began in 1866 and have become a widely used source, the Irish Returns preceded them by a clear twenty years.[3] For investigating output they provide acreages under crops, average crop yields, animal numbers which distinguish within animal groups those of different ages, and other 'vital' statistics. As the appendixes show, the bulk of these data are pretty well available annually from 1847 to the Great War. Armed with such riches it should be a simple task to combine these classes of data with the price material which is available. There are many prices to choose from, though many

---

[1]   A version of the new output series presented in table 4.2 in this chapter was first presented at the Workshop 'Agricultural Productivity and the European Economy in the Past' at the Rockefeller Study and Conference Center, Bellagio, 13–17 March 1989. My thanks to the participants at the workshop for their helpful comments. See M.E. Turner, 'Agricultural output and productivity in post-Famine Ireland', chapter 16 of B.M.A. Campbell and M. Overton (eds.), *Land, Labour, and Livestock: Historical Studies in European Agricultural Productivity* (Manchester, 1991), pp. 410–38 for the first version of the output estimates. The present series is a reconstruction and now supersedes all previous versions.

[2]   M.J. Bonn, 'The psychological aspect of land reform in Ireland', *Economic Journal*, 19 (1909), 377.

[3]   For example, the English returns have been heavily used in general by J.T. Coppock, *An Agricultural Atlas of England and Wales* (London, 1976), and specifically in output estimates by, amongst others, E.M. Ojala, *Agriculture and Economic Progress* (Oxford, 1952), especially pp. 191–217.

of them are Dublin based,[4] yet making the correct choice of so-called 'standard prices' is not easy.[5]

In the simple equation of per unit factor productivity based upon output divided by unit factor inputs (land, labour etc.), it has been precisely the absence of measures of output which encouraged historians to look for indirect measurements of productivity, such as yields per acre. The fulsome nature of the nineteenth-century Irish data allow a more certain, though not an absolutely certain, estimation of output to be made, and this is the concern of this chapter. It is from such measures or estimates of output that agricultural productivity changes may be identified, analysed and understood, and this will be part of the function of chapter 5.

More can be said: the Irish data are of such quality that an annual series can be contemplated and hence the subtle annual fluctuations examined.[6] Heretofore, estimators have concentrated on output estimates

[4]   See T. Barrington, 'Review of Irish agricultural prices', *Journal of the Statistical and Social Inquiry Society of Ireland*, 15 (1927), 249–80; R.M. Barrington, 'The prices of some agricultural produce and the cost of farm labour for the last fifty years', *Journal of the Statistical and Social Inquiry Society of Ireland*, 9 (1887), 137–53; Cowper Commission, *Report of the Royal Commission on Land Law (Ireland) Act, 1881, and the Purchase of Land (Ireland) Act, 1885, B(ritish) P(arliamentary) P(apers)* [C. 4969], vol. xxvi, 1887, pp. 960–7 for a summary of Dublin prices up to 1886. For official prices from the 1880s see, for example, *Agricultural Statistics, Ireland, 1907–8. Return of Prices, of Crops, Livestock and other Irish Agricultural Products, BPP* [Cd. 4437], vol. cxxi, 1908, pp. 24–5 for the period 1888–1907 and, *Agricultural Statistics, Ireland, 1915. Return of Prices of Crops, Livestock and other Irish Agricultural Products, BPP* [Cd. 8452], vol. xxxvi, 1917–18, pp. 14–15 for the overlapping period 1896–1915. On the use of the price data with the annual agricultural data see J.P. Huttman, 'Institutional Factors in the Development of Irish Agriculture, 1850–1915', Ph.D. Thesis, University of London, 1970, pp. 469–505; C. Ó Gráda, *Post-Famine Adjustment, Essays in Nineteenth Century Irish Economic History*, Ph.D. Dissertation, University of Columbia, 1973, especially pp. 128–43; Ó Gráda, 'Irish agricultural output before and after the Famine', *Journal of European Economic History*, 13 (1984), 149–65; Ó Gráda, *Ireland Before and After the Famine: Explorations in Economic History, 1800–1925* (Manchester, 1988), pp. 128–31, 149–50; B. Solow, *The Land Question and the Irish Economy, 1870–1903* (Cambridge, MA., 1971), pp. 170–3 and 213–17; W.E. Vaughan, 'Agricultural output, rents and wages in Ireland, 1850–1880', in L.M. Cullen and F. Furet (eds.), *Ireland and France 17th to 20th Centuries: Towards a Comparative Study of Rural History* (Paris, 1980), pp. 85–97; Vaughan, 'Potatoes and agricultural output', *Irish Economic and Social History*, 17 (1990), 79–92; M.E. Turner, 'Towards an agricultural prices index for Ireland 1850–1914', *The Economic and Social Review*, 18 (1987), 123–36; Turner 'Output and productivity in Irish agriculture from the Famine to the Great War', *Irish Economic and Social History*, 17 (1990), 62–78.

[5]   *Royal Commission on the Financial Relations between Great Britain and Ireland, BPP* [C. 7720 and C. 7721], vol. xxxvi, 1895, appendix VII, 'Statement by Dr Grimshaw with regard to the values of output of Irish agriculture for certain periods', p. 451, re 'standard prices'.

[6]   See the complete, but unpublished annual series in Huttman, *Institutional factors*, pp. 471–505; an annual output series by volume is given in graph form in H. Staehle, 'Statistical notes on the economic history of Irish agriculture, 1847–1913', *Journal of the*

from cross-sectional data on which they have based their long-term analyses. Therefore one reason for constructing annual estimates is to avoid the possible pitfalls that those other estimators unwittingly may have fallen into if they happen to have chosen what turned out to be unusual years as start or end points in those longitudinal studies. The problems encountered with studies based on cross-sectional data include changes in census enumeration practices, such as changes in the dates at which animals were enumerated in the 1850s, or the use of years with unusually high or low average product prices. Indeed any number of factors could single out specific years as somehow unusual within a trend. The construction of an annual series will not solve such problems but it will more readily identify that there is a problem, and what the nature of the problem is. One treatment or solution might be to take running averages rather than measurements in individual years. The Famine itself was the most far-reaching economic fluctuation in Irish history, but there were others in subsequent years and these can be more readily identified in an annual agricultural output series, and their significance mapped against the broad trends. They can also be mapped against other influences, one of which was the land tenure changes which occurred after 1870 and which themselves were not unconnected with the political moves towards Irish independence which followed.

Ideally the final output series should be capable of comparison with other UK series in order to net out the Irish component and leave the GB-only estimates of output. There are a number of possible ways to do this. One is to employ the UK series constructed by J.R. Bellerby, though to do so would imply a high level of confidence in his procedures relative to those employed in this chapter. A reading of Bellerby's published and unpublished estimates, and their methods of construction, indicates that this is a possibility. Confidence in such a comparison is enhanced when it is realised that the better quality Irish data encouraged Bellerby to use them to interpolate for equivalent missing data for the other countries of the UK.[7] In particular this was the case for crop yields. English crop yields equivalent to the Irish ones were not available until the 1880s.[8]

Statistical and Social Inquiry Society of Ireland, 18 (1950–1), p. 445; W.E. Vaughan's annual estimates for the period 1851–83 are reported in T.W. Moody, *Davitt and Irish Revolution 1846–82* (Oxford, 1981), p. 569.

[7] J.R. Bellerby, 'The distribution of farm income in the UK 1867–1938', reprinted and revised in W.E. Minchinton (ed), *Essays in Agrarian History*, II (Newton Abbot, 1968), 261–79, with the original manuscript estimates in Bellerby MSS, Rural History Centre, University of Reading, D/84/8/1–24, especially /17. See also M.E. Turner, 'Output and prices in UK agriculture, 1867–1914, and the Great Depression reconsidered', *Agricultural History Review*, 40 (1992), 38–51.

[8] For the period before the 1880s J.B. Lawes and J.H. Gilbert made estimates of wheat yields which have become the standard source for agricultural historians of the mid to

C.H. Feinstein's UK output series is another which might be employed to net out the Irish component and thus allow an Irish/British comparison to be made, though Feinstein himself was heavily influenced by Bellerby's estimates. A third method would be to employ E.M. Ojala's UK estimates from which it is possible to compare aspects of the Irish and British agricultural trends, though there are limitations in such an exercise related to the unusual grouping of years which Ojala employed.[9] In the end a comparison with Bellerby will be employed.

This chapter proceeds by reviewing some of the problems associated with estimating output; it continues by outlining the methods by which a new annual output series has been constructed. This series is presented in several ways: in nominal annual output terms; in real output or constant volume terms; in the light of the distributional contribution made by the main products in the series which will be related to Irish agricultural structural change; and finally in terms of short-run fluctuations, where the special problems of the depression of 1859–64 are paramount, and from which some obvious implications for the output series as a whole will be highlighted.

Estimation problems

With information available on total physical agricultural output and product prices it should be a short and simple step to obtain the value of that output for what was the main source of Irish income in the nineteenth century. It is short in terms of archival research, but as a number of commentators have discovered, it is not simple. The main problem lies in the definition of crop, livestock, and livestock product output, that is, the *final* output which theoretically reached the market place and which distinguished it from the *total* output. For example, a proportion of seed from a previous year's harvest was retained and not marketed because it was required to grow the current year's crop: in that sense a part of each crop did not realise its value until a succeeding year. Of this initial equation certain elements are known, such as the crop acreages and the yields per acre, but what was the seeding ratio? In 1908 wheat seed for the following year was reckoned at 5 per cent of the value of current output, oat seed at 13 per cent, barley seed at 9 per cent, and

late nineteenth century, 'Home produce, imports, consumption and price of wheat, over forty harvest years, 1852–3 to 1891–2', *Journal of the Royal Agricultural Society of England*, 3rd series, 4 (1893), 77–133.

9   A comparison with Ojala's UK estimates was made as a part of the first annual estimation of Irish output, this is now superseded by the comparisons which follow in chapter 5 below. Turner, 'Agricultural output', 422–8.

potato seed at a massive 22 per cent. In 1912 the equivalent figures were 6, 10, 8 and 14 per cent.[10]

After the seed which was retained for future use was deducted there were other calls on the output of the major crops before the producers were left with that proportion which they sent to market. Much of their total crop output, for example, was retained on the farm for consumption by the animals and the farmers' own households. In one estimate of 1908 it was reckoned that 60 per cent of the oats and 46 per cent of the potatoes were consumed by the animals, and in another estimate of 1912 it was reckoned that 60 and 28 per cent of the same crops were consumed by animals.[11] That which was consumed by the farmer must be included as final output because it would otherwise have been purchased at the market, but that proportion consumed by the animals realised its value when those animals were milked, slaughtered, sold for fattening overseas or in other ways disposed of. To this extent there will be a tendency in certain analyses to overstate the importance of livestock husbandry relative to tillage production. Certainly much of the livestock production was only possible because of the substantial input from tillage, in the case of potatoes this was especially as a principal feed for pigs. This might have a particular importance for the relative labour productivities of livestock and tillage production. In the absence of a system where the farmer wholly bought in the feed inputs to the livestock sector from another party, some caution should be exercised in viewing agricultural production as a simple choice between animals and crops. All of the time, and perhaps sometimes for most of the people involved, the so-called two sectors were complementary to one another.

Nevertheless, for crops alone, there is still the question of defining the final produce, that is, distinguishing which part was sent to market and which was consumed by the farmer and his household. Within the animal sector there are similar problems of estimating the final output: some animals and their products were consumed on the farm and should rightly form part of that output; conversely, some products were recycled on the farm and should not be counted. For example, a certain amount of milk production was retained for rearing calves and some was fed to pigs. Finally, a certain proportion of the animals which were enumerated at the annual census never realised their full potential value because of unexpected mortality.

Many historians have tackled these problems of estimating final

---

[10] *The Agricultural Output of Ireland 1908* (Department of Agriculture and Technical Instruction for Ireland, Dublin, 1912) p. 21; and *Food Production in Ireland, 1912–13* [Cd. 8158], vol., 1916, p. 70.

[11] *Ibid.*, pp. 21 and 70–1 respectively.

output.[12] They have used estimating procedures which are sometimes similar to each other but also sometimes they are quite different.[13] They were almost always obliged to use the same basic agricultural data, but they often used different prices, and had individual ideas on the coefficients then to apply to those data. Table 4.1 gives some idea of these differences by reporting the estimating coefficients employed by various scholars in conjunction with the 'official' estimates for appropriate years. The resulting final output estimates which are produced and which are capable of comparison because of the convergence of example years indicate at times a considerable comparability across the estimates, but also sometimes divergencies, especially over the output from crops.[14] Where it is possible to compare the individual crop and livestock product components some of those divergencies are extreme.

The official method of estimating output as set out in the *Agricultural Output of Ireland 1908* has been important at least as a starting point from which many subsequent estimators have proceeded. For example, Barbara Solow's estimates of agricultural output for 1876, 1881 and 1886 were based on the 'official' coefficients employed in 1908. The 1908 report was part of the calculation towards the contemporary *Census of Production* in fulfilment of the *Census of Production Act of 1906*.[15] These coefficients were not precisely stated, but they can be calculated from the

[12] In addition to the Irish studies itemised in n. 4 above we might note L. Drescher, 'The development of agricultural production in Great Britain and Ireland from the early nineteenth century', *The Manchester School*, 23 (1955), 153–75, with a comment and critique by T.W. Fletcher, 176–83; O.J. Beilby, 'Changes in agricultural production in England and Wales', *Journal of the Royal Agricultural Society of England*, 100 (1939), 62–73; Ojala, *Agriculture and Economic Progress*, especially pp. 191–217; Bellerby, 'The distribution of farm income', especially pp. 264, 276–7. For Canadian studies see F. Lewis and R.M. McInnis, 'The efficiency of the French-Canadian farmer in the nineteenth century', *Journal of Economic History*, 40 (1980), 497–514. especially 504–7, for the discussion on estimating procedures; and R.M. McInnis, 'Output and productivity in Canadian agriculture, 1870–71 to 1926–27', in S.L. Engerman and R.E. Gallman (eds.), *Long-Term Factors in American Economic Growth* (Chicago, 1986). For a United States study see W.B. Rothenberg, 'A price index for rural Massachusetts, 1750–1855', *Journal of Economic History*, 39 (1979), 975–1001, especially 989–1001.

[13] Discussed in Vaughan, Solow, Ó Gráda and Turner in n. 4 above. See also, P.M. Solar, 'Agricultural productivity and economic development in Ireland and Scotland in the early nineteenth century', in T.M. Devine and D. Dickson (eds.), *Ireland and Scotland 1600–1850: Parallels and Contrasts in Economic and Social Development* (Edinburgh, 1983), pp. 70–88, especially 71–3.

[14] The aggregate crop, livestock and total estimates were compared in Turner, 'Agricultural output', 435, but as far as my own estimates are concerned these have now been superseded.

[15] *The Agricultural Output of Ireland 1908, passim.* See for example Barrington, 'A review of prices', 259–60 as an early example. See also, Ó Gráda, *Post-Famine adjustment*, 139, and Vaughan compares his own estimates for the year 1908 with those produced by the official report. This comparison showed, for crops at least, a close approximation in results, 'Agricultural output', 95.

various deductions which were made from total output to arrive at final output. A subsequent 'official' estimate of agricultural output in 1912–13 also indicated the final output proportions which were employed, from which the coefficients can be calculated. One obvious problem in adopting such coefficients in the estimate of a long-run series is that the passage of time may have been accompanied by technical change and other adjustments which render the final output proportions of 1908, for example, inappropriate to the agricultural realities of the mid or late nineteenth century. In this respect therefore, appendix 5 will be less an explanation of technical change, agricultural adjustment and adaptation, and more a review of the coefficients employed by official estimators and historians alike, by way of justification of the choices made in the current new estimates.

One major source of difference in the estimates made by historians when apparently covering the same specific years or groups of years will be perfectly apparent from the tabulations in table 4.1. Another source of difference is the exclusion of certain products altogether. W.E. Vaughan, for example, did not include an assessment of the hay, straw and potato crops which formed final output. In the case of potatoes he was certainly not neglectful, but rather he explained that their disposal and ultimate value was too complicated to include.[16] In the case of all these products – hay, straw and potatoes – the greater part of total output was consumed by animals, though a large minority of the potatoes either went to the market or must be counted as final output because they constituted the farmers' own consumption. In addition, some of the hay and straw constituted final output since proportions of both were consumed by off-farm horses either as food or as bedding. In 1908 estimated sales of hay and straw were said to account for 5.9 per cent of total hay and straw output, and that proportion of the potato crop which was consumed by humans was said to account for 32 per cent of the total potato crop.[17] In earlier periods the human consumption of potatoes surely was considerably higher.[18] Large or small however, there is an enormous problem in valuing potatoes: few of those which were consumed in the home actually came via the market, and as Cormac Ó Gráda has indicated, the market price which historians use to value these home consumed potatoes may reflect the better quality ones which were brought to sale. In addition, they are a dense bulky commodity and a large proportion of their market

[16] Vaughan, 'Agricultural output', 94, though he did assess the total production of potatoes in Moody, *Davitt*, p. 570.
[17] These are the 1908 estimates which were subsequently used by Barrington, 'A review of prices', 259–60.
[18] For which see Ó Gráda, *Post-Famine adjustment*, 139, and 'Irish agricultural output', 162.

Table 4.1   *Comparison of coefficients in the estimation of final output* (in percentages unless otherwise stated)

Estimator	Wheat	Barley	Oats	Potatoes	Flax	Hay
Vaughan 1854, 1874	100[a]	90[a]	33[a]	[b]	100	
Ó Gráda 1854	90	80	60	50	100	[c]
Solar 1856/60	92	91	57	41		
Solow 1876, 1881, 1886 [d]						
Ag. Output 1908 [e]	92	71	27	32	100	6
Food Production 1912 [e]	94	84	27	59	100	3.5
Turner	92	92–70	60–28	45–32	100	[f]

	Wool lbs/ head	Milk gals/ cow	Butter cwt/ cow	Eggs per bird	Pork cwt/ pig	Mutton lbs/ sheep
Vaughan 1854, 1874	6		1	48	2	
Ó Gráda 1854	5–6	350[g]	[g]	60	1.2	60
Solar 1856–60	5	304	1	60	1.25	60
Grimshaw 1885	4.5			100[h]		
Ag. Output 1908		400		100		
Food Production 1912				86		
Huttman 1850–1914	6	350	[i]		1	112
Turner	5	350–400		50[j]		

Notes:
a   Less unspecified seeding rate
b   3 tons per farmer
c   2.4 tons per off-farm horse
d   Follows the coefficients used in the 1908 Census of Agricultural Output
e   In the cases of 1908 and 1912/13 these are not stated coefficients, but they are in reality the proportions so allocated to final output
f   See Appendix 4 below
g   3 gallons of milk per lb of butter. But see appendix 5 for a discussion of the deductions to be made for recycling a proportion to calves and pigs
h   For 25 per cent of the birds
i   2.5 gallons of milk per lb of butter
j   For 95 per cent of the birds

Sources:
The Agricultural Output of Ireland 1908 (in connection with the Census of Production Act, 1906) (Department of Agriculture and Technical Instruction for Ireland, Dublin, 1912), derived from data on pp. 14, 21; *Food Production in Ireland (1912–13)*, *BPP* [Cd. 8158], vol. v, 1914–16, derived from data on pp. 70–1; T. Grimshaw, *Royal Commission on the Financial Relations between Great Britain and Ireland*, *BPP* [C. 7720 and C. 7721], vol. xxxvi, 1895, appendix VII 'Statement by Dr Grimshaw with regard to the values of output of Irish agriculture for certain periods', p. 455; J.P. Huttman, 'Institutional factors in the development of Irish agriculture, 1850-1915', Ph.D. Thesis, University of London, 1970, pp. 376, 378; C. Ó Gráda, 'Irish agricultural output before and after the Famine', *Journal of*

European Economic History, 13 (1984), 159–64; P.M. Solar, Growth and Distribution in Irish Agriculture Before the Famine, Ph.D. Dissertation, Stanford, 1987, p. 359; B. Solow, The Land Question and the Irish Economy, 1870–1903 (Cambridge, MA, 1971), p. 171; W.E. Vaughan, 'Agricultural output, rents and wages in Ireland, 1850-1880', in L.M. Cullen and F. Furet (eds.), Ireland and France 17th to 20th Centuries: Towards a Comparative Study of Rural History (Paris, 1980), pp. 94–5; Michael Turner, appendix 5 below.

value derived from large transport costs. Using market prices therefore tends to exaggerate the value of potatoes consumed on farms.[19] Similar problems arise for those products like potatoes and hay whose prices were set locally, and were dependent on annual yield and produce. In contrast, the prices for cereals, flax and many animal products were governed more by external international circumstances than domestic variations.[20]

Yet since potatoes were such a central item in the agricultural economy, both as an input to animals and one at least of the human staple foods, an exaggeration of potato prices has important repercussions in the valuation of final agricultural output. To this extent W.E. Vaughan's criticisms of my earlier published estimates are well taken, and indeed apply in some instances to other products.[21] The fault in those earlier estimates was the very high base price which was used for potatoes, and which was then applied to a potato prices index. If the base is wrong it does great damage to the whole series. In the light of this important criticism the estimates presented in this chapter are revisions. However, R.C. Geary once remarked that while 'Only a small fraction of potatoes produced in Ireland are marketed, i.e. are priced' nevertheless 'the price is a faithful reflector of the total production of potatoes'. In other words it is unlikely that the use of exaggerated prices will distort the *trend* of potato production. By the same reasoning it should be possible to use the existing potato price index, as I do, though now with more care over the base price which is employed (see appendix 4 below).[22]

[19] Ó Gráda, 'Irish agricultural output', 161.
[20] Barrington, 'A review of prices', 258–9.
[21] Vaughan, 'Potatoes and agricultural output', 79–92.
[22] R.C. Geary, commenting on Staehle, 'Statistical notes', 465. Vaughan used exactly the same procedures with regard to prices as I originally adopted in his own original valuations of agricultural output, including an assessment of the potato output. W.E. Vaughan, 'A study of landlord and tenant relations between the Famine and the Land War, 1850–78', Ph.D. Thesis, University of Ireland, 1973, especially pp. 337–8, and in Moody, *Davitt*, p. 570. In his critique of my work he should surely have pointed this out. It is an easy and elementary mistake to make. We might both take comfort from Geary's view, though again I think Dr Vaughan has already realised this, 'if average prices exaggerate the value of production in one period, they will exaggerate it in other periods and comparisons are possible', Vaughan, 'A study of landlord and tenant relations', p. 338.

There is another problem in estimating output; this is the selection of years or periods for study by those historians who seek a long-term view but without establishing an annual series. The choice of years is crucial, and subsequent comparison of one estimate from one estimator with those from another where different years or periods have been employed is not as helpful as at first sight might seem to be the case. In his original work Ó Gráda isolated three periods for study, 1840–5, 1852–4, and 1869–71. He later estimated an output total for 1854 in a comparison with a pre-Famine datum, and in another study chose 1912 in comparison with later years. Peter Solar made an estimate based on the annual average for 1856–60; Barbara Solow made estimates for 1876, 1881 and 1886; and in one study W.E. Vaughan isolated 1854 and 1874 as poles in a long-term study.[23] With so many estimates for the mid-1850s it might seem reasonable to compare them all to identify and check on common coefficients. This might be hazardous however. The estimation is based on simple arithmetic, but it involves volatile elements – annual acreages and yields, both of which fluctuated, and prices which in their own turn were partly responsive to acres and yields. There were also external factors. Vaughan has demonstrated this. He estimated the value of tillage in 1851 at £7.3 million (excluding potatoes and hay). It fell to £6.9 million in 1852, but then leapt to £11.3 million in 1853, £11.9 million in 1854 and £12.9 million in 1855, before falling to £8.7 million in 1856.[24] In output terms this was the most volatile five or six years in the long-run period from 1851 to 1883; grain supply problems to Western Europe during the 1850s related to the disruptions caused by the Crimean War had important consequences for market prices.[25] Both Vaughan and Ó Gráda use the period 1852–4 or the year 1854 (the former as a base, the latter as a terminus) in comparisons with other periods. In some respects this is a sensible period to choose: it is far enough removed from the Famine for the repercussions of that disaster to have worked their way through the system; and it was also during the mid-1850s that the annual returns became more regularised. In other ways there were extraordinary internal and external factors which make these years untypical of those which surround them. There was the Crimean War,

[23] C.f. items by Ó Gráda, Solow and Vaughan in note 4 above and also Peter Solar, *Growth and Distribution in Irish Agriculture before the Famine*, Ph.D. Thesis, University of Stanford, 1987, p. 360.

[24] Vaughan's estimates as reported in Moody, *Davitt*, p. 569.

[25] See S.H. Cousens, 'Emigration and demographic change in Ireland', *Economic History Review*, 14 (1961), 277, who identifies the effect of the Crimean War on wheat prices. See also C.S. Orwin and E.H. Whetham, *History of British Agriculture 1846–1914* (Newton Abbot, 1971), p. 96 and the effect of a recurrence of potato blight on British supply prices in 1853 and 1854. In Ireland in these years potato prices approached those of the Famine years, for which see Barrington, 'A review of prices', 251.

but also during 1853–5 there were outstanding crop yields.[26] These high yields did not have the effect of reducing price, as they might in most circumstances. This was partly because of the wartime disruptions, and partly because of the related fact that it was the British rather than the Irish market which determined price. Even before the Famine, let alone afterwards, prices were determined in the larger British economy.[27] Thus there was a period of high prices, which combined with high yields inevitably produced high output values. Peter Solar's judicious selection of the years 1856–60 for specific study might produce a better base or terminal point for subsequent comparison.

Similar problems are encountered when valuing livestock and live-stock products. The problems have added complications because the question of weighting or parameterising the different products on this side of agriculture is subject to more guesswork. How large was a pig, or rather how much bacon did it produce, and did that vary over time; how large was a sheep and how big was the wool clip per head: how many head of cattle of different ages were sold and for what purpose; were they slaughtered and therefore equated with beef prices, and how big was the carcass; or were they sold on the hoof and therefore related to livestock prices; how much milk did a cow produce, how much was marketed, how much was fed to calves and how much to pigs? Then there are subsequent questions; the estimate of milk output must then be translated into a butter equivalent because it is long-run butter prices rather than milk prices which are available in greatest quantities, and besides, butter constituted most of the final output for milk. Arguably, taking a butter price distorts the estimates because it embraces a transformation of the original product, it adds value to it in a produc-tion process, and this is something which is not picked up for other agricultural products in this study when the outputs are more closely related to farms. This was a problem also confronted by Patrick O'Brien and Leandro Prados de la Escosura who chose to exclude

---

[26] In the whole chronology 1850–1914 the wheat and barley yields of 1853–5 were not surpassed until 1874, the oats yield of 1854 was not exceeded until 1891 and the average potato yields of 1853 and 1855 were never surpassed in the chronology, and only once equalled, in 1913. See Appendix 2 below. See also P.M.A. Bourke, 'The average yields of food crops in Ireland on the eve of the Great Famine', *Journal of the Department of Agriculture*, 66 (1969), 29. In turn these high yields combined with prevailing high prices produced bumper product values. In the four years 1852–5 Vaughan estimated that the total value of potatoes in £s million was 21.5, 27.7, 33.2, and 28.8. The next highest year was 1857 at £19.3 million. Vaughan in Moody, *Davitt*, p. 370. These calculations use exaggerated potato prices which Vaughan would no longer defend, but the relative values from year to year are probably of the right order. See note 22 above.

[27] See P.M. Solar, 'Harvest fluctuations in pre-Famine Ireland; Evidence from Belfast and Waterford newspapers', *Agricultural History Review*, 37 (1989), 157–8.

manufactured dairy products from their analysis of agricultural production in Europe from 1890 to 1980. Instead they remained as faithful as possible to farm gate prices and not to the values resulting from secondary production.[28] Thus, the greater complications and uncertainties surrounding estimates for the livestock side of the industry have been well demonstrated by historians, and also by nineteenth-century commentators, in England at least.[29]

New estimates of agricultural output

It is difficult to have precise rules about the weights or coefficients to apply to products in estimating agricultural output, but in spite of differences in emphasis between historians there seems to be a fair measure of agreement over the eventual size of final agricultural output in different periods.[30] What is missing is an annual assessment either of total output or of the components which make up that output.[31] Vaughan's telling truism that the methods (of calculation) 'are based in the end, on more or less arbitrary estimates of the value of the commodities sold or consumed by Irish farmers', is ever present, but not a deterrent to having a shot at making that annual estimate.[32]

Thus the empirical base of this, and much of the following chapter, is a series of new output estimates for the long-run period 1850–1914. The new estimates which result are given in table 4.2, and the coefficients employed for each product are discussed in

[28] P.K. O'Brien and Leandro Prados de la Escosura 'Agricultural productivity and European industrialisation, 1890–1980', *Economic History Review*, 2nd Series, 45 (1992), 514–36, especially 515–16.

[29] For example, on the frailty of milk yield estimates see D. Taylor, 'The English dairy industry, 1860–1930', *Economic History Review*, 2nd Series, 29 (1976), 585–601. On the same subject see also T.W. Fletcher, 'The economic development of agriculture in East Lancashire, 1870–1939', M.Sc. Leeds University, 1954, appendix III. For contemporary debate and estimates on milk yields, carcass sizes, and wool clips see P.G. Craigie, 'Statistics of agricultural production', *Journal of the Royal Statistical Society*, 46 (1883), 26–30; for a discussion of carcass weights see R.H. Rew, 'Production and consumption of meat and milk: second report', *Journal of the Royal Statistical Society*, 67 (1904), 374–78; and for a discussion of milk yields see Rew, 'Production and consumption of meat and milk: third report', in *ibid.*, 385–93.

[30] *C.f.*, note 14 above.

[31] Though see the partial breakdown of output for the period 1851–1883 by Vaughan in Moody, *Davitt*, pp. 369–70. See also Vaughan, 'A study of landlord and tenant relations', especially pp. 33–5 and 336–58. See also Huttman, 'Institutional factors', pp. 471–6, 483–505 for a very detailed product by product breakdown. This last reference is in an unpublished form but it does cover the period which this book reviews.

[32] Vaughan, 'Agricultural output', 95.

appendix 5.[33] The raw crop and livestock data in terms of acres, yields and numbers are taken from the annual agricultural statistics. In the first presentation of these output estimates these basic statistics were combined with the indexes of product prices given in Thomas Barrington. These indexes are easily converted back into proper prices by reference to a base price. Therein, however, lay a major problem which my first attempts at estimation did not address. This has been rectified in appendix 4.

In broad terms the contribution of tillage fell from £16.5 million per annum in 1850 to £6 or £7 million by 1900 (with a low of £5 million in 1897), before turning up to £8 million by the start of the Great War. This long-term trend in tillage output was disrupted early on, especially in the 1850s when output rose to £26 million in both 1854 and 1855. This period, above all others, was the most dramatic of the short-term fluctuations. The volatility of the years 1852–6 can be clearly seen in table 4.2. Notwithstanding the possibility of insecure estimation in the early years of the annual census, when the enumerators were honing their skills, there are nevertheless clear consequences in terms of supply and supply prices arising from the disruptions of the Crimean War.

The comparable trend in livestock and livestock product output rose from £11 million in 1850 to an initial peak of £33 to £36 million in the mid-1870s.[34] There was then a fall to £26 million in the mid-1880s, a rise to a plateau of £29 million in the mid-1890s, before a recovery to a level of output which was on or over £40 million by the outbreak of the Great War.[35] There was a steep rise through the 1850s, but unlike the equivalent rise in tillage output this can be viewed as part of a long-term trend and not as a short-term aberration. It can be viewed as a significant post-Famine adjustment in agricultural production out of tillage and into pastoral activities, an adjustment which came early in the chronology and which persisted up to the Great War. The body which eventually became responsible for administering the annual census enumeration, the

[33] The enumeration of milch cows as distinct from all cattle over 2 years of age was not made by the census enumerators until 1854, but over the period 1854–9 milch cows as a proportion of all cattle over 2 years of age was fairly constant. In those years it varied from a low of 66.7 per cent in 1857 to a high of 68.4 per cent in 1854, giving a mean over six years of 67.5 per cent. Therefore it is assumed that this proportion prevailed earlier in the decade.

[34] By my methods of estimation this includes zero output for cattle less than 2 years of age up to 1854. Underenumeration may have been a problem but also there was a period of replenishment of stocks in the wake of the Famine. This is not to say that there were not any cattle sales, but rather that the method of calculation reveals 'negative' sales.

[35] I have used store cattle prices throughout, but as an alternative I made the assumption that all cattle were fat cattle and applied beef prices to them using certain assumptions about carcass sizes. The final estimates for the contribution of cattle were not much different one way or the other.

Table 4.2 *The final value of Irish agricultural output, 1850–1914* (in £000s)

	Crops	Animals	Total		Crops	Animals	Total
1850	16,520	11,161	27,681	1890	6,578	28,822	35,401
1851	15,402	13,099	28,501	1891	9,094	28,800	37,894
1852	15,643	12,118	27,761	1892	7,286	28,399	35,685
1853	22,965	15,001	37,966	1893	8,089	29,795	37,884
1854	26,175	19,020	45,195	1894	7,058	29,041	36,099
1855	26,168	20,546	46,714	1895	7,516	29,258	36,774
1856	15,726	20,937	36,663	1896	5,590	27,931	33,521
1857	16,051	23,410	39,462	1897	4,937	29,185	34,122
1858	14,208	22,385	36,592	1898	7,186	28,981	36,168
1859	14,835	23,718	38,553	1899	6,253	30,379	36,631
1860	15,707	25,596	41,303	1900	5,970	31,125	37,096
1861	10,598	22,963	33,562	1901	7,693	32,400	40,093
1862	9,447	23,460	32,907	1902	7,005	32,484	39,489
1863	12,104	22,539	34,642	1903	6,239	36,368	42,607
1864	13,183	22,962	36,145	1904	6,655	33,159	39,814
1865	14,571	23,961	38,531	1905	7,288	34,427	41,715
1866	14,054	24,873	38,927	1906	6,652	34,871	41,524
1867	14,375	17,991	32,367	1907	6,958	35,989	42,947
1868	15,948	32,697	48,645	1908	7,468	35,252	42,721
1869	12,724	26,022	38,746	1909	7,238	38,760	45,997
1870	13,745	29,919	43,664	1910	7,114	40,035	47,148
1871	11,259	28,943	40,203	1911	8,305	40,527	48,832
1872	10,671	33,196	43,866	1912	7,666	39,457	47,122
1873	10,676	32,026	42,701	1913	8,730	41,090	49,819
1874	10,360	35,662	46,022	1914	8,046	42,391	50,437
1875	12,379	33,037	45,417				
1876	12,196	36,196	48,391				
1877	9,896	34,085	43,982				
1878	10,221	32,149	42,370				
1879	8,149	29,043	37,192				
1880	9,757	32,404	42,161				
1881	10,661	29,406	40,066				
1882	8,101	32,074	40,175				
1883	9,311	29,892	39,203				
1884	8,001	28,931	36,932				
1885	7,831	25,704	33,535				
1886	7,102	25,838	32,940				
1887	6,564	25,964	32,527				
1888	6,709	29,331	36,041				
1889	7,329	30,698	38,027				

Source: Derived from the methodology outlined in this chapter and Appendixes 4 and 5.

Department of Agriculture and Technical Instruction for Ireland, attributed this trend in its report of 1902 to a labour supply constraint directly
arising from the Famine and subsequently from the rate of emigration.
This explanation is explored further in chapter 6.[36]

The long-run output series can be summarised in terms of thirteen half
decades, as in table 4.3. In conjunction with figure 4.1 it shows that there
were three phases of roughly equal duration: a phase of generally rising
output down to the mid-1870s, held in check by the declining contribution from cash crops; a middle phase to the mid-1890s of downturn in
output; and finally a phase to the Great War of pronounced upturn in
output.

In Ireland, over the whole period, the contribution of off-farm crop
output fell by nearly 60 per cent, animal output increased by nearly 190
per cent giving an overall increase of nearly 45 per cent. If there are any
doubts about the completeness of the early census returns then a solution
might be a comparison of the annual average output in the half decade
1855–9 with the annual average output in 1910–14. Over this period there
was a fall of 54 per cent in crop output and increases of 83 per cent and
23 per cent for livestock and the combined final output.

In current price terms total agricultural output stood at a little over
£28 million in 1850; it rose to a peak of £45/6 million in the mid-1850s (a
result largely of the exceptional tillage values of the Crimean War
period); stabilised in the medium term around the mid to late £30
millions until the late 1860s; reached a peak in the mid to upper £40
millions in the middle 1870s; before falling to the end of the century to
the mid £30 million, and a final recovery to a new high of £50 million on
the eve of the Great War.

These new estimates compare rather well in rounded terms (crops and
livestock) with the estimates of others (table 4.4). And even if at times
one or other estimator has apparently overvalued the crops or the
livestock, the totals square up reasonably well, and just as importantly
there is no mistaking the basic balance between crops and livestock, or
the trend over time. The comparative figures in table 4.4 have been
adjusted as far as possible so that the same crops and livestock products
are included. In the case of the official estimates of 1908 and 1912–13 a
number of minor crops and animal products, and also some major ones
such as the horse trade, have been subtracted for comparability. These
are items which most modern estimators have excluded in direct terms
but have included by adding a small mark-up of something like 5 per
cent to cover their exclusion. As the footnote to table 4.4 shows, while 90

[36] *Agricultural Statistics of Ireland . . . . .1901, BPP* [Cd. 1170], vol. cxvi, 1902, p. viii.

Table 4.3   *Annual average final agricultural output, 1850–1854 to 1910–1914*

	Crops	Animals	Total
In £s million			
1850–4	19.3	14.1	33.4
1855–9	17.4	22.2	39.6
1860–4	12.2	23.5	35.7
1865–9	14.3	25.1	39.4
1870–4	11.3	31.9	43.3
1875–9	10.6	32.9	43.5
1880–4	9.2	30.5	39.7
1885–9	7.1	27.5	34.6
1890–4	7.6	29.0	36.6
1895–9	6.3	29.1	35.4
1900–4	6.7	33.1	39.8
1905–9	7.1	35.9	43.0
1910–14	8.0	40.7	48.7
Percentage changes			
1850–4 to 1870–4	−41.5	126.2	29.6
1870–4 to 1890–4	−32.7	−9.1	−15.5
1890–4 to 1910–14	5.3	40.3	33.1
1850–4 to 1910–14	−58.5	188.7	45.8
1855–9 to 1910–14	−54.0	83.3	23.0

Source: Table 4.2

per cent of the items covered in 1908 and 1912–13 are included in my own estimates, a mark-up of around 5 per cent may be an underestimate. No matter, because in the case of the present estimates a fixed mark-up over time will not adjust the trends identified at all. Although the various global estimates are close to one another, the discussion in appendix 4 reveals that the use of different product prices by different estimators uncovers some large differences in the contribution of those different products to final output. The encouraging comparisons in table 4.4 therefore, partly arise from arithmetical accident rather than certitude.

It is possible to engage in a certain amount of sensitivity analysis to see to what extent the application of different prices or different coefficient weights alter the final output profiles. Such a sensitivity analysis regarding prices is carried out in appendix 4. A glance at table 4.1 indicates that for the early part of the period, for the main crops, the only major difference in the estimates concerns oats, where Vaughan suggests that only one-third of the output from that crop was actually

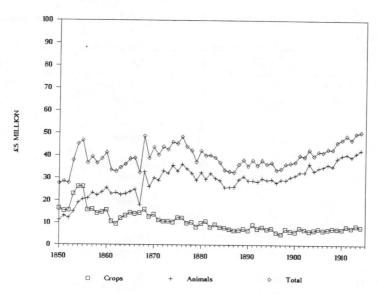

4.1   Final agricultural output in Ireland, 1850–1914

marketed, as compared with nearly two-thirds according to the judgement of others. For the later period the major anomaly relates to the high assessment for potatoes in the *Food Production* survey of 1912–13. It seems unreal to weight potatoes at a higher level than prevailed sixty years earlier. As for the animals, the major sensitivity in final output seems to attach to the milk yields. My choice of 350 gallons per cow stands uneasily besides Solar's 304 gallons per cow (a difference of 13 per cent). Conversely, it seems to agree with Ó Gráda's estimate of 350 gallons per cow, but his is a gross measure before a deduction for recycling a proportion as feed to pigs and calves, a deduction which is in line with the official estimates made in 1908. I still use 350 gallons because I have no separate estimate for the trade in buttermilk, which was a by-product of butter production which was partly fed to animals, but was principally fed to humans in mid-century. This omission may counterbalance the lower milk yield figure used by Solar.[37] Nevertheless, a reduction in my butter estimates of 10–15 per cent to bring me into line with other estimators does not do great damage to the monetary estimate of final output; it reduces the proportion of milk and butter in final

[37]   My thanks to Peter Solar for pointing out the importance of buttermilk.

Table 4.4 *Comparison of different estimates of final agricultural output* (in £s million)

	Crops	Livestock	Total	Source
1854[a]	20.8	24.2	45.0	Ó Gráda (1)
1854	26.2	19.0	45.2	Turner
1856–60[a]	15.0	25.4	40.4	Solar
1856–60	15.3	23.2	38.5	Turner
1869–71	15.5	29.1	44.6	Ó Gráda (2)
1869–71	12.6	28.3	40.9	Turner
1876[a]	11.9	35.2	47.1	Solow
1876	12.2	36.2	48.4	Turner
1881[a]	9.4	29.0	38.4	Solow
1881	10.7	29.4	40.1	Turner
1886[a]	6.1	24.2	30.3	Solow
1886	7.1	25.8	32.9	Turner
1908[b]	5.5	36.3	41.8	Agricultural output 1908
1908	7.4	35.3	42.7	Turner
1912[c]	10.9	41.2	52.1	Food production 1912
1912	7.7	39.4	47.1	Turner

Notes: [a] For comparability, unspecified 'other' items have been excluded. This usually meant a mark up of around 5 per cent.
[b] In the case of 1908 the total quoted here equals 89 per cent of the actual final output. This sub-total excludes horses, poultry and certain minor animal products, and minor crops, fruit and timber.
[c] In the case of 1912 the total quoted here equals 91 per cent of the actual final output. This sub-total excludes horses, poultry and certain minor animal products, and minor crops, fruit and timber.

Source: The Agricultural Output of Ireland 1908 (in connection with the Census of Production Act, 1906) (Department of Agriculture and Technical Instruction for Ireland, Dublin, 1912), p. 18; *Food Production in Ireland (1912–13)*, *BPP* [Cd. 8158], vol. v, 1914–16, pp. 70–1; C. Ó Gráda (1), *Ireland Before and After the Famine* (Manchester, 1988), p. 68; C. Ó Gráda (2), *Post-Famine Adjustment, Essays in Nineteenth Century Irish Economic History*, Ph.D. Dissertation, University of Columbia, 1973, p. 137; P.M. Solar, *Growth and Distribution in Irish Agriculture Before the Famine*, Ph.D. Dissertation, Stanford, 1987, p. 359; B. Solow, *The Land Question and the Irish Economy, 1870–1903* (Cambridge, MA, 1971), p. 171; Michael Turner, *This chapter.*

output by less than 2 per cent, and it reduces the overall proportion of animal products by 1 per cent at most; the profiles of the various products discussed in the next section hardly change at all, and nor does the relative distribution, broadly speaking, of the crops and the animals.

The distribution of output

The distribution of output over time between its different components throws up some interesting patterns. At the macro level the tillage sector represented nearly 60 per cent of final output in the early 1850s, but it slumped more or less progressively thereafter to a level of under 20 per cent by the late 1890s (with an upturn in the mid-1860s to about 40 per cent), at which level it remained up to the Great War. Thus the share of the livestock and livestock product sector rose from a little over 40 per cent in the early 1850s to over 80 per cent by the close of the century and on to the Great War. *If* the transition from a crop-based to a livestock-based agricultural economy is an indication of the farmers' economic pragmatism, especially at a time when the international climate and terms of trade had swung towards livestock products, then Ireland surely had few rivals. In about 1910 Irish livestock contributed 84 per cent to final output. In the UK as a whole it was something like 75 per cent, and in Germany 66 per cent, but in France it dipped below the half-share mark at 45 per cent, and in both Italy and Spain it was only 32 per cent.[38] More than in all of these other countries, the Irish economy was dominated by meat production, whether it was on the hoof or not. In terms of its profile Ireland looked more like Germany and the rest of the UK than like other European countries, but then the Irish industry was very much directed by events which took place on the other side of the Irish Sea.

At the micro level in Ireland, the cash crops of wheat, barley and flax declined respectively from 8–9, 4–5 and 5–6 per cent of output in the early 1850s, to 1–2 per cent each, more or less, from the 1880s onwards. The brief interruption in this trend occurred in the 1860s when the contribution of flax increased to 10–11 per cent of final output. This came about as a result of a twin revival in flax production: from one direction in the wake of the Indian Mutiny and the increased demand for British army uniforms; but importantly from another direction with the increased demand for flax during the cotton famine induced by the American Civil War. This revival amounted to an increase in the flax acreage from 92,000 to 302,000 acres from 1858 to 1864, though it was an increase which was almost entirely confined to Ulster.

[38] O'Brien and Prados de la Escosura, 'Agricultural productivity', p. 525.

Potato output was as high as 20 or 25 per cent of final output in the early 1850s, reducing to about 10 per cent or less from 1860 onwards. The lowest it ever reached was about 5 per cent in 1897. To underplay potato output is to remove a substantial component of final output, especially early in the series. The real issue is not that so little of the crop actually reached the market, for sale, but rather that it was an essential home-grown crop for home consumption.[39] Although a large proportion of it was fed to animals, a large residual was home consumed and must be counted as final output.

The contribution of oats fell from nearly 15 per cent of final output in the early 1850s to less than 5 per cent from 1895 onwards. Finally, although the acreage under hay rose to become second only to pasture in terms of land use, nevertheless by current estimation procedures the final output of hay on only one occasion was greater than 2 per cent of final output (in 1892).[40]

Within the livestock sector young cattle seemed to make a negative contribution to total output up to 1854 because cattle stocks which had suffered during the Famine were still in a state of replenishment. Thereafter, the contribution of cattle rose from over 20 per cent in 1860 to over 30 per cent by the late 1870s, at which level it remained up to the Great War.[41] The contribution from milk in the guise of an estimate of butter production accounted for 15 per cent in the early 1850s, rising to over 20 per cent from the late 1850s to 1880, declining to 17 or 18 per cent thereafter.[42] Therefore the two components of 'cattle' output together contributed close to 40 per cent of final output in the 1860s rising to

---

[39] Though see Vaughan's total value estimates for this crop in Moody, *Davitt*, p. 570, demonstrating just how large the crop was in money terms.

[40] Rye was always at or below one-half of 1 per cent, and straw was the same.

[41] Reaching nearly 40 per cent at the end of the 1880s and again at the end of the 1890s with a peak of 42 per cent in 1903.

[42] For more detail on the butter trade see Peter Solar, 'The Irish butter trade in the nineteenth century: new estimates and their implications', *Studia Hibernica*, 25 (1989–90), 134–61. This really supersedes a section of one of the otherwise classical works in the historiography, J. O'Donovan, *The Economic History of Live Stock in Ireland* (Cork, 1940), pp. 301–31. For competition in the British market from large supplies of butter initially from Holland and France, then especially from Denmark from about 1880, and Australia from 1890 see C Ó Gráda, 'The beginnings of the Irish creamery system, 1880–1914', *Economic History Review*, 2nd series, 30 (1977), 286; O'Donovan, *Live Stock in Ireland*, pp. 306–8; J.S. Donnelly, 'Cork Market: its role in the nineteenth-century Irish butter trade', *Studia Hibernica*, 11 (1971), 146–7; The value of Irish milk/butter output fell from £9 or £10 million in the mid-1870s to £6 or £7 million in the mid-1880s, recovering to £9 million or thereabouts by the Great War. Some of this trend will have been a price effect, but some will have been due to this competition within the British market. From the early 1870s to the late 1880s the average price of first-quality Cork butter fell by 27 per cent, J.S. Donnelly, *The Land and the People of Nineteenth-Century Cork: The Rural Economy and the Land Question* (Cork, 1975), pp. 153, 157.

nearly 50 per cent by the early 1870s, and remaining at or several points over that level down to the Great War. There was a peak of 59 per cent in 1903. The contribution of pigs did not fluctuate very much except in the short term, starting at a level of about 10 per cent in the 1850s and rising to about 12 per cent by the 1860s, and on to 14 or 15 per cent by the 1880s. Wool was always less than 2 per cent of output except during the period 1860–75 when in every year except 1868 and 1870, it rose to over 2 per cent.[43] It received a boost from the cotton famine arising from the disrupted cotton supplies during the American Civil War. The other component of the sheep industry, sheep and lambs, rose from less than 2 per cent of final output in the early 1850s to on or about 5 per cent from 1860, 7 per cent from 1870 to the early 1880s, reducing to less than 5 per cent by the Great War.

Finally, the output from eggs contributed less than 2 per cent of final output in the 1850s rising steadily as the poultry population rose to reach about 5–6 per cent in the 1880s and 1890s and leaping to 9 per cent from 1907 to 1914.[44] On the basis of official sources Joanna Bourke reports that Irish egg exports to Britain in 1910 and 1911 amounted to a value of £4 million per annum, and this was an increase of £1 million over the 1904 level of exports.[45] On the basis of current estimates it looks as though a massive 90 per cent of net egg output was exported.

This all leads rather naturally to the dramatic conclusion vividly brought to life by Ó Gráda and with which chapter 2 opened.[46] The general point about the shifting balance of agricultural output is well made both in terms of land use and now in terms of product values. In simplified terms under main headings, this shift is given in table 4.5. There was a clear, almost dramatic switch in agricultural concentration away from tillage and towards livestock output, which by the Great War produced an agricultural economy which was a livestock production specialist.

The 'real' volume of final agricultural output

The analysis so far has been presented in terms of current prices. There is some comparability between the trends of final output and the

---

[43] This was nearly 3 per cent in two years, 1864 and 1871, though as with flax this may have been due to the American Civil War and the associated famine of textiles' raw materials.

[44] This may be artificially high coincident with the great leap in poultry numbers in 1906–7. An adjustment of pre-1907 poultry output upwards, or post-1906 poultry output downwards will necessarily affect the proportions attributed to all other categories.

[45] J. Bourke, 'Women and poultry in Ireland, 1891–1914', *Irish Historical Studies*, 25 (1987), 308.

[46] Ó Gráda, 'Irish agricultural output', 154.

Table 4.5 *Distribution of final agricultural output, 1850–1914* (in percentages)

	Cash crops (1)	Oats	Potatoes	Cattle	Butter	Pigs	Sheep (2)	Eggs	Total crops (3)	Total live
1850–4	19.1	14.3	23.3	12.7	15.3	10.1	2.5	1.6	57.7	42.3
1855–9	15.8	10.9	15.4	16.1	21.6	11.6	5.4	1.9	43.4	56.6
1860–4	13.9	9.3	9.6	22.9	21.3	12.2	7.2	2.5	33.9	66.1
1865–9	16.4	9.2	10.1	16.8	22.6	13.4	8.0	2.5	36.8	63.2
1870–4	9.0	7.0	9.1	26.3	23.0	12.6	9.0	2.8	26.3	73.7
1875–9	8.7	6.8	7.5	29.2	21.3	13.3	8.4	3.5	24.2	75.8
1880–4	7.1	6.6	8.2	32.4	18.7	13.5	7.7	4.6	23.1	76.9
1885–9	5.9	5.6	7.9	34.2	18.0	14.8	6.9	5.5	20.6	79.4
1890–4	5.6	6.1	7.2	32.7	19.1	14.3	7.5	5.7	20.8	79.2
1895–9	3.9	4.6	7.7	36.6	18.6	13.7	7.1	6.4	17.7	82.3
1900–4	4.0	4.2	7.2	37.7	18.3	14.2	6.7	6.2	16.9	83.1
1905–9	3.9	3.9	7.3	36.0	18.8	13.7	6.4	8.4	16.6	83.4
1910–14	3.9	3.5	7.6	35.7	18.1	14.4	5.7	9.6	16.4	83.6

Notes: (1) Wheat, barley, rye and flax (2) Includes wool (3) Includes hay and straw

Source: Derived from the methodology outlined in Appendixes 4 and 5, and table 4.2.

general course of late nineteenth-century prices. The international prices peak of the early to mid-1870s is more or less mirrored in animal and total final output terms, and the prices depression which followed down to the mid-1890s again follows the ambient trend. There was a progressive recovery of prices thereafter.[47] This is not a perfect mirror of British agricultural prices, but then price is only one variable in the output equation. What might be useful now is to compare this computed trend of output values with the general price history of the age. One way which has been employed by other historians is to deflate the trend of output, or simply compare the trend of output with a wholesale prices index. The usual one which has been employed has been the Statist–Sauerbeck price index.[48] In a crude way this procedure

[47] Though with a greater absolute fall in the mid-1880s.
[48] As used by Barrington in a comparison with agricultural prices, 'A review of prices', 251–3; Vaughan in a section on living standards, 'Agricultural output', 85; Turner, 'Agricultural output', 426–7. See also Huttman, 'Institutional factors', 370–1. K.T. Hoppen, *Ireland Since 1800: Conflict and Conformity* (London, 1989), pp. 93, 100 said he used the index 'because with all its faults, it probably remains the most useful and has been widely accepted as such'. This is not the same as saying that it is correct to use it.

might define a terms of trade index between the agricultural sector and all other production sectors. Yet even if the agricultural component is netted out of the Sauerbeck index, and the question of whether others have done this is by no means certain, the remaining components of the index are based in the main on their unweighted contributions.[49] In addition, the kinds of products other than agriculture which make up the Sauerbeck and against which the agricultural component might be compared are not necessarily appropriate to a relatively non-industrialised, agriculturally dominant economy such as Ireland. Indeed, such a deflating exercise produces an unrealistically optimistic impression of the long-term trend of Irish agricultural output.[50]

The correct way to investigate real output is by looking at the agricultural industry from within, so to speak. The value of final output can be compared with the price trend of its component agricultural products. This could be achieved by constructing a composite agricultural product price index as the deflator.[51] However, since we have available a full range of the approriate data, the yields of crops, the crop acreages, the numbers of animals and estimates of milk yields and wool fleeces, and so on, as well as the comparable product prices, we can proceed directly to an estimate of constant agricultural output using the standard Laspeyres quantity index of the form.[52]

$$\frac{\sum P_0 Q_1}{\sum P_0 Q_0} \quad \text{------} \quad \frac{\sum P_0 Q_n}{\sum P_0 Q_0}$$

in which P_0 is based on the 5 year average of product prices from 1848 to 1852, and Q_0 to Q_n are the annual individual agricultural product outputs from 1850 to 1914. In 1850 Q_0 and Q_1 are the same. The results of this exercise are given in table 4.6 and figure 4.2. While taking a 5 year average of product prices for the base in this calculation might counter the problem of using a set of product base prices which were untypical of their broad period, it was still the case that the country was once

[49] An improved version of the Sauerbeck has been calculated, J.T. Klovland, 'Zooming in on Sauerbeck: monthly wholesale prices in Britain 1845–1890', *Explorations in Economic History*, 30 (1993), 195–228.

[50] For which see Turner, 'Agricultural output', 426–7. My thanks to Professor Charles Feinstein for convincing me that my continued use of the Sauerbeck was not only wrong but would likely confuse others in their appreciation of my analysis.

[51] See also Rothenberg, 'A price index', 985–6.

[52] As for example in R.G.D. Allen, *Statistics for Economists* (London, 1972 reprint), p. 105. My thanks to Professor Charles Feinstein for his advice over the construction of this index. My thanks also to my colleagues Steve Trotter and Gerry Makepeace.

dominated by arable production, but then switched to livestock production. Such a structural change may not properly be captured in the Laspeyres quantity index which uses as its denominator the value sum of outputs in the base period. The structural changes should be accounted for in the changing composition of the quantity enumerator. However, the structural change was quite radical. A standard method of adjusting the bases to accommodate this problem is to use a 'chained' Laspeyres index.[53] This takes the form:

$$\frac{\sum P_0 Q_1}{\sum P_0 Q_0} * \frac{\sum P_1 Q_2}{\sum P_1 Q_1} * \frac{\sum P_2 Q_3}{\sum P_2 Q_2} \cdots * \frac{\sum P_{n-1} Q_n}{\sum P_{n-1} Q_{n-1}}$$

Where P and $Q_{1,2,3...n}$ represent different points along the trend of output which signal the appropriateness of constructing a new base for the Laspeyres index. In some indexes, such as those associated with retail prices where different commodities enter or leave the basket of goods, it is not unusual to rebase the index annually. The Irish material is rich enough to undertake a similar procedure here. Thus, all of the P_0 to P_{n-1} are annual prices based on 5 year averages, and all the Q_0 to Q_{n-1} are the annual agricultural product outputs. The resulting index is also reported in table 4.6 and figure 4.2. It is different, but not significantly different from the same index derived from a constant base.

Evidently, the real volume of agricultural output did not improve over the full course of the period. There was in fact a decline in that real volume down to the early 1860s, a rise to about 1870, and a further decline to about 1880. To this extent the recently developed and new popular view of a long boom in the rural economy in the three decades after the Famine is in need of qualification.[54] Such views have slipped easily into those modern texts which report only the nominal value of output.[55] This new orthodoxy portrays a tide of prosperity accruing to Irish agriculture, especially in the first two to three decades after the Famine (an extended discussion of this issue is reported in chapter 7). Liam Kennedy captures this view when he states that 'the value of Irish agricultural output rose by over 40 per

[53] See for example Allen, *Statistics*, p. 109, though by common consent there is a printing error in the formula which is itemised. See also W.R. Crowe, *Index Numbers: Theory and Applications* (London, 1965), p. 186; D.A. Leabo, *Basic Statistics* (5th edn, Homewood, IL, 1976) pp. 365–9; F.G. Forsyth and R.F. Fowler, 'The theory and practice of chain price index numbers', *Journal of the Royal Statistical Society, Series A*, 144 (1981), 224–46.

[54] See the bald statement of this in J.J. Lee, 'Patterns of rural unrest in nineteenth-century Ireland: a preliminary survey', in Cullen and Furet (eds.), *Ireland and France*, 230.

[55] M.J. Winstanley, *Ireland and the Land Question 1800–1922* (London, 1984), p. 9.

Table 4.6 *The constant volume of Irish agricultural output, 1850–1914*

	Standard Laspeyres quantity index	'Chained' Laspeyres quantity index		Standard Laspeyres quantity index	'Chained' Laspeyres quantity index
1850	100.0	100.0	1884	99.2	101.8
1851	109.4	109.4	1885	99.6	101.2
1852	106.9	106.8	1886	102.0	105.8
1853	117.4	116.8	1887	103.5	105.3
1854	119.2	119.8	1888	99.6	103.3
1855	125.0	123.6	1889	101.3	104.1
1856	107.7	108.1	1890	95.8	99.8
1857	107.2	109.0	1891	100.7	102.7
1858	117.2	116.9	1892	97.0	100.7
1859	109.9	110.5	1893	104.5	108.2
1860	107.8	111.0	1894	102.5	108.0
1861	91.6	95.5	1895	106.2	109.2
1862	94.6	98.2	1896	100.3	104.5
1863	103.8	104.3	1897	93.5	99.8
1864	106.1	104.8	1898	101.5	105.9
1865	99.6	98.5	1899	103.0	107.5
1866	96.8	98.2	1900	96.2	101.4
1867	89.9	91.2	1901	105.1	108.4
1868	122.7	119.2	1902	103.9	108.2
1869	102.1	101.6	1903	106.7	114.0
1870	114.7	112.6	1904	103.3	109.0
1871	100.6	101.0	1905	106.4	111.1
1872	97.5	100.4	1906	103.7	109.5
1873	101.0	102.6	1907	102.3	109.2
1874	105.9	105.7	1908	103.7	108.1
1875	110.4	109.7	1909	109.5	115.6
1876	115.9	114.0	1910	106.1	112.5
1877	100.2	104.5	1911	111.6	116.9
1878	102.4	104.8	1912	104.4	110.2
1879	89.6	94.9	1913	108.5	112.7
1880	104.4	107.5	1914	107.9	112.4
1881	102.6	103.2			
1882	98.6	103.0			
1883	101.1	102.0			

Note: In the case of the standard Laspeyres quantity index the base year in 1850 uses a 5 year average of product prices for 1848–52. In the case of the 'chained' version it uses 5 year average prices throughout for each year in the chain.

Source: See text.

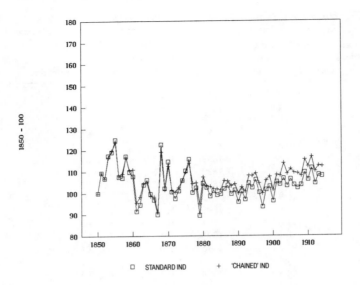

4.2   Irish agricultural output, 1850–1914, constant volume index

cent between 1851/55 and 1871/75'.[56] While the value of that output
rose it did so against a backcloth of favourably rising prices, but on
the whole the constant volume of Irish agricultural output barely
improved at all except in short bursts. The downturn in real output in
1879 should be noted. In one of the real volume indexes this was the
lowest point reached in the whole period, a low point also replicated in
Britain.[57] The coincidence here with J.S. Donnelly's theory of 'rising
expectations' which took a severe jolt in the late 1870s is appealing,
though those rising expectations from the mid-1860s look less impressive in trend terms with one or two outstanding years at odd
intervals (see discussion in chapter 7). Nevertheless, 1879 was an
outstandingly poor year in volume terms and perhaps tenants were
particularly mindful of the poor yields which were readily observable in
that year, against a backcloth of relatively fixed land (rent) and labour
(wages) costs. After about 1880 there was almost certainly a levelling-
off in real output.

[56]   L. Kennedy, 'The rural economy, 1820–1914', in L. Kennedy and P. Ollerenshaw, *An
Economic History of Ulster 1820–1940* (Manchester, 1985), p. 41.
[57]   See Turner, 'Output and prices', especially 48–50.

Output and the agricultural depression of 1859–1864

Special mention should be made of the agricultural depression of 1859–64, not least of all because it is sometimes seen as a crossroads in the structure of post-Famine Irish agriculture. It shows up in the trend of livestock output, but not particularly dramatically, though the value of the livestock sector during this depression has surely been exaggerated.[58] There must have been an incalculable loss of pigs probably through a higher than average rate of mortality arising from a relative fodder famine which must have accompanied the depression, and this was probably also the case with other animals. In general the animals may have been slaughtered earlier than usual at lower carcass weights than have been employed in estimation. Milk yields were probably less than the otherwise contemporary average. Ó Gráda says that 'Dairy farmers particularly were in a bad way', with yields of butter per cow at two-thirds to one-half of the normal average.[59] The current estimates are incapable of capturing this without making assumptions about milk yields during identifiable fodder crisis years. In addition, the repeal of a customs duty on foreign butter in 1859, which had given a measure of protection to Ireland, resulted in a large increase in foreign imports into Britain which had the effect of lowering prices.[60] For Irish producers, therefore, there were diminished milk yields and also diminished prices, and therefore greatly diminished output. These effects cannot be, or have not all been captured properly. Similar problems apply to the later depression of 1879–82.[61] In addition, in the 1860s the output from wool and flax was up on the trend of the previous decade as a result of the disruption of cotton supplies to Britain during the American Civil War.

The derived demand for livestock products came principally from Britain; 58 per cent of the value of final livestock output came from exports by 1908.[62] Prices for animal products, therefore, were not greatly affected by such local crises as the depression of 1859–64. However, if

[58] See chapter 2 above and J.S. Donnelly, 'The Irish agricultural depression of 1859–64', *Irish Economic and Social History*, 3 (1976), 33–54.

[59] C. Ó Gráda, *The Great Irish Famine* (London, 1989), p. 68; Donnelly, 'Irish agricultural depression', 42–3.

[60] Donnelly, 'Irish agricultural depression', 44. See also Donnelly, 'Cork Market', 154–7. See this trend in Solar, 'The Irish butter trade', 156, 159 where the high point of Irish butter exports between 1850 and 1914 was 1853 but followed closely by 1859, and then 1860 began the massive increase in UK imports.

[61] Donnelly, *The Land and People of Cork*, pp. 150–3.

[62] *The Agricultural Output of Ireland 1908*, p. 6; export figures are reported in *Agricultural Statistics of Ireland ... ..1891, BPP* [C. 6777], vol. lxxxviii, 1892, p. 23. See also R. Perren, *The Meat Trade in Britain 1840–1914* (London, 1978), pp. 96–7 for 1865–89 and Huttman, 'Institutional factors', pp. 538–40 for 1850–1915. See also a digest of export figures in chapter 2 above, table 2.3.

product prices were relatively unaffected, fodder supplies certainly were affected: there was a fodder famine, and this had an important effect on the value of output.[63] It meant that to maintain that output, to pay rents, and to offset short supplies of fodder, the farmers, relatively speaking, flooded the market with animals. Cattle over 2 years of age 'disappeared' in large numbers; 500,000 in 1855 rising to 580,000 in 1858, 550,000 in 1859 – the beginning of the crisis – and 712,000 in 1860, 634,000 in 1861, 725,000 in 1862 and 630,000 in 1863. This depleted the animal stocks and necessarily led to a reduction in disappearances in subsequent years; to 552,000 in 1864, 477,000 in 1865, 459,000 in 1866 and only 207,000 in 1867.

It is difficult at this stage to distinguish between sales of stock and excessive mortality during these years. Since the estimation of output employs fixed mortality levels over time it almost certainly results in an underestimation of the severity of the depression and hides the fact that there must have been a downturn in livestock output in the first half of the 1860s.[64] In addition, in the face of a fodder famine, it is likely that a smaller proportion of the tillage output reached the market. Therefore the value of tillage output from the late 1850s to the early 1860s was probably smaller than has been estimated. In actual fact, therefore, the total value of agricultural output was probably lower than has been demonstrated. Conversely, the post-depression recovery was probably much more impressive and hence the sense of rising expectations into the 1870s equally more impressive.

The downturn in livestock 'disappearances' is nicely represented in the livestock output figures even though there is the suspicion of under-estimation. It was as if the depression and its aftermath finally turned the country into a livestock producer. In 1867 there was a sharp decline in livestock output (and therefore agricultural output), but in the following year 1.2 million cattle over 2 years of age 'disappeared', followed by 626,000 in 1869, and thereafter a fairly steady rise up to the 1 million mark by the Great War.

The fluctuations in sheep numbers mirrored the cattle trends in and around the 1859–64 depression, though on a less dramatic scale. Yet an appraisal of the numbers of livestock disguises the important repercussions on the likely quality of the sheep flocks during the depression. Adverse weather conditions damaged fodder crops, but they also encouraged the snails which were responsible for the parasitic liver-fluke in sheep, and the induction of sheep rot.[65] The general deterioration in the

---

[63] Donnelly, 'Irish agricultural depression', 37.
[64] Ibid., 33, 37–8.
[65] Ibid., 39–40.

quality of livestock therefore goes beyond mortality, but cannot readily be quantified. Nevertheless, the coincidence of the depression and the cotton famine spilling over from the American Civil War made wool prices rise by over 50 per cent from 1859 to 1864, and the output of wool rose from £0.7 million to over £1 million over those years. A fixed wool clip size has been employed in these estimates, and yet with a suspicion of a reduction in animal quality, once again it is likely that the value of this product has been overestimated. Note also that although flax production was given a boost during the cotton famine, most of the addition to the flax acreage was to the benefit of Ulster producers. The bad effects of the depression were more or less national, the few good effects were localised.

There was an upturn in some prices during the depression, but these were mainly short-term movements in pork prices and a longer-term movement in sheep prices.[66] In the immediate wake of the depression sheep numbers recovered dramatically to reach an all-time peak of nearly 5 million in each of the three years 1867–9. Much the same can be said of pigs, the population of which turned down during the depression but recovered progressively after 1864 to record a peak of 1.4 million pigs of less than 1 year of age in 1871. Pig and sheep populations fluctuated dramatically in the short term, as is known from the downturn of pig numbers during the Famine, but recovery rates were rapid with either multiple litters and/or rapid animal maturity.[67] In contrast, cattle numbers fluctuated less wildly. Nevertheless they did turn down during the depression, but farmers were more reluctant to deplete their stocks because the replacement rate for cattle was low, and the animals took longer to reach maturity, unless they were offered to the market in the form of calves for fattening or for veal.[68]

The way the crisis affected market prices was probably greatest not at the sale end, but rather at the restocking end of the chain. Graziers were confronted with high prices for replacement stock from 1863 onwards, partly because of shortages in market supplies and partly because of slow in-farm replacement. Dublin lamb prices ranged from 20–28 shillings in 1859 but rose to over 20–30 shillings in each year from 1860 to 1862, with the upper range prices well in excess of 30 or even 40 shillings until

[66] It has been suggested that there was a close and sensitive relationship between sheep and pig numbers and their respective product prices, Staehle, 'Statistical notes', especially 448–54.

[67] For a local study of this for Down and Antrim see M.E. Turner, 'Livestock in the agrarian economy of counties Down and Antrim from 1803 to the Famine', *Irish Economic and Social History*, 11 (1984), 19–43.

[68] There was a trade in veal but the current estimates, as indeed those of others, ignore it because of lack of adequate data.

1867.[69] One repercussion of the high price of replacement stock was either a reduction in other investment, or indeed a spate of lettings appearing in the press.[70]

The dramatic fall in the value of cattle output in 1867 and the more dramatic recovery in 1868 partly reflected the smaller numbers which reached the market in the wake of the depression, and then the great surge once the depression had passed. The numbers of young cattle which 'disappeared' (cattle less than 1 year) also rose during the depression, from 4,000 in 1859 to 78,000 in 1860. This trend probably reflected the severity of the depression, the lack of fodder, the need to pay rents in the face of declining incomes, and so on. In subsequent years the sale of cattle less than 1 year in age fluctuated and did not settle down to much of a pattern until the mid-1870s. In 1864 these sales were as low as 12,000 animals, the third lowest year in the long-run period 1854–1914. Over the course of the depression the number of milch cows declined by 10 per cent, and this reduction was reflected in reduced butter sales which declined from £9.5 million in 1859 to £7.1 and £7.4 million in 1863 and 1864. At times milch cows had to be disposed of as if they were beef cattle, and in addition there was an immeasurable decline in average milk yields to reflect the fodder crisis. Finally, egg prices began their long upward trend during the depression, to reach ten times their 1840 price by 1918.

Conclusion

This chapter, and the discussions in appendixes 4 and 5 show that there are profound difficulties in estimating agricultural output, whether of selected years, of groups of years, or as this study shows, of a long-run annual series. These problems principally involve the weighting procedures employed in netting out mortality, seeding, and the recycling of individual products from the total output, thus to arrive at what is called final output. This is an estimate of the output which was traded in markets or consumed in farmers' households. Nevertheless, with the guidance of the official government inquiries of the early twentieth century, combined with the accumulated knowledge acquired by historians who have made estimates for isolated periods but often with local and more detailed price material than that which is available on a larger scale, it is possible to arrive at an annual estimate of Irish agricultural final output.

By comparison with other historians' estimates based on individual years or groups of years, the annual series bears up well. At the individual product level there remain some differences of opinion, but the

---

[69] Cowper Commission, *Royal Commission on Land Law*, pp. 962–3.
[70] Donnelly, 'Irish agricultural depression', 40–1.

long-run history of Irish agricultural change shines through. The country became greener, but then we discovered this easily in chapter 2. The reason for estimating output, therefore, is not to confirm a trend which was already understood, but rather to open up different questions related to performance. This chapter has shown that there are important questions regarding that performance. A nominal review of the value of output indicates a long-term rise and therefore *apparent* prosperity. This is true notwithstanding years and periods of depression, such as 1879, one of the worst single years, and 1859–64, one of the worst sustained periods of depression. A more detailed review of the long run indicates a cyclical pattern of output, but this does not hide the fact that the nominal value was higher, by some margin, at the end of the period than at the beginning.

In addition, the long-run series has identified a problem which not all historians may have appreciated, and this is the importance attached to choosing the correct periods or years in cross-sectional studies. This begs a question regarding 'correctness', correct for what? One favoured individual year, or favoured group of years has been variously located in the early to middle 1850s. This was a time remote enough from the Famine to suppose that the main repercussions from that traumatic event had passed, thus allowing a comparison with the pre-Famine economy. It was also far enough in advance of the next major structural change which affected the agricultural economy, the depression of the late 1870s and early 1880s, thus to allow a discussion of long-run post-Famine recovery. Yet the annual series shows that the early to mid-1850s was a time when agricultural output was at a peak, a peak which in some important respects was quite artificial. To an extent it was created by exogenous factors which caused high prices. An annual series is only partly a solution to such a problem, in that at least it shows it up, because that annual series is determined by relatively fixed attitudes – we give those attitudes technical terms like weights or coefficients – and these in turn cannot respond to environmental shocks such as occurred during the depression of 1859–64. Indeed there are a number of environmental and regional factors which may point to the futility of making the estimate in the first place. This is a national income accounting issue with a recognised methodology. It is related to the identification of trends and not always, or necessarily, to absolute truths. That trend of output seems well established, but when it is related to other trends, such as the general agricultural price movements, the real volume of Irish agricultural output seemed at best to remain level except in short sharp periods of a few years. Initially at least, this suggests some confusion in the performance of Irish agriculture, a confusion which will be further explored in chapters 5–7.

5   The performance of agriculture

Introduction

In chapter 4 the broad trends in Irish agricultural output were identified. The assessment of nominal value of that output when compared with the volume of output left a question mark over the general performance of Irish agriculture. This analysis is now taken several stages further: by putting the performance of Irish agriculture into a wider UK context; by investigating the productivity of the industry in its own right; by an alternative estimate of the physical volume of output; and in conclusion by widening the comparative lens to include some European comparisons.

Relative performance of Irish and British agriculture

The two main components of Irish agriculture in the nineteenth century formed part of agricultural output in the United Kingdom as a whole, and if Ireland and Irish agriculture can be viewed as a region of the UK it may be instructive to compare the performance within that region in the context of the whole nation. In 1956 J.R. Bellerby published estimates of UK agricultural output. His method of construction was not far removed from the current exercise for Ireland, though he did not have as complete a data set.[1] For example, official crop yields for England, Wales and Scotland were not available before 1884. On this issue, as on a number of others, he was informed partly by the Irish data. In the

---

[1]   J.R. Bellerby, 'The distribution of farm income in the UK 1867–1938', reprinted and revised in W.E. Minchinton (ed.), *Essays in Agrarian History*, II (Newton Abbot, 1968), 261–79, with the original manuscript estimates in Bellerby MSS, Rural History Centre, University of Reading, D/84/8/1–24, especially /17. See also M.E. Turner, 'Output and prices in UK Agriculture, 1867–1914, and the Great Depression reconsidered', *Agricultural History Review*, 40 (1992), 38–51. The estimates employed in comparisons in this chapter are taken from the Bellerby manuscripts, though as I illustrated in Turner, 'Output and prices' (especially p. 41), the difference between the original manuscript estimates and the final published ones is very small.

Table 5.1 *United Kingdom and Irish final agricultural output, 1867–1871 to 1909–1914*

	Irish crops	Irish live	Irish total	UK crops	UK live	UK total
Annual averages in £s million						
1867–71	11.5	27.1	38.6	89.5	89.5	179.0
1869–73	10.2	30.0	40.3	82.4	92.6	175.0
1874–8	9.5	34.2	43.7	83.3	103.4	186.7
1879–83	7.9	30.6	38.4	62.8	96.8	159.6
1884–8	6.3	27.2	33.4	63.6	90.0	153.6
1889–93	6.9	29.3	36.2	59.4	90.8	150.2
1894–8	5.9	28.9	34.8	53.7	91.0	144.8
1899–1903	6.1	32.6	38.7	52.9	105.7	158.5
1904–8	6.4	34.7	41.2	49.8	107.4	157.2
1909–14	7.2	40.4	47.6	48.2	116.7	164.9
	Irish crops as % of UK	Irish live as % of UK	Irish total as % of UK			
1867–71	12.8	30.3	21.6			
1869–73	12.4	32.4	23.0			
1874–8	11.3	33.1	23.4			
1879–83	12.5	31.6	24.1			
1884–8	9.9	30.2	21.8			
1889–93	11.6	32.3	24.1			
1894–8	11.0	31.7	24.0			
1899–1903	11.5	30.8	24.4			
1904–8	12.9	32.4	26.2			
1909–14	14.9	34.6	28.9			

Sources: Derived from table 4.2 but with the exclusion of flax output in the crop and total output estimates in order to compare directly with Bellerby's estimates, and J.R. Bellerby's UK estimates outlined in M.E. Turner, 'Output and prices in UK agriculture, 1867–1914, the Great Agricultural Depression reconsidered', *Agricultural History Review*, 40 (1992), 38–51.

absence of a regional breakdown of his results into the component countries of the UK it is impossible to check accurately the Irish component of his estimates with the new ones. Nevertheless, because the methods of construction were similar it is not an outrageous presumption to attempt to disaggregate the Irish component. Table 5.1 compares UK output (including Bellerby's built-in estimate for Ireland) with the new Irish estimates. The annual British agricultural returns were not begun until 1866.

From the onset of the Great Depression in British agriculture in the 1870s the share of UK tillage output coming from Ireland was about 12 per cent and eventually rose to about 15 per cent by value. This seems impressive, but since Irish tillage acreage was in absolute decline over the period, the estimates more properly reflect the very poor performance of mainly English tillage output from about 1870 to 1913. The contribution of Irish livestock remained fairly steady at about 30 per cent, or a little above, from 1870 to the Great War. Overall therefore, Irish agriculture contributed about one-fifth of UK agricultural output in the 1860s rising to one-quarter or better on the eve of the Great War. This seems to have been a substantial achievement, though it may demonstrate more the straitened times which faced British agriculture in general.

With the removal of protection in the middle of the century and the more general opening up of the British economy to world trade, it comes as no surprise to see that the British performance retreated in comparative terms. For Ireland, however, the removal of protection in her most essential market opened up the virtues of improved performance in what became a more competitive local Western European situation.

The relative outcomes of the two industries are demonstrated in table 5.2. These are not true measures of income, but rather measures of the per capita contribution of net agricultural output values.[2] Annual average output from both the new Irish estimates and from Bellerby's UK estimates are reworked on the population census years 1851, 1861, down to 1911. In addition, the Irish estimates have been deducted from Bellerby's UK estimates to leave GB estimates. From data contained in the census reports for England and Wales, and Scotland and estimates for Ireland, an attempt is made to isolate the agricultural workforce from the total population.[3] This is achieved in two ways: by isolating the agricultural labour force; and by employing J.P. Huttman's estimate of the farm population, which in a sense attempts to isolate that part of the population which was directly dependent on agriculture. Whichever estimate is chosen, the results hold up a light to the relatively improving position of Irish agricultural performance relative to Britain. These are not intended as productivity estimates, though there is a hint of

[2] For an earlier comparison with E.M. Ojala's UK estimates see, M.E. Turner, 'Agricultural output and productivity in post-Famine Ireland', chapter 16, pp. 410–38 of B.M.S. Campbell and M. Overton (eds.), *Land, Labour, and Livestock: Historical Studies in European Agricultural Productivity* (Manchester, 1991), especially pp. 422–28.

[3] Using standard Census sources, and for Ireland see J.P. Huttman, 'Institutional Factors in the development of Irish agriculture, 1850–1915', Ph.D. Thesis, University of London, 1970, pp. 384–7, 414. In all of this the warning given by David Fitzpatrick is readily accepted that it is a difficult task to disentangle the size of the agricultural labour force, separating the overlapping classes of wage labour, farmer's own labour, and the labour of the farmers' relatives. D. Fitzpatrick, 'The disappearance of the Irish agricultural labourer, 1841–1912', *Irish Economic and Social History*, 7 (1980), 66–92.

Table 5.2   *Agricultural output in Ireland, the UK and Britain* (in £s)

	IRELAND		UK	GB
	(1)	(2)	(1)	(1)
1851	16.8	5.3		
1861	27.2	8.0		
1871	36.2	10.7	60.8	76.2
1881	38.1	11.2	60.4	74.2
1891	39.2	12.2	62.0	75.9
1901	46.3	14.5	70.2	84.1
1911	59.7	19.1	70.2	75.5

Notes: Final output per head based on 5-year averages centred on population census years.
(1) Output per agricultural worker.
(2) Output per head of the estimated Irish farm population.

Sources: Same as in table 5.1 and J.P.Huttman, 'Institutional factors in the development of Irish agriculture, 1850–1915', Ph.D. Thesis, University of London, 1970, pp. 384–7 and 414; B.R. Mitchell, *Abstract of British Historical Statistics* (Cambridge, 1962), pp. 6, 60.

comparative labour productivity in them; rather they are measures of the values of *home grown* agricultural output per unit of labour input.

As a crude income indicator, Irish 'agricultural incomes' increased up to the mid-1860s (from £17 to £27 per agricultural worker), and then remained roughly level during the 'Great Agricultural Depression' to the mid-1890s, and improved dramatically thereafter. In the rest of the UK, however, although the nominal level of crude income per capita of the agricultural workforce was higher than in Ireland, by 100 per cent early on and still by over 30 per cent on the eve of the Great War, there was in fact a marginal decline in that income during the depression. Output per Irish agricultural worker was 47 per cent of the same measure in Britain in about 1871, but nearly 80 per cent by the Great War. In the post-depression period of price recovery in Britain average agricultural incomes had still not reached pre-depression levels by 1914–18. Set against the British experience alone therefore, Irish agriculture was performing well. In wider circumstances however, in some of the major economies of north-west Europe such as France, Germany, Denmark and Holland, there was also a general narrowing of the rural income gap from 1890 to 1910 between those economies and a UK base, especially in Denmark and Germany.[4]

[4]   P. K. O'Brien and Leandro Prados de la Escosura, 'Agricultural productivity and

The estimated nominal incomes of the British agricultural labourers were always higher than those of their Irish counterparts, but the gap between Ireland and Britain narrowed over time.[5] In 1871 the output per Irish worker was about 60 per cent of the output of the overall UK worker and 48 per cent of those in Great Britain, but by about 1891 his output was 63 and 52 per cent of the equivalent in the UK and GB, and by about 1911 it was 85 and 79 per cent (table 5.2, columns labelled 1).

Partial factor productivity performance in Irish agriculture

Given that the amount of land under cultivation did not change very much – the acreage under crops and hay for example varied from a high point of 5.97 million acres (1860) to a low of 4.583 million acres (1909) over the whole period – land productivity in terms of the value of output followed the course of that output. Give or take short-term fluctuations, Irish output per acre (cultivated land *less* grass) rose modestly from about £5 per acre in 1850 to a little over £8 per acre in the mid-1870s, then fell back to £7 per acre by 1890, before rising steeply to over £10 per acre by the Great War. Thus there was an advance in crude land productivity, but in volume terms, as we have seen, there was a less encouraging pre-war trend.

The analysis of land productivity can be extended a stage further, and at the same time it can include a combined labour productivity measure. The number of occupiers fell from something over 550,000 in the early 1860s to 537,000 by the end of the decade, and hence at a slower rate to a low of 521,000 by 1884. Their numbers then recovered to over 550,000 by 1905 and remained at that level until 1914. The combination of this trend with the trend of output values (ignoring short-term fluctuations) meant that output per occupier rose from something close to £60 in the early 1860s to about £90 in 1876: it then declined to about £60 by 1887; rose modestly to £70 in the early 1890s; fell back again to the mid £60s in the late 1890s; and finally expanded almost continuously to over £90 by 1914. This last period of expansion coincided in part with changes in tenure and the land law and a general move towards owner-occupancy and greater peasant independence.[6] In nominal terms occupiers managed to improve the gross value of their output for the first half of the period

European industrialization, 1890–1980', *Economic History Review*, 45 (1992), pp. 518, 531.

5   For an historiographical summary of the rise of average incomes in Ireland see T. Guinnane, 'Economics, history, and the path of demographic adjustment: Ireland after the Famine', *Research in Economic History*, 13 (1991), 154–5.

6   For a brief, recent summary see K. T. Hoppen, *Ireland Since 1800: Conflict and Conformity* (London, 1989), pp. 94–8.

up to the mid-1870s, and then again from the mid-1890s, from which time they took part in a strong movement towards greater independence. The coincidence of this apparent productivity advance with the independence movement is seductive as a general explanation of agricultural change during this last period and will be reinvestigated in a later chapter.

If real output is measured by volume, then, apart from isolated outstanding yearly improvements in the 1860s and mid-1870s, in fact at best it only remained level. It marginally declined from the mid-1870s to the Great War. Thus from 1860 to the mid-1880s, whilst output per occupier rose and then declined fairly evenly, real output by volume rose sharply at first and then declined to approximately the same level as existed in about 1860. Thereafter, although the number of occupiers increased and nominal output per occupier also increased, real output per occupier by volume declined, or at best stood still. If the number of farming units (in our case occupancy units) is a productivity parameter, then, although the nominal monetary improvements are evident, that productivity has to be questioned, generally speaking, in real volume terms.

Total factor productivity in Irish agriculture

There is enough fragmentary information to offer an estimate of total factor productivity. Table 5.3 summarises the data in terms of annual averages by decade. Four cross-sections are taken which roughly represent the traditional turning points of late-nineteenth-century UK agriculture: the 1850s represents the onset of High Farming; the mid-1870s is the high point in Western European price history on the eve of the Great Depression; the bottom of that depression came in the 1890s; and thereafter there was the semblance of a recovery down to the Great War. Such turning points are not inappropriate to Ireland, not least because of open economy considerations, but also because post-Famine recovery was certainly under way; the onset of the Land War period came in the late 1870s; there were concerted moves towards peasant ownership from the 1890s; and the period culminated in the Great War and a high degree of owner-farming and agricultural independence. The total factor productivity model assumes an agricultural production function which is linear in character, where land, labour and capital are the main inputs with no account taken of purchased inputs.

The real output values are simply annual averages based on the volume index produced in chapter 4; the labour input is based on Huttman's measure of the agricultural labour force (substituting alternative figures

Table 5.3   *Agricultural output and factor inputs in Irish agriculture, annual averages for the 1850s, 1870s, 1890s, and 1910–1914*

	Output Q Current prices £s millions	Output Q Constant prices £s millions	Output Q Constant prices £s millions	Labour L Agricultural labour force millions	Land N Acres millions	Capital K LUEs millions
(A) Output and factor inputs						
1850s	36.509	36.509	36.509	1.575	15.015	4.734
1870s	40.986	33.844	34.204	1.112	15.598	5.182
1890s	56.278	32.762	34.078	0.923	15.165	5.340
1910s	54.283	35.115	36.780	0.797	14.681	5.803
	ΔQ	ΔQ	ΔQ	ΔQ	ΔN	ΔK
(B) Annual rate of change Δ						
1850s–70s	0.580	−0.378	−0.326	−1.725	0.191	0.453
1870s–90s	−0.968	−0.162	−0.018	−0.927	−0.141	0.150
1890s–1910s	1.517	0.347	0.382	−0.731	−0.162	0.416
1850s–1910s	0.663	−0.065	0.012	−1.129	−0.037	0.340

Notes: LUEs refers to Livestock Unit Equivalents.
The two constant output series are based first on the standard Laspeyres quantity index and second on a 'chained' version of that index, for which see tables 4.2 and 4.6.

Sources: This chapter, and tables 4.2 and 4.6.

from the census makes little difference to the outcome of the estimates), and Huttman's estimated farm population is also included for reasons explained below; the land input is based on the annual average cultivated area. Ideally, variations in soil quality should be accounted for, though Ó Gráda suggests that before the Famine it was the *supply* of land rather than its quality which was the important element in productivity differences, in association with other factors, particularly labour.[7] The capital input is a surrogate for an unknown true capital factor. One estimate suggests that landlords ploughed back only 3–5 per cent of their rental income on improvements. The most enlightened landlords may have allowed 8–10 per cent, and Vaughan's research on a number of estates suggests 11 per cent, though his sample of estates included those

[7]   C. Ó Gráda, 'Poverty, population, and agriculture, 1801–45' in W.E. Vaughan (ed.), *A New History of Ireland, V, Ireland Under the Union, I, 1801–70* (Oxford, 1989), pp. 124–5.

'where high expenditure on improvements was most likely to occur'. A more plausible level of investment was 4–5 per cent. Whatever the landlords' contribution was, it was reckoned to have been chronologically uneven, and with a tendency to lump in periods of crisis or depression when it acted as a device to avoid rent reductions.[8] The main adjustment in Irish agriculture from the Famine down to the Great War has been identified as a switch in emphasis out of tillage and into livestock and livestock products. The capital factor measure used here attempts to capture the capital implications of that adjustment. It is a measure of the livestock component based on livestock unit equivalents.[9]

These estimates, or rather surrogates for the true annual capital inputs necessarily exclude the cost of providing specialist equipment for the main changes in farming enterprise into livestock production, or for other improvements. These would have included fixed costs in the provision of specialist buildings, especially for the output of milk products, and also drainage or other land improvements, and they would also have included variable costs. However, in Ireland before the Great War the level of artificial fertilisers and non-human or non-animal energy such as fuel was so small as to be negligible. While recycled feeding stuffs in agriculture accounted perhaps for as much as 80 per cent of all farm inputs in 1912/13, the information on the application of inorganic, bought in, fertiliser, remains largely a mystery. Fertilisers as a whole may have accounted for one-eighth of inputs.[10] The major technical and capital change was out of spade cultivation and into ploughs, but since the country became greener and the tillage acreage declined the impact of such a switch on the measure of capital inputs was necessarily reduced. Yet there was an impact which is very difficult to capture. J.P. Huttman attempted to do it through an investigation of the annual statistics. For the machinery associated with the arable economy, the reapers, threshers, sowers and winnowers, there was a peak in about

---

[8] C. Ó Gráda, 'The investment behaviour of Irish landlords 1850–75: Some preliminary findings', *Agricultural History Review*, 23 (1975), 151–4. W.E. Vaughan, *Landlords and Tenants in Mid-Victorian Ireland* (Oxford, 1994), pp. 122, 218, 277–8, and in general on the low levels of capital expenditure see pp. 124–130. See also W.E. Vaughan, 'An assessment of the economic performance of Irish landlords, 1851–81', in F.S.L. Lyons and R.A.J. Hawkins (eds.), *Ireland under the Union: Varieties of Tension: Essays in Honour of T.W. Moody* (Oxford, 1980), especially p. 192.

[9] There is encouragement to equate livestock directly with the capital factor by J.P. Huttman, 'The impact of land reform on agricultural production in Ireland', *Agricultural History*, 46 (1972), 366. If a milk cow is weighted by a value of one, then cattle greater than two years of age are worth 0.75, adult sheep 0.2 and so on. Crudely it measures the feeding requirements of all the animals. See J.T. Coppock, *An Agricultural Geography of Great Britain* (London, 1971), p. 150, for a full list of weights.

[10] R. O'Connor and C. Guiomard, 'Agricultural output in the Irish Free State area before and after Independence', *Irish Economic and Social History*, 12 (1985), 94.

1881, followed by a serious downturn to 1890, and then a recovery by 1912 (for threshing machines and winnowers this recovery was by a huge amount, but then these were the only machines for which the figures extended to 1912). For machines related to hay production there was a similar peak either in 1881 or by 1875 (noting that Huttman's figures relate to 1865, 1875, 1881, 1886, 1890 and 1912), followed similarly by a downturn. The numbers of steam ploughs recorded was negligible and declined from 1875.[11] The lack of hard data conspires against a more detailed appraisal of agricultural machinery. In the allied processing industries, for example, Cormac Ó Gráda has documented the rise of the creamery industry which involved the use of specialised equipment. It took off in the late 1880s, and the number of creameries increased eightfold from the mid-1890s to the Great War. This has to be compared however, with the earlier rise in the Danish industry and the much larger numbers associated with it, though to come second to Denmark was no mean feat.[12] More generally, Ó Gráda saw the late, but large, emergence of Irish mechanisation in the transition from the nineteenth to the twentieth centuries.[13]

Whilst there are statistics on the adoption of some agricultural machinery at intervals from 1865,[14] there are no estimates of ploughs prior to 1908, and yet in terms of factor substitution the transition from the spade to the plough, from labour to capital, was one of the most important of the period. It would be expected that this would have a negative correlation with output and productivity since it is generally accepted that spade cultivation produced higher yields per acre than plough cultivation, though at higher costs.[15] This is a negative view. Rather, there was an adaptation to factor costs, in this case a response to higher labour costs, and this was no more pronounced than in Ulster. L. Kennedy refers to a revolution in the use of farm machinery by the turn of the century. The items he lists for 1912 were for tillage production in the main.[16] Since the data for these measures are the least complete, the present calculation is based on a series of compromises in which both the size of capital as an input, and its share as an output, may well be underestimated. At least on the input side the method used here might

[11] Huttman, 'Institutional factors', p. 524.
[12] C. Ó Gráda, 'The beginnings of the Irish creamery system, 1880–1914', *Economic History Review*, 30 (1977), 284–305, especially 296.
[13] C. Ó Gráda, *Ireland Before and After the Famine: Explorations in Economic History, 1800–1925* (Manchester, 1988) p. 141.
[14] T. Barrington, 'The yields of Irish tillage crops since the year 1847', *Journal of the Department of Agriculture*, 21 (1921), 222–3.
[15] *Ibid.*, 223, and see the discussion on yields in Chapter 2 above.
[16] L. Kennedy, 'The rural economy, 1820–1914', L. Kennedy and P. Ollerenshaw (eds.), *An Economic History of Ulster, 1820–1939* (Manchester, 1985), p. 24.

capture the trend and therefore act as a good approximation to the rate of growth of capital input, which in a total factor productivity model is the important thing. It is not unusual for the construction of a capital series to be the most problematical in productivity exercises.[17] In their wider study of European agricultural productivity O'Brien and Prados de la Escosura fell back upon inferential rather than empirical indicators.[18]

The question of factor shares, which is that share of income which returns to the factors of production, is another rather tricky issue. At least there are some ideas on wages and rents and hence on the order of magnitude of the returns to land and wage labour, but not all labour enjoyed a wage (farmers' and farmers' relatives own labour, for example), and the returns to capital are very problematical. Ó Gráda has suggested a variation in the share of capital from agricultural income from 6 per cent in about 1854, rising to 7, 8 and 10 per cent in 1876, 1908 and 1926.[19] In the first instance a similar linear variation from 6 to 9 per cent will be adopted for the four cross-sections, thus leaving between 91 and 94 per cent of agricultural income to be divided between land and labour. If the nominal labour bill, that is wages alone, varied from a little over £9 million in the early 1850s, to £10.6 million in the early 1870s, to £11 million in the early 1880s, before falling back to £10.6 million in the late 1900s, it has been suggested that the rent bill varied from £10 to £12 to £11.5 and £8 million over the same period.[20] This suggests a rough and ready equality in factor returns or shares until the period of greatest movement towards peasant independence in the early twentieth century

---

[17] See R.M. McInnis, 'Output and productivity in Canadian agriculture, 1870–71 to 1926–27', in S.L. Engerman and R.E. Gallman (eds.), *Long-Term Factors in American Economic Growth* (Chicago, 1986), pp. 758–9, 762. C. Ó Gráda has pointed out to me in correspondence that there are two possible biases in my measure of capital: crops were part of the capital stock but their value fell over time; but inputs of machinery are likely to have risen more rapidly than livestock rose. To a degree these may be compensating biases. The sort of additions would include the £7–8 million expenditure between 1850–75 estimated by C Ó Gráda, 'Agricultural head rents, pre-Famine and post-Famine', *The Economic and Social Review*, 5 (1973–4), 390, which in disaggregated form included £3.1 million on drainage from 1847–80 under the 'Land Improvement' legislation, and £712,000 on farm buildings, labourers' cottages and scutching mills, all as a result of loans. The productivity benefits of some of this expenditure, however, is open to some doubt. See Ó Gráda, 'The investment behaviour', 146–50.

[18] O'Brien and Prados de la Escosura, 'Agricultural productivity', 515.

[19] Ó Gráda, *Ireland Before and After the Famine*, p. 130.

[20] Hoppen, *Ireland Since 1800*, p. 100, but the time periods are, in fact, 1852–4, 1872–4, 1882–4 and 1905–10. The trends might suggest that the annual rent figures reported and meant to represent the 1890s may be too high, and by 1910–14 the direct rental costs almost certainly would have been lower given the move to owner-occupancy. Minor adjustments to these figures in fact make little difference to the final total factor productivity estimates.

when rent was converted into self-ownership. Yet the wage bill does not capture the whole return to labour because it does not include the returns to represent the farmer's own labour and the labour of his family. The ratio between Huttman's agricultural labour force and his estimated farm population as a whole varied narrowly between 1:3.1 and 1:3.3. In contrast David Fitzpatrick's estimate of the ratio of farm workers to farmers declined from 2.29:1 to 1.31:1 in a roughly linear fashion from 1851 to 1911.[21] This might suggest that the return to labour, through a re-estimated notional wage bill, might be anything from two to three times the size of the current estimate. Bearing in mind that some of this own-farm population-cum-labour comprised children, women, the old and the infirm, of varying capacities to contribute to output and thus of varying capacities to take a share of the agricultural income acquired, it might be reasonable, if not plausible to suggest a doubling of the wage bill as a surrogate return to labour. On the basis of the ratios of farm workers to farmers or labourers to the rural population this would still tend to underestimate the labour share relative to land, especially so if as Fitzpatrick suggests in relation to wage estimates 'the vast majority of Irish farm workers are by definition excluded from analysis'.[22] Doubling the nominal wage bill generates the putative factor shares in table 5.4.[23]

The standard total factor productivity equation has been applied to these factor share estimates and the data in table 5.3, namely:

$$\Delta\mathrm{TFP} = \Delta Q - (A_1 \Delta \mathrm{Lab}) - (A_2 \Delta \mathrm{Land}) - (A_3 \Delta \mathrm{Cap})$$

where the Δs represent annual rates of change of the factor inputs between time cross-sections, and the As represent the factor shares. For the purposes of computation the factor shares take two forms, those relevant to the start of a period and those relevant to the end.

Table 5.5 reports the total factor productivity estimates. The output has been adjusted in constant volume terms using the constant volume index from chapter 4. Three sets of estimates are included for each inter-

---

[21] Huttman, 'Institutional factors', p. 414; Fitzpatrick, 'The disappearance of the Irish agricultural labourer', 88.

[22] Fitzpatrick, 'The disappearance of the Irish agricultural labourers', 80.

[23] In my original attempts at estimating total factor productivity I used factor shares which varied from 52–43 per cent for labour, 39–51 per cent for land, and 1–9 per cent for capital. Peter Solar has commented privately that I surely underestimated the labour share because taking a simple national nominal wage bill neglects what is known as the farmer's incentive income, effectively that which was left after he paid his labour, land and capital costs. In this connection, and also for details about factor shares in Britain, see C. Ó Gráda, 'Agricultural Decline 1860–1914', in R. Floud and D.N. McCloskey (eds.), *The Economic History of Britain Since 1700, Part 2, 1860 to the 1970s* (Cambridge, 1981), pp. 178–9.

Table 5.4   *Factor shares in Irish agriculture*
1850s–1910s (in percentages)

	Labour	Land	Capital
1850s	61	33	6
1870s	59	34	7
1890s	60	32	8
1910s	66	25	9

Source: See text.

censal or period between cross-sections: those derived from factor shares relevant to the beginning of the period; those at the end; and an assumed constant factor share of 60:30:10, labour:land:capital. For Britain the distribution of the equivalent factor shares from 1870/2 to 1910/12 suggest in general a decrease in the share of the land component from 23 to 18 per cent, a rise in the labour share from 63 to 69 per cent, and a fairly even capital share of 12 or 13 per cent.[24] As the Irish results show, there is only marginal sensitivity to the differential factor shares.

It appears from table 5.5 that there was positive growth throughout the period, in rounded terms of 0.4–0.8 per cent per annum. However, there were variations across the three main sub-periods: on balance they identify a disjointed improvement over time from 0.6 per cent per annum up to the 1870s, a fall to 0.4 per cent to the 1890s, but a trend completed by an impressive rise to 0.8 per cent per annum up to the Great War. During the same period Ó Gráda has measured the equivalent British growth at 0.3 per cent per annum, and he reminds us that one per cent per annum was achieved in the emerging economy of Japan and the expanding economy of the USA.[25] It appears that while Ireland was by no means in the vanguard of agricultural development, she was also by no means faring badly, with an overall growth rate in the half century or so after the Famine of about 0.6 per cent per annum.

The combination of a *relatively* fixed area of land under cultivation which necessarily declined less fast than the decline of the labour force, in whatever way the size of that labour force is measured, and with small increases in capital, suggests overwhelmingly a labour productivity

[24] It was an anonymous referee who suggested the not implausible and more simple distribution of factor shares of 60:30:10, labour:land:capital, intimating that marginal variations from this are not greatly significant.
[25] Ó Gráda, 'Agricultural decline 1860–1914', p. 178. See also T. Weiss, 'Long-term changes in US agricultural output per worker, 1800–1900', *Economic History Review*, 46 (1993), 324–41.

Table 5.5 *Estimates of total factor productivity in Irish agriculture,*
1850s–1910s (in annual percentage growth rates)

	Using starting decades for factor shares	Using finishing decades for factor shares	Using fixed factor shares
In constant volumes derived from a standard Laspeyres Quantity Index			
1850s–70s	0.584	0.543	0.554
1870s–90s	0.422	0.427	0.421
1890s–1910s	0.805	0.833	0.793
1850s–1910s	0.616	0.659	0.590
In constant volumes derived from a 'chained' Laspeyres Quantity Index			
1850s–70s	0.637	0.596	0.607
1870s–90s	0.566	0.571	0.565
1890s–1910s	0.839	0.868	0.828
1850s–1910s	0.693	0.736	0.667

Note: Time periods refer to the annual averages for the decades shown, as in table 5.3.

Sources: Tables 5.3 and 5.4, and see text.

effect.[26] These results partially confirm those estimated by Ó Gráda. For
the period from 1854 to 1908 he found a productivity rate of about 0.6
per cent per annum of roughly two equal halves, from 1854 to 1876, and
1876 to 1908.[27] He further stated that the differences between the two
sub-periods, that is virtually no differences at all, 'raise doubts about the
strong emphasis in the work of Crotty and Solow on a turning point (for
the worse) in the fortunes of Irish agriculture around the 1870s.'[28] What
the new estimates add to this story, perhaps, is a period of retrenchment

---

[26] A series of simulation exercises is very revealing about the sensitivity of the labour factor
and hence of labour productivity. If total factor productivity is re-estimated on the basis
of plausible changes in the factor shares, it is the labour share which is the most sensitive
and the main determinant of the overall productivity. To this extent a greater refinement
of the capital input makes little difference to total factor productivity, although there is
considerable doubt about estimating the size of the labour force,(*cf.* Fitzpatrick, notes 3,
21, and 22 above, and accompanying text), nevertheless, by substituting the estimates in
Fitzpatrick which indicate the size of the male only component of the farm labour force
produces a TFP estimate which emphasises the growth of that TFP over the full time
period by reducing the productivity in the period before the 1870s to something close to
0.2 per cent per annum.

[27] Ó Gráda, *Ireland Before and After the Famine*, p. 130.

[28] *Ibid.*, p. 130.

from the 1870s followed by a period of *pronounced* growth after the 1890s. The significance of these new estimates relates to the relationship between greater productivity and the strong moves towards peasant independence, when a greater share of income returned to labour which formerly may have gone to the landlord. In this sense the TFP model, even with a weight of 0.66 as the labour factor share, may yet underestimate the returns to labour. Indeed, the growing share which accrued to labour may itself have acted as a spur or incentive to the now independent farmer to perform even better than before.

Physical output in Irish agriculture

The performance of agriculture can also be expressed in physical output terms, as distinct from the monetary value of output. The distinction is a real one with an established literature.[29] It is a more direct measure of output without the problems of data inconsistencies, such as choices over price. The method uses fixed conversion weights whose only subsequent control, for crops for example, is determined by crop yields. Once completed the exercise will furnish some estimates of land and labour productivity and it will indicate the degree to which Ireland was able not simply to feed itself, but also produce a surplus for export. One measure of the success of agriculture might well be the degree to which it could sustain its own population. Enthusiasm for such an indicator might be tempered with the knowledge that the population was falling drastically. A measure of physical output may more helpfully be evidence regarding labour productivity and adaptability.

The simplest method might be simply to add up the acreage of land under cultivation, and the animals in occupation of the land, and use these as indexes of production. Yet a definition of output is rather more complex. For example, an acre of wheat is not the same as an acre of hay, and a chicken is not a calf, nor is it a pig. Assigning monetary values has equated unlike products with one another, and by constructing a constant volume index the real volume of output has been derived, but there are other methods. The Irish Department of Industry and Commerce in 1930 in their digest of *Agricultural Statistics 1847–1926* produced a series of tables based on land use which reduced all the corn, root and green crops, and hay, to common units based on the number of

---

[29] See for example L. Drescher, 'The development of agricultural production in Great Britain and Ireland from the early nineteenth century', *The Manchester School*, 23 (1955), 153–83, including the critique by T.W. Fletcher, but especially 162–6, 170–5; O.J. Beilby, 'Changes in agricultural production in England and Wales', *Journal of the Royal Agricultural Society of England*, 100 (1939), 62–73.

starch lbs per gross ton each crop produced.[30] Wheat was valued at 1,605 lbs of starch per gross ton, oats at 1,335 lbs, potatoes at 399 lbs, and hay 531 lbs. The product of these weights and the acreages under the crops and estimates of their yields produced an estimated annual average output of 2.66 million starch tons for the years 1851–5. This was the high point in crop output. Thereafter it declined to 1.892 million tons in 1860–4, and then rose to a pre-1914 peak of 2.323 million tons in 1910–14. *The record year with 2.757 million tons was 1853* followed by 1855, 1851, 1854, 1852 and 1913 (2.512 million tons), in that order. The lowest year was 1879 with 1.527 million tons. Most of these dates are now familiar landmarks in Irish agricultural output. A similar exercise was not carried out on the animal side of production, and the estimates referred only to the geographical area of post-Partition Ireland.[31] Leo Drescher produced an annual index of the physical volume of production for the years 1866–1931 for Great Britain and Ireland. It included separate indexes for arable and livestock, and separate indexes also for Great Britain and Ireland, but as a production index it was much criticised.[32]

A new physical volume output index for Ireland is presented in this section. It is essentially an inventory, or stock-taking, of standing crops and feeding and grazing animals. In its construction there is the danger of introducing a certain amount of double counting: a proportion of potatoes for example, all roots, some hay and some oats were produced for animal consumption. To count these as standing crops and also to include them in an inventory of animals is to count them twice. Drescher recognised the problem but did not attempt to solve it. The solution adopted here is similar to that used in chapter 4 in deriving the monetary output estimates; the application of final output crop coefficients or weights.[33] A second problem is to decide which conversion factors to employ to change crops and animals into common units of starch equivalents. Arguably animals could be ignored altogether because they are accounted for in total crop production. Both animals and humans eat the products of the soil and therefore the sum of those products captures final output.[34] Yet this does not take into account value added in the

---

[30] Saorstát Éireann, *Agricultural Statistics 1847–1926. Reports and Tables* (Dublin, 1930), pp. 6–13.

[31] And for a similar exercise for the whole of Ireland, but for crops only, see, Barrington, 'The yields of Irish tillage crops', 210, 213, and 215.

[32] Drescher, 'The development of agricultural production', 174–5, and the accompanying critique by Fletcher.

[33] See also *The Agricultural Output of Ireland. 1908* (Department of Agriculture and Technical Instruction for Ireland, London, 1912), pp. 4–5.

[34] Huttman did this in his estimates of the physical output of Irish agriculture. He argued that animal feed was a very important means of consuming crop output in Ireland in the period. The nutritive value of the crops measured as feed is, therefore, apparently, a

conversion of the crop by the animal into animal fats, or other body substances, in both a physical and a monetary sense. Furthermore, it assumes the same conversion by animals of all shapes and different sizes (ages). Conversely, if the crop feed units already take this into account then those crops intended for non-animal consumption will have been overvalued in crop feed unit terms, or at least incorrectly valued for human consumption purposes.

Drescher used a set of conversion factors for both crops and animals which were based on official returns from *The Agricultural Output of England and Wales* in 1925 and 1930/1.[35] When O.J. Beilby estimated long-run agricultural production for England and Wales for the period 1885–1937 he used calorie conversion factors which were assembled from a variety of sources.[36] As a first approximation the starch conversion factors employed in the official Irish report of 1930 are adopted.[37] The resulting annual output index expressed in total crop starch equivalents is given in table 5.6 and figure 5.1. There was a decline in output in the 1850s to a low point in 1861, a decline accentuated by the depression of 1859–64, and there was a similar downturn in 1879, in the wake of the outstanding peak of 1876. Overall, however, the trend was not determined by those depressions. In a similar exercise for Ireland, Barrington produced a similarly shaped long-term profile but his yearly figures were generally below mine. For example, my figures never fall below 2,000 million tons of starch whereas his are always below 2,000 million tons from about 1860 to 1915 (with the exception of 1875).[38]

measure of crop production. He proceeded to base his calculations on crop feed units derived from estimating procedures appropriate to relatively modern Danish agriculture. Huttman, 'Institutional factors', pp. 366–7.

[35] Drescher, 'The development of agricultural production', 169–73.

[36] Beilby, 'Changes in agricultural production', 72. There is a general discussion of such conversion factors in more modern works such as W.B. Morgan and R.J.C. Munton, *Agricultural Geography* (London, 1971), pp. 106–10.

[37] There is not a conversion rate for all crops grown. The notable absentees are peas and beans, flax, and grass, and there is no provision for straw, unless it is assumed that the straw of the grain crops is counted in the crops themselves. In the present calculations this assumption is made, flax is ignored, bere is counted along with barley, all root and green crops not otherwise specified are counted the same as cabbage, and peas and beans are counted the same as wheat. There is some encouragement to do this by Drescher, 'The development of agricultural production', 172, who took the raw production estimates reported in *The Agricultural Output of England and Wales, 1925*, BPP [Cmd. 2815], vol. xxv, 1927, *passim*, and applied starch equivalent coefficients to them. The starch equivalent per ton of wheat was 0.701, for peas it was 0.686, and for beans it was 0.666. Taken together the problem crops cover such a small acreage as to have a negligible effect on the final index. The number of starch lbs per gross ton of each crop is, wheat and rye 1,605, oats 1,335, barley 1,525, potatoes 399, turnips 164, mangels 152, cabbage 148 and hay 531.

[38] Barrington, 'The yields of Irish tillage crops', 215; he failed to say what conversion factors he used, or indeed to indicate his sources.

Table 5.6   *Physical output of crops from Irish agriculture, 1847–1914* (in annual averages)

	Tons of starch (millions)	Index 1850–4 = 100	Tons per acre	Index 1850–4 = 100
1847	2.936	89.1	0.573	95.7
1848		Insufficient information		
1849	3.158	95.9	0.583	97.3
1850–4	3.295	100.0	0.599	100.0
1855–9	2.984	90.6	0.531	88.7
1860–4	2.424	73.6	0.440	73.5
1865–9	2.517	76.4	0.482	80.4
1870–4	2.555	77.5	0.489	81.6
1875–9	2.617	79.4	0.521	87.0
1880–4	2.563	77.8	0.530	88.5
1885–9	2.577	78.2	0.532	88.8
1890–4	2.671	81.1	0.566	94.5
1895–9	2.733	82.9	0.591	98.6
1900–4	2.804	85.1	0.622	103.8
1905–9	2.867	87.0	0.634	105.8
1910–14	2.960	89.8	0.632	105.4

Source: See text.

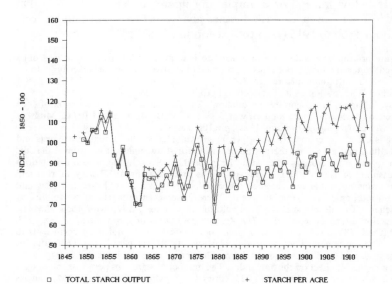

5.1   Physical output of Irish agriculture, 1847–1914 – Starch index

Table 5.7 *Starch output of crops in Ireland*

	Starch	Total population	Labour force	Tons per head of population	Tons per head of labour force
	mill tons	000s	000s	tons	tons
1849–54	3.27	6,552	1,575	0.50	2.08
1855–64	2.70	5,799	1,264	0.47	2.14
1865–74	2.54	5,412	1,112	0.47	2.28
1875–84	2.59	5,175	1,008	0.50	2.57
1885–94	2.62	4,705	923	0.56	2.84
1895–1904	2.77	4,459	834	0.62	3.32
1905–14	2.91	4,390	797	0.66	3.66

Source: See table 5.6, and for population see J.P. Huttman, 'Institutional factors in the development of Irish agriculture, 1850–1915', Ph.D., Thesis, University of London, 1970, pp. 384–7, 414.

The output equation is based on acres times yields, and this is sensitive to relative changes in acreage, and the state of the weather. The control of the annual yields was mainly outside the influence of the individual, aside from good husbandry practices, because the weather was a more important factor. Yet responses to environmental changes, as reflected in the distribution of crops, was controlled by the body of husbandmen in their decision-making processes. To this extent they did have some control over land productivity. The tillage acreage (including hay) fell from 5.8 million acres in 1850 to 4.58 million acres in 1909, but there was not a matching decline in physical output. There was, therefore, a productivity improvement in physical output per acre from about 1870, or even 1865 onwards. The grass acreage increased from 8.7 million acres in 1851 to 9.5 million acres by 1861, and remained on or around 10 million acres for the remainder of the period. Ordinarily grass is not a good producer of starch equivalent material, but when fertilisers are applied it becomes one of the most productive crops in terms of animal feed.[39] In nineteenth-century Ireland, a country of mainly small farmers and low levels of capital, it is assumed that grass was not a good producer of feed compared with other products. In spite of its large acreage it will not have had a large impact on the *trend* of physical output.

A more important factor to consider might be the population. It declined in size, as did the size of the labour force. Table 5.7 summarises some crude population and labour productivity estimates. The decline in starch output per capita in the 1850s and 1860s in the face of a

[39] See Coppock, *An Agricultural Geography*, p. 136.

population decline appears less sharp than the fall in output per acre in table 5.6, but the rise in labour productivity from about 1860 looks stronger than it did in unit acre terms. Output per capita of the labour force grew throughout the period, with a maximum improvement of nearly 17 per cent between 1885/94 and 1895/1904. Once again there is a coincidence in this later period with the growth of greater farmer independence in the decade or so before the Great War.

The calculation still makes no allowance for the conversion of vegetable matter into animals.[40] Beilby's method of assessing agricultural production is adopted for this, in association with the same weighting coefficients employed in chapter 4 and set out in appendix 5 to calculate the final proportions of output from each product. Thus the final output of crops for human consumption is combined with the final output of animals.[41] The results are given in table 5.8 and the resulting annual index in figure 5.2.

Crop calories for human consumption were nearly always greater than the calories derived from animals, but from about 1880 onwards (or even as early as 1860 if the wild fluctuations of the 1860s and 1870s are ignored) there was an irreducible minimum of non-fodder crops below which the Irish industry did not or could not fall. Beginning at nearly 6 billion calories in the 1850s, it fell to under 4 billion, and as low as 2.6 billion in 1879. Thereafter it never rose above 4 billion calories again. The short-term extremes in the trend are the function of the weather and other major influences on the harvest, but the long-term trend is unmistakable. In contrast, the contribution of animals increased steadily

---

[40]   R.C. Geary adopted the Saorstát Éireann calculation of starch equivalent output, but included the pasture element (presumably as a proxy for animal output). 'At a guess' he took the yield of pasture per acre as equal to one quarter of the yield of crops (including hay). Thus the following emerged:

Date	Rural pop. millions	Crops	Pasture	total	Total per head of rural pop. in tons
		(millions of starch tons)			
1861	3.424	1.9	0.9	2.8	0.8
1881	2.946	1.9	1.1	3.0	1.0
1911	2.207	2.3	1.4	3.7	1.7

Interestingly, Geary offered this table, which includes estimates for 1841 and 1926, 'instead of "output" which it would be impossible to estimate for the remote past.' R.C. Geary, 'The future population of Saorstát Éireann and some observations on population statistics', *Journal of the Statistical and Social Inquiry Society of Ireland*, 15 (1935), 15–35, especially 29.

[41]   Using weights in Beilby, 'Changes in agricultural production', 72, and the procedures adopted in appendix 5 below with additional information from *The Agricultural Output of Northern Ireland, 1925*, BPP [Cmd. 87], 1928, especially data on carcass weights, pp. 25–32 since for the purposes of calculation I have converted all animals into dead weight.

Table 5.8   *Index of the agricultural output of Ireland, 1850–1854 to 1910–1914*

	Crops	Index	Animals	Index	Total	Index
1850–4	5,710	100.0	2,050	100.0	7,760	100.0
1855–9	5,258	92.1	2,593	126.5	7,851	101.2
1860–4	3,890	68.1	2,565	125.1	6,456	83.2
1865–9	3,968	69.5	2,569	125.3	6,537	84.2
1870–4	3,761	65.9	2,796	136.4	6,557	84.5
1875–9	3,587	62.8	2,871	140.1	6,459	83.2
1880–4	3,535	61.9	2,730	133.2	6,265	80.7
1885–9	3,377	59.1	2,822	137.6	6,198	79.9
1890–4	3,328	58.3	2,944	143.6	6,272	80.8
1895–9	3,305	57.9	3,039	148.3	6,344	81.7
1900–4	3,289	57.6	3,166	154.4	6,455	83.2
1905-9	3,480	61.0	3,227	157.4	6,708	86.4
1910–14	3,628	63.5	3,389	165.3	7,017	90.4

Note: Net crops and net livestock in 000s million calories.
Index 1850–54 = 100.

Source: See text.

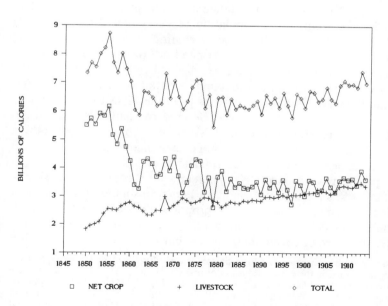

5.2   Physical output of Irish agriculture, 1850–1914 – in calories

from about 2 billion calories in about 1850 to over 3 billion by 1900, with very little year-to-year fluctuations. There was a low level of output from animals in the early 1850s when the process of post-Famine replenishment of stocks was under way. Apart from odd years, the contribution of animals never equalled the contribution of crops, and from about 1890 or 1895 they were both on a rising trend. In aggregate there was an uneven downturn to about 1880, but a more even upturn thereafter. The degree of movement away from these trends was a function of the harvest more than anything else.[42]

There is a fear, which must be guarded against, that these estimates may be used to suggest questions and provide answers about Irish agricultural production which they are incapable of doing. After all, Ireland was *not* a closed economy and, for many products, many if not most of these calories were exported. In contrast, there were other calorie equivalent materials which were imported either as animal or human feedstuffs. However, the estimates may indicate, though not rightly speaking measure, productivity trends. Two interesting views of Irish agricultural performance could emerge: the estimates could demonstrate per capita output for the agricultural population; and they could be used in a comparison with the parallel performance in England and Wales.

Table 5.9 is the estimated decadal annual average agricultural production in Ireland, centred on population census years. Per head of the agricultural labour force there was a continuous increase amounting overall to about 70 per cent. The right-hand side of the upper part of the table shows a variety of measures per diem (dividing through by 365). There was an overall rise in production per head per diem for the total population from 3,200 to 4,300 calories from the 1850s to 1914, with an identifiable upturn centred on 1871. By modern standards of nutrition this was easily enough to provide basic needs, though it says nothing about product mix and the other important ingredients derived from food, such as protein, fat, calcium, iron and so on.

Two problems are immediately apparent when it comes to a consideration of what was retained in Ireland for home consumption: the needs of men and women are different; and a proportion of output was exported. The differences between men and women are basically those of bodily size and therefore of consumption needs, but also of occupation, with a

42   To the extent that there are existing estimates of agricultural production see H. Staehle, 'Statistical notes on the economic history of Irish agriculture, 1847–1913', *Journal of the Statistical and Social Inquiry Society of Ireland*, 18 (1950–51), 458–62, 466–7 for isolated years, and 445 for a long-term graph; B. Solow, *The Land Question and the Irish Economy, 1870–1903* (Cambridge, MA,1971), pp. 100–1; Huttman, 'Institutional factors', pp. 451–61 in terms of crop feed units.

Table 5.9 *Agricultural output in Ireland in calories*

| | Total calories | Total Population | Agricultural labour | | Calories per head | | |
| | | | | Population | Population per day | Labour force | Labour force per day |
	000s millions	000s	000s	millions		millions	
1850–4	7,760	6,552	1,575	1.18	3,245	4.93	13,498
1855–64	7,153	5,799	1,264	1.23	3,379	5.66	15,504
1865–74	6,547	5,412	1,112	1.21	3,314	5.89	16,130
1875–84	6,362	5,175	1,008	1.23	3,368	6.31	17,291
1885–94	6,235	4,705	923	1.33	3,631	6.76	18,508
1895–04	6,399	4,459	834	1.44	3,932	7.67	21,022
1905–14	6,862	4,390	797	1.56	4,283	8.61	23,590

	With a gender allowance[a]	With an export allowance[b]	With both gender & export allowance[c]	With an export allowance[d]	With both gender & export allowance[e]

(Calories per head of total population per day)

1850–54	3,678	3,123	3,540	2,583	2,928
1855–64	3,829	3,189	3,614	2,466	2,794
1865–74	3,756	3,084	3,496	2,332	2,643
1875–84	3,816	3,089	3,500	2,330	2,640
1885–94	4,110	3,301	3,736	2,484	2,812
1895–04	4,451	3,552	4,020	2,662	3,014
1905–14	4,840	3,904	4,412	2,937	3,319

Notes: [a] Using a conversion factor of males:females of 1:0.77, the origin for which see text and accompanying notes.
[b] After deducting estimated meat calories exported, according to the percentages explained in the text.
[c] Based on both [a] and [b] above.
[d] Based on deducting meat exports and all milk calories.
[e] Based on [d] above and also the gender division in [a].

Sources: Tables 5.7 and 5.8 and explained in text.

variation according to manual or non-manual, sedentary or active occupations. This last consideration would also have affected different people of the same sex. In addition, pregnant women, or women with children at the breast require different intakes of calories to other women. On the basis of recent estimates of the calorie requirements of men and women a simple bifurcation of the population is made, with the women weighted against men in the ratio 0.77:1.[43] A norm for pre-Famine Ireland suggests a ratio of 1:0.8 men to women based on a purely potato diet.[44]

The second consideration is the amount of output which was exported. There are other problems associated with this estimate. For example, there is concern over the definition of the enumeration year compared with the export year, and with the distinction between store cattle and fat cattle. Roughly speaking about 15 per cent of all Irish cattle enumerated annually were exported to Great Britain, 18 per cent of the sheep and 35 per cent of the swine. As a proportion of the number of animals which 'disappeared' each year it turns out that the majority of the cattle, about half the sheep, and perhaps one third of the pigs were exported.[45] For example, in all years from 1869 onwards 60 per cent of all cattle which 'disappeared' were exported (except in 1877 when the proportion was 39.5 per cent). The average export level for all the years for which there are data was 76.6 per cent. For argument's sake, let exports account for three-quarters of all cattle 'disappearances'. Therefore, one quarter of the cattle calories were retained in Ireland for consumption. The range of sheep exports suggests that 38.5 per cent of 'disappearances' in 1865 were exported, but 85.9 per cent in 1854, with no discernible trend in either direction from about 1870 onwards and an annual average of 51.8 per cent. In 1913 as few as 12.9 per cent of pig 'disappearances' were exported but in 1902 as many as 45.7 per cent were exported. The annual average of exports was 31.2 per cent.

For present purposes it is assumed that 75 per cent of the cattle, 50 per cent of the sheep and 30 per cent of the pigs were exported, and so were their calories, leaving the residual 25, 50 and 75 per cent for home consumption. This analysis is only partial because there was a net trade position – there was the importation of calorie equivalent materials. For example, we know from Solar's meticulous estimation of the trade in certain animals and animal products that although there were imports as well as exports of ham and bacon, the net outflow was massive until the late 1870s but from then onwards it declined. There was indeed a

---

43 Differences between adults and children are ignored, as are other considerations of different age-groups. 'Reference man' requires 3,000 and 'reference woman' 2,300 calories per day giving the ratio of 0.77 (see note 56 below).

44 P.M.A. Bourke, 'The use of the potato crop in pre-Famine Ireland', *Journal of the Statistical and Social Inquiry Society of Ireland*, 21 (1967–8), 77. A weighted estimate to allow for the inclusion of children would inevitably increase the calories available per head of population. For example, in one study children of 11–15 years have been counted equal to women, and children under 11 have been counted in the ratio of 1:0.35 with respect to men, *ibid.*, 77.

45 All along, as in chapter 4 and appendix 5 it is assumed that the mortality rate for cattle was 5 per cent and for sheep was 12 per cent.

narrowing in the gap between exports and imports as the century proceeded, but it was always a position of net exports. In other areas of trade, although there was a growth in both butter and cheese imports from the 1860s down to the Great War, together they were massively overwhelmed by the counterpart export of butter, and later the export of condensed milk.[46]

The allowance for gender and export considerations is clearly seen in the lower half of table 5.9. This still reveals an overall increase in calories per head, with still a rise centred on 1871, and an estimate which is in excess of 3,000 or even 4,000 calories per day depending on whether the gender and/or export allowances are considered. There are still some unrevealed exports in these estimates; in particular there was the exported portion of the butter trade (i.e. milk).[47] Even allowing for *all* milk output to be exported, as in the final parts of table 5.9, the estimates suggest that there was still always well over 2,000 calories per head per day available for home consumption from home production.

This is an exercise in agricultural output; it is *not* an exercise in standard of living or about the diet of nineteenth-century Irishmen, although these estimates may allow other historians to form some tentative observations on these related subjects. For example, it may be possible to follow up R.C. Geary's study of 1935 based on physical output and wage movements in which he pronounced emphatically that, 'There can be little doubt that during the last century the average standard of living in this country increased from being one of the lowest to one of the highest in Europe.'[48] His evidence was not robust, though the estimates from the present chapter and from what has been said about performance with respect to England and Wales suggests that whatever gap existed across the Irish Sea in mid-century, it was closed substantially before the Great War.

These estimates of the calories per day *available* to Irishmen from Irish sources bear up to contemporary or near-contemporary food intake norms. Carol Shammas has reconstructed the calorific equivalent diets of English labourers during the food supply crises of the 1790s. Her study is based on the limited evidence from the few budget surveys which were constructed at the time. Labourers consumed an estimated 2,100 and

---

[46] P. Solar, *Growth and Distribution in Irish Agriculture before the Famine*, Ph.D. Thesis, University of Stanford, 1987, pp. 155–8, 198.

[47] Milk output was about 16 per cent of total calorie output in the 1850s, always over 20 per cent from 1856 with a maximum of 27 per cent in 1879 and an average over the period of 22 per cent. It was 60 per cent of animal output alone in the 1850s, over 50 per cent to the mid 1880s, always 45 per cent thereafter, with an average over the whole period of 52 per cent.

[48] Geary, 'The future population', 31.

2,800 calories per day in the south and north of England respectively, which after allowances for likely food consumption which went unrecorded in the budget surveys Shammas raised to upper bounds of 2,500 to 3,200 calories per day.[49] D.J. Oddy reports the nutrient intake of a small sample of cotton operatives in Lancashire in the 1860s. The diets of seven 19–20 year old unemployed females during the cotton famine amounted to energy values of 4,160 calories per head per day under 'normal' circumstances, which reduced to 2,555 calories in the winter of 1862. The 'normal' average family diet was 2,745 calories per head per day and 1,955 in the winter of 1862. Thus even for normally active young women their diets might be considered as lavish, or even 'certainly lavish'.[50] Furthermore, diets which provided energy values of between 2,000 to 2,500 calories per day 'were certainly not starvation diets'.[51] The daily energy intake of contemporary (i.e. 1860s) low paid indoor workers, like silkweavers and shoemakers, was 2,190 calories per day.[52]

These estimates are not directly comparable with the current Irish estimates which are based on converting a crop into calories which is then assumed to be the available food before further processing. The diets, in contrast, are based on the conversion of estimated food intake into calories. Energy gains and losses can take place in the cooking process.[53] However, at face value, without yet converting crops into food, it looks as though Irish agriculture was able to produce calorie equivalent output on a par with, or better than, current estimates of calorie intakes of labourers in different trades from different parts of England in the 1860s, and still have a substantial surplus for export. We can go further with the comparison. The Irish net production estimates can be compared with Robert Fogel's estimates of the calorie intake in eighteenth-century France and England. He has re-estimated Toutain's original estimate of the average caloric consumption in France in about 1806 of 1,846 calories per capita. This figure does not standardise for differential age and sex distributions across the population, but when it does so it reveals a figure of 2,400 calories per *consuming unit* (defined as

---

[49] C. Shammas, 'The eighteenth-century English diet and economic change', *Explorations in Economic History*, 21 (1984), 256–7.

[50] D.J. Oddy, 'Urban famine in nineteenth-century Britain: the effect of the Lancashire cotton famine on working-class diet and health', *Economic History Review*, 36 (1983), 78–9, based on T.C. Barker, D.J. Oddy and J. Yudkin, *The Dietary Surveys of Dr Edward Smith 1862–3*, (Department of Nutrition, Queen Elizabeth College, University of London, Occasional Papers No. 1, 1970), especially pp. 35–50.

[51] Oddy, 'Urban famine', 79.

[52] *Ibid.*, 80.

[53] There is a large literature on this, as a sample see A. Bender, 'The nutritional importance of fruit and vegetables' in D.J. Oddy and D.S. Miller (eds.), *Diet and Health in Modern Britain* (London, 1985), pp. 274–95.

an adult male equivalent aged 20–39 years). An equivalent estimate for 1785 suggested a daily consumption of 2,290 Kcalories per consumer unit. For comparison, Fogel's estimate of the Englishman's daily consumption per consuming unit for 1790 is 2,700 Kcalories.[54] For Belgium G. Bekaert has estimated that the per capita caloric consumption per male equivalent adult was something a little over 2,700 calories per day in 1812, at which level it remained up to the mid-century.[55]

In the second half of the nineteenth century in Ireland, on the basis of the *availability* of *home produced* calories, and even allowing for deductions for exports, it looks as though Irish agriculture was producing a decent stock of calories. This kind of exercise, from the standpoint of agricultural productivity, only has real value when comparing closed economies. The present exercise is only a partial productivity exercise; if it was extended it could embrace wider questions regarding standards of living. Thus, from the point of view of living standards this is *before* any allowance is made for food imports. This has two strands to it: a proportion of whatever food was brought in would have entered normal consumption through the Irish diet; but a proportion would have entered as a supplement to livestock fodder or have acted as crop seed for a future year. That which went into human consumption would have raised the caloric intake of Irishmen, but that which was fed to animals would influence any exercise concerned with productivity; it would be an input and would need to be deducted from total output to arrive at net output.

In the Irish economy of the late nineteenth century farmers produced high money value products, such as meat and dairy products, but with low caloric value. In contrast they imported low money value food, but with high caloric values, like cereals. An accurate and true measure of productivity for the whole economy would surely need to capture these influences. From the point of view of the current exercise we have more narrowly identified that the total caloric output of Irish output was high, and by only partially taking into account the influence of trade (on the export side) we have still created a large residual which was retained in Ireland. *If* we could add to that an estimate of the calorie

[54] R.W. Fogel, 'Second thoughts on the European escape from hunger; famines, price elasticities, entitlements, chronic malnutrition, and mortality rates', *Working Paper Series on Historical Factors in Long-Run Growth*, 1 (National Bureau of Economic Research, 1989), 37–43. See also Fogel, 'New sources and new techniques for the study of secular trends in nutritional status, health, mortality, and the process of aging', *Historical Methods: A Journal of Quantitative and Interdisciplinary History*, 26 (1993), especially 9–13.

[55] G. Bekaert, 'Calorific consumption in industrializing Belgium', *Journal of Economic History*, 51 (1991), 633–55, especially 635, 639.

equivalent imports, it would further enhance the welfare implications for Irishmen. There are two separate issues here: one relates to welfare and the other to production; and it is the production (meaning performance) of the agricultural economy which is the main concern of this study.

Some useful international comparisons can be made. In modern times 'reference man' is said to require 3,000 calories per day and 'reference woman' 2,300.[56] In Ireland in 1839 the daily energy value of actual potato diets has been calculated at 5,720 calories. This is way above the estimates presented here of calories *available* for home consumption except for the highest estimates for the decade before the Great War and *before* an allowance is made for exports. What this demonstrates is the large energy intake and nutritional qualities of the pre-Famine potato diet. In the decade before the Second World War Ireland apparently had the second highest per capita caloric intake of all European countries, at 3,400 calories per day, second only to Denmark at 3,450, followed by Norway, Switzerland, Sweden, the UK and Germany all with over 3,000 calories. Modern Canadian data suggests that the recommended intake for active males in the age group 25–49 is 2,700 calories, which is in line with other modern estimates.[57] This whole exercise is limited in value however since diet can be determined environmentally and culturally, and the use of the diet to translate into energy for work depends not only on the work involved but the environment in which the work takes place – hot or cold climates, living at sea level or altitude, active or sedentary occupations all have a bearing on consumption needs. The best that can be said is that Irish agriculture in the nineteenth and early twentieth centuries seemed to perform quite well.

A more detailed comparison can be made with the contemporary output from England and Wales. Beilby's estimates of the caloric value of agricultural production in England and Wales in the period 1885–1937 are compared with the Irish estimates in table 5.10. The comparison is *not* a perfect one. Beilby took into account the conversion of crop output into fodder, and he also deducted inputs, in his case imports of stock from overseas (including therefore some of the Irish exports), and finally he included an estimate of the calorie production by animals derived

---

[56] See D.S. Miller, 'Man's demand for energy' in Oddy and Miller (eds.), *Diet and Health*, pp. 278–9 for a definition of 'reference' man and woman and examples of actual energy intake in modern times of less than, and sometimes much less than 2,000 calories per day.

[57] P. Lamartine Yates, *Food, Land and Manpower in Western Europe* (London, 1960), p. 49; C. Ó Gráda, *The Great Irish Famine* (London, 1989), p. 27.

Table 5.10 *Gross calorific output from agriculture in Ireland, and England and Wales* (in 000s million calories)

	Crops		Animals		Total	
(A) Ireland						
1885–9	3,377		2,822		6,198	
1890–4	3,328		2,944		6,272	
1895–9	3,305		3,039		6,344	
1900–4	3,289		3,166		6,455	
1905–9	3,480		3,227		6,708	
1910–14	3,628		3,389		7,017	
(B) England and Wales						
		(1)	(2)	(1)		(2)
1885–9	9,630	5,109	3,522	14,739		13,152
1890–4	9,054	5,363	3,613	14,417		12,667
1895–9	8,943	5,352	3,190	14,295		12,133
1900–4	8,166	5,516	3,235	13,682		11,401
1905–9	8,634	5,845	3,634	14,479		12,268
1910–14	8,323	5,969	3,636	14,292		11,959

Notes: (1) Includes livestock derived from imported feed and imported stores.
(2) Less livestock derived from imported feed and imported stores.

Sources: Table 5.8 and O.J. Beilby, 'Changes in agricultural production in England and Wales', *Journal of the Royal Agricultural Society of England*, vol. 100, (1939), p. 64.

from imported feedstuffs. England and Wales were therefore modelled as receivers of inputs, while Ireland was a supplier.

The time periods in table 5.10 are those employed by Beilby. For England and Wales there are two estimates of the gross livestock contribution, the first includes that portion derived from imported feed and imported store cattle, but the second then deducts these categories. Depending on which columns are compared, therefore, there is some overlap between Irish output (which includes exports) and the measure of output from England and Wales which includes (Irish) imports. The contribution of crops for human consumption in England and Wales quite considerably exceeded the contribution of livestock. In Ireland the gap was narrow, and narrowing over time. In general, Irish output improved steadily *vis-à-vis* England and Wales: by most measures in 1914 it was over one half the size of the output from England and

Wales, while in the late 1880s it had been mostly well under one half.[58]

Conclusion – the search for perspective

Great caution must be exercised regarding the conclusions which can be derived from this study of the performance of Irish agriculturalists. The precise numbers which have been generated are less important than the trends and distributions of one sector of agriculture compared with another, and of one country compared with another. Nevertheless, as a region of the UK, during a time of considerable change in Western European agriculture in general, let alone the specific upsets and changes in Ireland itself, the Irish agricultural industry seems at times to have made a successful adjustment to those changes. This was the case in absolute terms when measured in nominal monetary values, and by *some* measures in real volume terms as well. Certainly with respect to the rest of the UK, Irish agriculture was contributing a larger share of output as time proceeded, and was enjoying productivity gains compared with the other side of the Irish Sea. In terms of physical output two things can be established: there was a large disposable energy output from Irish agriculture for home consumption after deducting for exports and fodder consumption; and there was an improvement in the size of that output per capita over time. In a comparison with England and Wales, it looks as though farmers in Ireland acted both more positively, and more quickly, to changing agricultural situations in the late nineteenth century.

Yet this optimistic gloss is precisely because Ireland can be treated as an independent region within the UK industry as a whole, in which case it may say more about the retarding state of that wider economy than it does about Ireland itself. If the idea of a nation state can be relaxed even further, and if the UK is regarded itself as a region, or part of a wider region in Europe as a whole, it might encourage us to spread the net of comparison of European agricultural performance a little wider and yet still retain the focus of attention on Ireland. According to Van Zanden's

---

[58] A comparison of total crop output equivalents reveals an Irish output in the 1880s of 18–19 billion calories, rising to 20–21 billion by the Great War. This was a rise from 30 + to 40 + per cent of the equivalent gross crop output in England and Wales, as follows,

Billions of crop calories (including crops destined for animal fodder)

	1885/9	1890/4	1895/9	1900/4	1905/9	1910/14
England & Wales	55.3	52.0	50.3	50.3	52.6	49.1
Ireland	18.7	19.4	19.8	20.5	21.0	20.9
Ireland as % of England	33.8	37.3	39.4	40.8	39.9	42.6

and 1910, Britain had one of the poorest agricultural growth and productivity performances over this period. In comparison, while Ireland was not in the vanguard of land and labour productivity in 1870, neither was it in the rear, but by 1910 it was one of the worst performers broadly in Western Europe by both of these measures. Land productivity suffered because of the failure to add fertilisers and other additives to the soil, but at that same time there was no substitution effect by increasing labour. In all sources, therefore, there was a neglect of factor inputs. Thus while there was an increase in both land and labour productivity, this was at a slower rate than almost everywhere else except Britain for both measures, and Italy and France for labour productivity.[59] These comparisons, however, are subject to important influences. The common element in the output equation is the prices of commodities, and these experienced similar trends in what were generally speaking internationally determined price situations. The only adjustment in these trends, at least in their magnitudes, was with respect to the presence or absence of protection. Thus in France, generally speaking, the internal prices of many products did not fall as much as in an open economy such as prevailed in the UK. In France the value of agricultural output rose modestly by 3.6 per cent from 1870 to 1880, but then fell equally modestly by 4.4 per cent and then 2.2 per cent over the next two decades, before recovering by 33 per cent before the Great War.[60] Such considerations make some European comparisons difficult to construct and equally difficult to interpret.

Given the growing importance of livestock production in Ireland at the time, and within it the continued importance (if somewhat diminishing over time) of dairy-based products, the classic alternative 'region' of Europe against which to make comparisons with Ireland is the Danish agricultural economy. Several references have already been made to this country, a foremost agricultural reformer and performer, if still only a small economy.

Danish data become good enough to compare with their Irish equivalent from about 1861, and they reveal that the output performance of Ireland, by some measures, was under a considerable Danish shadow. While the volume of Irish cereal production declined (total production by weight for example), Danish production expanded. From 1861 to 1909 a decline of over 20 per cent in one economy was matched by an

[59] J.L. Van Zanden, 'The first green revolution: the growth of production and productivity in European agriculture, 1870–1914', *Economic History Review*, 44, (1991), especially 229–30, 235.
[60] M. Tracy, *Agriculture in Western Europe* (London, 1982), p. 76, based on J.C. Toutain, *Le Produit de L'Agriculture Française de 1700 à 1958* (Paris, 1961).

Table 5.11:   *Distribution of agricultural output in Ireland and Denmark*

	Crops		Livestock	
	Ireland	Denmark	Ireland	Denmark
1851	58		42	
1861	34		66	
1871	26	58	74	42
1881	23	47	77	53
1891	21	27	79	73
1901	17	15	83	85
1911	16	16	84	84

Note: Percentage distribution according to off-farm sales by value
Based on 5-year averages.
Includes all of the main items, but not horses.

Sources: Table 4.5 above and adapted from E. Jensen, *Danish Agriculture: Its Economic Development* (Copenhagen, 1937), p. 262.

expansion of nearly 50 per cent in the other.[61] The differences in the two economies was even more profound in terms of livestock production. There was a rise of nearly 50 per cent in Ireland, but this was easily matched by a rise of nearly 400 per cent in Denmark. Overall therefore there was only a 30 per cent rise in Irish output but an increase of 150 per cent in Denmark. If Denmark produced 60–70 per cent of the size of the physical output of Ireland in 1861, by 1909 she produced about 130 per cent.[62] This came about through the combination in Denmark of an already large cereal sector which became larger, and the growth of its livestock from a relatively low position with respect to Ireland, such that by 1909 Denmark was a considerable rival to Ireland in terms of physical output.[63] This is not without significance, though it might cast Ireland in a false shadow. From tables in chapter 4 above we gained an appreciation of final output distributions between crops and animals. Table 5.11 compares these data with equivalent data for Denmark.

In terms of the distribution of the value of output, the Danish industry of 1871 mirrored post-Famine Ireland, and it was not until the twentieth century that the distributions of agricultural incomes across the two economies looked about the same. It looks as though Ireland was in the vanguard of change, but eventually she was caught up and overtaken.

---

[61] This comparison is made by Staehle, 'Statistical notes', 458–9, in which case it refers to the territory of what became the Irish Free State and later the Republic.
[62] *Ibid.*, 459.   [63] *Ibid.*, 459–60.

Even so, by 1911 cereal production still represented as much as 13 per cent of output in Denmark, but only 4 per cent in Ireland, but in contrast, butter, which was less than 10 per cent of output value in 1871 in Denmark had risen to 44 per cent in 1911. This compares with the equivalent figures in Ireland of 23 per cent and 18 per cent. Comparably in France, cereals represented 25 per cent of the value of final output in c. 1870, and still about 20 per cent on the eve of the Great War, while butter and milk represented 8 per cent and 9.5 per cent respectively.[64] The other great growth in Denmark was in pig production which rose from 12 to 24 per cent of output. In comparison *total* meat production in France represented 12 per cent in c. 1870 and still only 13 per cent in c. 1910.[65]

Therefore a more accurate description of Irish agricultural development would be to say that it became a different, rather than a less productive economy compared with Denmark, and certainly compared with France: cattle, in all of its non-butter senses began to dominate output (rising from 26 per cent in 1871 to 36 per cent in 1911), while in Denmark it was butter production which dominated, and in France it remained cereal production. Undoubtedly Ireland enjoyed a period in the third and fourth quarters of the nineteenth century when her product range and her anticipation of the changed international terms of trade was ahead of other rivals, but it was a short-lived experience, and whilst one biographer of the Danish experience could boast that the shift from grain to livestock production in that country 'was well under way before the years of low agricultural prices began in the seventies', this was decidedly an introspective judgement. The same observation could have been made about Ireland, but with more assurance.[66] Like Ireland, Denmark was an open economy and became a major exporter of animals, especially pork and dairy products, and, in at least this one respect shared a common export market in Britain. Given a policy of British free trade the only unnatural advantage the Irish had in the livestock trade was through the provisions of the contagious diseases legislation. This required meat which entered the British market to be slaughtered outside territorial waters, but it did not apply to Ireland which could therefore gain an advantage in the store cattle trade and in live animals in general. In such circumstances, and with the prevailing environmental conditions which ensured a generally wet Ireland, it was

[64] Tracy, *Agriculture*, p. 76.
[65] *Ibid.*, p. 76.
[66] E. Jensen, *Danish Agriculture: Its Economic Development* (Copenhagen, 1937), pp. vii and 60. See also the plea for a greater appreciation of the Irish performance in Staehle, 'Statistical notes', 460.

Table 5.12:   *Trends in agricultural output in Ireland and Denmark*

	Crops		Livestock		Total	
	Ireland	Denmark	Ireland	Denmark	Ireland	Denmark
1871	100	100	100	100	100	100
1881	81.4	89.9	95.6	137.8	91.9	110.0
1891	67.3	57.7	90.9	214.5	84.7	123.5
1901	59.3	39.6	103.8	315.2	92.1	155.4
1911	70.8	66.4	127.6	488.9	112.7	244.0

Note: Based on 5-year averages centred on population census years, 1871 = 100.

Sources: See table 4.2 and source cited in table 5.11.

Table 5.13   *Irish/Danish price relativities*

	wheat	barley	oats	butter	pork
1861	100	112	108	103	
1871	100	100	100	100	100
1881	94	98	111	77	100
1891	96	104	114	100	97
1901	105	115	110	80	101
1911	101	103	111	78	100

Sources: E. Jensen, *Danish Agriculture: Its Economic Development* (Copenhagen, 1937), pp. 373–82 and the Irish price indexes used in this book.

natural that the country should have specialised in animals. A simple comparison in agricultural output and distribution between Ireland and Denmark is shown in table 5.12.

When translated into index numbers the trends in output in the two countries can be compared and they show the large development in livestock production. Compared to Ireland, which was already an important livestock and livestock product producer before the 1870s, Denmark was a late developer, but thereafter she became a livestock producer with a vengeance. One reason for this was undoubtedly the 'speed and width of diffusion of the centrifugal separator in Denmark and Ireland [which] shows the latter in relatively poor light'.[67] While the combination of a co-operative creamery system and the latest technology had a major impact on Denmark, it is also possible, if not probable 'that

[67]   Ó Gráda, 'The beginnings of the Irish creamery system', 298.

Table 5.14 *Index of agricultural output per unit of labour input*

	Denmark	Ireland
1871	100	100
1881	94	105
1891	110	110
1901	239	138
1911	388	178

Sources: Labour statistics taken from B.R. Mitchell, *European Historical Statistics 1750–1975* (London, 1981 edn), p. 162 for Denmark, and J.P. Huttman, 'Institutional factors in the development of Irish agriculture, 1850–1915', Ph.D. Thesis, University of London, 1970, p. 414 for Ireland. The output data come from previous tables.

the innovation had spread as far as was viable in the Irish context by the 1910s'.[68] Such international comparisons can only be taken so far.

One reason for the dramatic advance in Danish agriculture relative to Ireland might have been due to discordant price movements. The combination of cereals, butter and pork in the agricultural output of Denmark always accounted for between 76 and 82 per cent of final output. Therefore if adjustments in relative prices explain the dramatic rise in Danish output, it is in these products that such adjustments should have occurred. Table 5.13 shows that there was some discordance, but it was hardly dramatic enough to aid our explanations. The cereal price indexes were not so discordant early on when cereals were important to Danish output, and later on such discordance was no longer so influential because the generally higher relative prices in Ireland had an impact on a much diminished sector. The pork price relativities remained remarkably level. The big influence was the butter index after about 1890 from when the trend was towards a lowering of butter prices in Denmark. Denmark therefore made her impact through volume sales and not through increased unit values.

The only redemption left for the relatively poorer performance of Irish agricultural production is the fact that the Irish population and labour force was falling. But this is a short-lived glow of comfort for although the Danish population was rising through this period, there was in fact a

[68] *Ibid.*, 299.

decline in the labour force and the ultimate result was a much improved labour productivity in Denmark relative to Ireland, as in table 5.14.

Thus, while the labour productivity of Irish agriculture looked good when compared with England and Wales, and even good, or as good when set against a more appropriate comparison with Denmark before about 1890, when the Danish agricultural economy took off from the 1890s the progress in Ireland looked somewhat pedestrian. Nevertheless, this chapter has uncovered an important improvement in labour productivity in Ireland in the sixty years after the Famine and it is specifically to these labour considerations to which we turn in chapter 6.

6   Labour and the working of the land

Introduction

When the literature on post-Famine Ireland is reviewed to seek a link between labour supply and changing agricultural production, it reveals a series of essentially conflicting stories and interpretations. The basic choices are of two sorts: was there a shortage of labour as a consequence of the mid-century crisis which determined changes in production out of tillage and into livestock; or was there a change in the terms of trade between tillage and livestock production which enhanced livestock production, independent initially of labour supply issues, but which then necessarily was labour saving, and therefore in turn led to people leaving the land; or was there a combination of the two? In this chapter this literature is reviewed, before the empirical exercise which confronts some of these issues is presented.

Conflicting models: labour constraint or market forces?

The demonstrable switch in emphasis in Irish agriculture from tillage to pasture after the Famine has been directly associated in official papers with a decline in labour supply. The clearest statement of this is contained in the comment by William Coyne, the Superintendent of Statistics at the Department of Agriculture, in his introduction to the 1901 *Agricultural Statistics*.

Adverting to the continuing contraction of the arable land of the country – this is, of course, the outstanding feature of these returns, especially when the process is regarded, not from year to year, but as an historical tendency. Land is being laid down to grass in Ireland, as in England, partly no doubt from the fact that stock raising and dairying are, at present prices, more profitable than corn growing, but far more largely (though this is not altogether a distinct cause from that just mentioned) because labour is scarce and dear.

The report continued:

There are fewer agricultural labourers in the country, and those who are left are said to be not the most efficient. Pasture involves much less outlay than any form of cultivation, the great saving being of course in the labour bill. On the other hand, a certain proportion of the land going into grass in Ireland is essentially unsuitable for pasture, while a net decrease of 29,615 acres in the area under cereals in a single year, coinciding with the decrease of 18,758 acres in the area under potatoes, 8,100 acres in the area under turnips, and 1,606 acres in the area under other green crops represents a shrinkage which can hardly be explained, in its entirety, by reference to market conditions.[1]

This seems to indicate the choices which faced Irish farmers, and some of the constraints which were in play. While recognising that market conditions *favoured* the major switch to stock and dairying as in England, it was in fact, or by implication, a declining labour supply which was the major *cause*. The comparison with England is not altogether a justifiable one when it is remembered from analysis in earlier chapters that Irish farmers adjusted their production and output to livestock some time in advance of English farmers. The official view also recognised that the cost of pasture cultivation was less than the cultivation of the land for other crops, with a particular emphasis on the saving of the labour bill. At face value that seems correct. The implication is that after the initial great loss of labour (through death and emigration), in spite of agricultural developments, a labour supply shortage persisted. For late-nineteenth-century English agriculture there is a suggestion that the official inquiries on agricultural depression in the 1890s demonstrated this developing shortage of labour: in their case labour-saving methods resulted because of a dearth of labour, and were not intended to shed labour; instead those labour-saving methods became a solution to an existing problem.[2]

The possibility that agricultural adjustments pushed labour off the land is not usually considered, and yet in Ireland there were other problems in place which may lend support to such an idea. For example, occupiers had difficulty in finding by-employments to supplement their livelihood; there was a reduction in the numbers of the smaller holdings, though not of the smallest; and the general movement up the agricultural holdings ladder resulted in a modest increase in average farm sizes. In the short term the economies of scale on larger holdings offset the need to hire an equivalent labour force as when the holdings were more numerous and in smaller units, but under the same acreage. In the long

---

[1] *Agricultural Statistics of Ireland ... 1901, B[ritish] P[arliamentary] P[apers]* [Cd. 1170], vol. cxvi, 1902, p. vii.

[2] F.M.L. Thompson, 'An anatomy of English agriculture, 1870–1914', in B.A. Holderness and M.E. Turner (eds.), *Land, Labour and Agriculture, 1700–1920: Essays for Gordon Mingay* (London, 1991), p. 217.

term this would have pushed labour off the land, and at the same time possibly sowed the seed of a long-term labour shortage.[3] For northern Ireland it has been suggested that, 'the principal cause of the depopulation of the rural districts ... has been the reduction in the number of small farms'.[4] This seems contrary to the labour shortage model.

There were many factors at play: there was the rate of decline in the agricultural labour force which was greater than the rate of decline in the rural population; in addition, the decline in tillage reduced the opportunities for agricultural employment. However, while the switch to large-scale farms economised on labour as a whole, it increased the demand for non-family labour, that is, for hired labourers.[5] There are implied, and sometimes circular, causal explanations in all of this, but no clear analysis. In addition, although there was a decrease in male agricultural workers, there was an increase in women workers on the farm. This resulted partly from the increase in the poultry sector, those animals now raised for sale rather than simply for household consumption. This had the effect of increasing the employment for wives and daughters.[6]

Irish emigration is an important element in any discussion of labour supply. R.C. Geary emphatically stated that emigration since the Famine resulted from an attraction from abroad rather than a repulsion from within the country.[7] He pointed out that the decline in the rural population of Saorstát Éireann was roughly the same across counties whereas the changes in agriculture were markedly different. In other words, the relative changes in the county rural populations were not necessarily equal to the relative changes in the emphasis on cattle and tillage. If the changes in the rural population were poorly correlated with basic agricultural change, then by implication so also was labour supply poorly correlated with agricultural change. From 1881 to 1911 cattle numbers increased by 21 per cent and the rural population declined by 25 per cent. Initially this argues for a close trade-off in labour usage, but great regional variations diminish the explanatory powers of these mean values. While the decrease in rural population varied only narrowly between 23.4 and 27.6 per cent, the increase in cattle numbers oscillated

[3] *The Agricultural Output of Northern Ireland, 1925, BPP* [Cmd. 87], 1928, p. 44.
[4] *Ibid.*, p. 48.   [5] *Ibid.*, p. 48.
[6] *Ibid.*, p. 49; B. Solow, *The Land Question and the Irish Economy, 1870–1903* (Cambridge, MA, 1971), p. 96; J. Bourke, 'Women and poultry in Ireland, 1891–1914', *Irish Historical Studies*, 25 (1987), 293. See also J.P. Huttman, 'The impact of land reform on agricultural production in Ireland', *Agricultural History*, 46 (1972), 361.
[7] R.C. Geary, 'The future population of Saorstát Éireann and some observations on population statistics', *Journal of the Statistical and Social Inquiry Society of Ireland*, 15 (1935), 23–4.

widely and wildly between 8.7 and 37.7 per cent.[8] In addition, emigration was greater by 14 per cent from 1881 to 1911 than it had been in the crisis years of 1846–9, but this was now at a time when agriculture was recovering (the measurement Geary uses for recovery is crop yields). Thus, Geary contends, the pull from abroad outweighed the push from home as the determinant of emigration. Nevertheless, since the correlation between population change and agricultural change was at least consistent, that is, a decline in population was correlated, however weakly, with a decline in tillage and an increase in cattle numbers, a relationship with labour supply cannot be entirely discounted. The big question remains, was agricultural change a response to a shortage of labour or a cause of its decline?

Geary's views have been challenged. It has been suggested forcefully that 'The reason for emigration from Ireland was obviously poverty; the people went to America because they had no work at home.'[9] This proposition poses several unasked questions: was the country overcrowded by the mid-nineteenth century? Did the Famine act as a catalyst for increased emigration and lead to yet more emigration because more information became available? Did post-Famine changes in agricultural production push people off the land? The market may have determined those changes, but a declining labour requirement produced the flow of migrants.

Yet another interpretation is offered by P. McGregor in his analysis of the impact of the potato blight on the rural economy. He suggests that the rural economy was hit in two ways. The potato blight devastated the economy of the poorer producers (cottiers and those on conacre plots). Initially this increased the supply of labour for hire, but, coincidentally, demand for labour fell as the income of the next strata of landholders, the tenants, fell. It fell because they were also wholly or partially dependent on potatoes for personal consumption, and for feeding their pigs and poultry. For these tenants there was a more attractive proposition to staying on the land, and this was to emigrate. The potato blight was part of a more widespread harvest problem in Europe at the time which had the effect of raising prices in general. Through real wage movements, this led in turn to a rise in the efficiency wage floor. The

[8]   *Ibid.*, 24. A recent analysis of the impact of the potato blight on the rural economy argues strongly for the opposite, and a strong association between agricultural change and labour supply. It suggests 'that increased tillage intensity was a substitution of labour for livestock'. By implication there was a close association between livestock changes and labour changes, or rather between livestock and the price of labour, P. McGregor, 'The impact of the blight upon the pre-Famine rural economy in Ireland', *The Economic and Social Review*, 15 (1984), 289–303, especially 292.

[9]   Geary, 'The future population', 34.

combination of a rise of wages and a decline in demand for labour was to increase the pool of unemployed labour.[10] There then entered two different forces: the relief mechanisms provided relief for the unemployed poor; but emigration became the salvation of the peasants. Those peasants who stayed on the land reacted to the relative increase in the price of labour by substituting livestock production for tillage.[11] This was a short-term solution, but it gave way to a longer-term permanent change. The argument continues that there was a withdrawal of many peasants from production altogether because the blight ruined them, or so diminished their status that emigration became a better option. This is a combination of push and pull factors, with the push preceding the pull. Those who remained on the land rationalised their situation and countered the higher costs of labour-intensive tillage production by switching to livestock production.

In his appreciation of the labour question Raymond Crotty exercised caution and allowed for a combination of market forces and labour supply issues to interact. Nevertheless, he suggested that *the order* of responses was initially from a position of inadequate labour supply, followed by an acceleration in agricultural change because of changes in market prices. Agriculture in Ireland therefore adjusted to a steep rise in rural labour costs as a result of labour constraints (the Famine and emigration), and this resulted in the move from labour-intensive tillage production to less labour-intensive grassland farming. In turn, higher rents could be had from the relative improvement of cattle prices over corn prices, but those rents could not be borne by tillage-derived incomes. To this extent, Irish farmers responded to the secondary rather than the primary forces of the market. They could not see the demand patterns for produce, and therefore anticipate them, but they could react to them with a delay. In any event, the further implication was that labour saving resulted from the changes in agricultural production.

Seemingly Crotty believed the ball was set in motion by a labour supply constraint.[12] If this interpretation is correct then it is the message which also emerges from J.S. Donnelly's study of Cork, since he seems

---

[10] Or as Peter Solar has said, potato was the main wage good, and therefore a fall in yields after the Famine made labour more expensive, thus making labour-intensive cultivation more costly. In turn this should have encouraged Irish farmers to seek ways to reduce labour costs by shifting away from labour-intensive production and reducing the demand for labour. P. Solar, The Great Famine was no ordinary subsistence crisis', in E.M. Crawford (ed.), *Famine the Irish Experience 900–1900: Subsistence Crises and Famines in Ireland* (Edinburgh, 1989), p. 128

[11] McGregor, 'The impact of the blight', 297–8.

[12] R.D. Crotty, *Irish Agricultural Production: Its Volume and Structure* (Cork, 1966), pp. 66–8.

simply to follow the Crotty line.[13] Similarly this is also Mary Daly's summary of events. Thus, as a result of the Famine and the immediate post-Famine disruptions to labour supply through death, emigration, destitution, eviction *et al.*, the primary problem that emerged was a labour supply constraint. The ready expression of this was the much reduced tillage acreage in 1847. The *secondary* reaction was the increase in the price of labour which brought to an end, or at least forced an adaptation of old labour-intensive practices. This adaptation included a reduction in the application of manure, generally reducing the impact of the good but labour-intensive husbandry practices, and the general economising on labour.[14] Thus, what began as a labour constraint because of labour disruption, became labour adaptation with less hired labour and relatively more family labour.

Necessarily there were short cuts in good husbandry practice when compared with the pre-Famine period, and the evidential consequences of this were reflected through crop yields and the volume of output.[15] This is a short or medium-term effect which commentators generally seem to ignore, but it allows the long-term responses to the market to take over at the appropriate time, and allows an explanation which seems to reconcile all or most views. Barbara Solow is one of the few to take this double-sided view of post-Famine labour issues and their relationship with economic change.[16] Donnelly, in taking a long-term view from the 1850s to the 1890s, makes a similar point, though starting from the position of emigration. The more emigrants there were, the more there was a labour constraint, and the more natural it was to switch to grassland farming. In Cork, those rural areas with the highest emigration rates experienced the largest decreases in ploughed land, and vice versa. Yet Donnelly has it both ways because the continuous shift to pasture farming was a clear response to the British market, to relative price movements and to the shift in the arable:livestock terms of trade thus magnifying the attractiveness of livestock production over tillage.[17]

The short-term explanation certainly has its attractions in economic, or shall we say, pragmatic terms. The dislocations which followed the Famine, including the deaths and emigration, gave many estates a 'breathing space', as Daly describes it. Now that they had vacant

[13] J.S. Donnelly, *The Land and the People of Nineteenth-Century Cork: The Rural Economy and the Land Question* (London, 1975), p. 133, and generally pp. 132–5 with the labour constraint provoking the introduction of threshing machines in the late 1850s.
[14] M. Daly, *The Famine in Ireland* (Dublin, 1986), pp. 63–4.
[15] See also T. Barrington, 'The yields of Irish tillage crops since the year 1847', *Journal of the Department of Agriculture*, 21 (1921), 223, 225–6.
[16] Solow, *The Land Question*, pp. 93–4.
[17] Donnelly, *The Land and the People*, p. 233.

holdings the landlords could indulge in some restructuring, probably for the first time for perhaps a century. Yet a proper restructuring probably required a greater volume of vacant holdings than that which the Famine created: it required vacancy at such a high level that it could only come about by eviction. Thus, 'evictions added a measure of compulsion to be resorted to by landlords who felt that natural wastage had not been sufficient'.[18] Additionally, it was not necessarily rent arrears which provoked eviction, but rather a whole package of financial problems including arrears and the local taxation which was introduced to help finance Famine relief schemes. Evictions may have added to that tax burden by putting yet more pressure on those schemes.[19] Apart from any sense of altruism they might have felt, along with the fulfilment of their local duty, the landlords did derive some tangible rewards, because the situation afforded them the opportunity to restructure. There are no obvious correlations between levels of eviction and rent arrears, and so perhaps too much is read into Daly's summary of events. However, her claim is supported by reference to pre-Famine secret societies and the attendant agrarian crimes which stultified earlier attempts at restructuring. These organisations were weakened by the Famine, and restructuring could proceed.[20]

Historians should guard against exaggerating the total level of eviction, or indeed the pervasiveness of evictions.[21] The seeds of a major change, however, may have been planted, and a change in the structure and production of agriculture was about to take place. This was the tendency towards livestock farming in the 1850s which was accelerated by the sharp fall in labour available. The economic advantages of abandoning labour-intensive methods proved sound in the long run, because in the post-restructuring phase it was the obvious, and economically most rational move to make as the international terms of trade moved against grains.[22] Kevin O'Rourke has tested this and a number of other models with respect to post-Famine rural depopulation. He poses a familiar question. Was the depopulation caused by push or pull factors: was it the push from the land because of agricultural restructuring in the wake of the Famine which responded to a price shock as the terms of

[18] Daly, *The Famine*, p. 112. Eviction in the long-term was not ever-present nor always the method of replacing tenants except when they fell into rent arrears. See the Bessborough Commission evidence related to the decade before the 1880s quoted by Solow, *The Land Question*, p. 53. It so happens that evictions lumped in the few years after the Famine. See Solow, p. 55, and W.E. Vaughan, *Landlords and Tenants in Ireland 1848–1904* (Dublin, 1984), pp. 15–17, 23–24.

[19] Daly, *The Famine*, pp. 109–10.   [20] *Ibid.*, p. 111.

[21] For a clear statistical history of evictions from the 1850s to 1880s see W.E. Vaughan, *Landlords and Tenants in Mid-Victorian Ireland* (Oxford, 1994), pp. 229–31.

[22] Daly, *The Famine*, p. 119.

trade favoured livestock over tillage production; or was it a pull from abroad allowing Irish labour to filter into the international labour market? His main conclusion, which is in the same vein as that of Geary over fifty years earlier, is that the pull from abroad outweighed the push from home, though this is not to dismiss entirely the push from home. Whilst relative price changes or shocks did help to restructure agriculture – and since those price shocks were largely determined internationally, they were present to a certain degree even if the Famine had not occurred – they did not bring about the long-term observable decline in population. Of course the Famine and immediate Famine years (1845–9) were a shock to the rural population, but thereafter it was the foreign labour market which attracted labour to leave Ireland. O'Rourke wins on both accounts however because the upwards pressure on Irish wages after the Famine further reinforced the agricultural changes which were underway, and these were inherently labour saving. The pull and the push seemed to work together. The ultimate conclusion is that a labour constraint existed, and it helped to mould agricultural change coincidental with adjustments in the terms of trade, which in turn encouraged the switch from grain to tillage production. In another study he states rather more firmly that the Famine was 'a major watershed' in hastening the switch from tillage to pasture and thus reducing long-run agricultural employment. This seems to argue more for the internal push than the external pull, or rather it allows for the push to precede what became the (greater) pull.[23]

This line of argument regarding push or pull factors is given greater impetus by Jeffrey Williamson's work on international labour markets, and in his collaborations with Tim Hatton. Williamson's new estimates point to the rise in real wages which McGregor (cited above) suggested occurred. For the present purposes Williamson establishes one great point about the evolution of a global labour market, and specifically one important point regarding Ireland. In the late nineteenth century there was a convergence in the average wage gap between the New World and the Old leading to a convergence of, though not an equalisation of, real average wages. This has implications for emigration. It had its impact on the push factors as a result of the local Irish crisis in mid-century, and also on the pull factor arising from the attraction of trans-Atlantic migration in particular. 'Real wages in Ireland started a dramatic

---

[23]   K.H. O'Rourke, *Agricultural Change and Rural Depopulation: Ireland, 1845–1876*, Ph.D. Dissertation, Harvard, 1989, especially chapter 4 (the substantial argument) and chapter 7 (the conclusion). See also his 'Rural depopulation in a small open economy: Ireland 1856–1876', *Explorations in Economic History*, 28 (1991), 409–32; O'Rourke, 'Did the Great Irish Famine matter?', *Journal of Economic History*, 51 (1991), 1–22.

convergence on the UK during the 1850s.' In 1845 they were 54 per cent
of the UK real wage, and in 1853 they were still only 53 per cent (with
one abnormal year in 1850 of 61 per cent), but in 1854 they reached 68
per cent, thence to 71 per cent by 1870, 79 per cent by 1890 and 83 per
cent by 1913 (with a maximum of 92 per cent in 1905). In contrast, the
US real wage converged in the opposite direction, though not consis-
tently, and at best it remained level relative to the UK from 1870. Thus
'The Irish convergence towards real wages in America must have been
even more dramatic since relative real wages were falling in America
during most of this period.'[24] This narrowing of the wage gap, it is
contended, explains the secular fall in the emigration rate from Ireland,
along with adjustments in the size of the family and the acquisition of a
place on the landholding ladder, though as we emphasise below, this last
effect was disabled by the restrictions on property inheritance which
developed, and which worked against splitting up family farms.[25] Un-
deniably there was a push from the Famine, though Ireland also had the
highest emigration rate in Europe *before* the Famine. Equally evidently,
there was a decline in the pull from America, the main receiving centre
for displaced Irishmen. This line of argument, based on real wage effects,
however, may be self-fulfilling and may distort the importance of those
effects. The bulk of the population decline at the Famine, after all, was
concentrated amongst the poorest sectors of the population. Naturally
their loss would boost the wage floor, as McGregor suggested.[26]

This line of argument increasingly stresses improvements in living
standards in Ireland in the second half of the nineteenth century and the
ability of such a condition to slow down, though not reverse, the
emigration of Irishmen.[27] At the same time there developed adjustments
in local Irish economic circumstances which made it desirable to resist
the move overseas, but made it necessary also to have in possession the
vital factor of production, land. Changes in land inheritance after the
Famine, when the division of family small-holdings ceased and the
average size of farms increased, was the encouragement to stay put, but

---

[24] J.G. Williamson, 'The evolution of global labor markets in the first and second world
since 1830: background evidence and hypotheses', *Working Paper Series on Historical
Factors in Long Run Growth*, 36 (National Bureau of Economic Research, 1992), 36,
30–31, figures 13 and 14, and appendix 2. See also Williamson, 'Economic convergence:
placing post-Famine Ireland in comparative perspective', *Irish Economic and Social
History*, 21 (1994), 5–27.
[25] T.J. Hatton and J.G. Williamson, 'After the famine: emigration from Ireland, 1850–
1913', *Journal of Economic History*, 53 (1993), 575–600.
[26] See also K.A. Kennedy, T. Giblin, and D. McHugh, *The Economic Development of
Ireland in the Twentieth Century* (London, 1988), pp. 20–1.
[27] See also J.D. Gould, 'European international emigration: the role of "diffusion" and
"feedback"', *Journal of European Economic History*, 9 (1980), 281.

not everyone could inherit in such a situation. Thus was created a constant pool of frustrated would-be landholders. From 1870, though more importantly from the early twentieth century, the moves towards Irish peasant independence aided and abetted this desire to stay in Ireland. Yet in so doing, unless there was a complete halt on procreation, it necessarily created a surplus supply of an otherwise would-be landholding class. This begs a question, did the move into pasture-related husbandry which was less labour intensive and which therefore further restricted entry onto the land, guarantee that this situation developed and assumed a dynamic of its own?

This chapter began with an 'official' explanation of the Irish labour market, and this must be questioned. It equated a labour constraint with a specific course of agricultural change. It suggested that the solution to the constraint was to turn to pasture farming *because* it was labour saving compared with arable cultivation. This is a common assumption and is upheld with some force in Thomas Grimshaw's comments of 1895. He was the Registrar General of Ireland. With enormous certainty he proclaimed that: 'We know that the labour expended in stock-raising is much less than the labour expended on tillage, and that all other things being equal, the cost of working an acre for tillage purposes is much greater than that of working an acre for the growth of live stock.'[28] The demand for arable products is derived from the needs of the farmers' own table and from the market. The demand for pasture, in its broadest sense, is derived from the need to supply fodder and hay to animals. The labour component in raising grass therefore is inseparable from the labour supply in milking cows and tending sheep and cattle. Furthermore, much arable production is highly seasonal, but dairying demands a daily labour input. Finally, much arable was actually destined for fodder, and so the labour input to that arable should rightly be equated with livestock and livestock product output. When the labour supply is related to different land uses, the simplistic model which looks at the simple adjustment out of arable and into pasture is not enough. The implications for local populations and labour usage of such adjustments, in reality, was much more complicated. Indeed labour usage as distinct from labour supply is given very little consideration in such a model.

One popular method of investigating labour in relation to land use is the simple construction of land to labour ratios. For the nine historic counties of Ulster Liam Kennedy has established an increase in the land

---

[28] *Royal Commission on the Financial Relations between Great Britain and Ireland, BPP* [C. 7720 and C. 7721], vol. xxxvi, 1895, appendix VII, 'Statement by Dr Grimshaw with regard to the values of output of Irish agriculture for certain periods', p. 453.

to labour ratio from 10:1 to 12:1 from 1851 to 1876, and then to 16:1 by 1913.[29] Within fairly narrow limits the area under cultivation in Ulster (defined as pasture plus tillage), did not change much. It increased from 3.956 to 3.976 million acres from 1851 to 1913, or an increase of only 0.5 per cent, but the increase in the land to labour ratio was of the order of 60 per cent. Therefore the outcome of the ratio was driven entirely by labour adjustments. By about 1900 the active male labour force was farming more acres per head than in almost all other European countries bar Britain, Denmark, Greece, the Baltic States and European Russia.[30] In view of the economic diversity of these countries the use of land to labour ratios as a measure of productivity change is surely very misleading. At best they indicate the allocation of resources where labour is abundant relative to capital, though they do also indicate the importance of shifting labour out of primary production for more general economic development. It is this area of resources allocation which starkly links agricultural productivity with economic development.[31] In the Irish case it still seems more sensible to treat the country as a region of the UK rather than as a wholly separate economy.

Kennedy points to one very interesting feature. The tillage acreage in Ireland as a whole did not change appreciably in the decade following the Famine, whether hay is counted as tillage or not. From 1850 to 1860 it increased from 5.758 to 5.97 million acres with hay, or decreased from 4.558 to 4.376 million acres without hay. A 4 per cent swing one way or the other establishes that to all intents and purposes the tillage acreage was stable. Yet this took place alongside a huge decline in population, and therefore Kennedy concludes that 'a labour shortage was not a serious restraining factor on labour-intensive forms of agriculture'.[32] He may be confusing the declining population with a decline in the labour force of those who *actually worked for others*, as distinct from being self-employed. To explain this we invoke Crotty who pointed out that the Gregory or 'quarter-acre' clause in the Irish poor law disqualified occupiers of holdings above one quarter of an acre from qualifying for poor relief. It was those close to the quarter acre, and perhaps up to 5 acres, who lived closest to the margin of survival. Numerically they were numerous, and their position was made precarious at the time of the Famine. 'The loss of these holdings

[29]   L. Kennedy, 'The rural economy, 1820–1914', in L. Kennedy and P. Ollerenshaw (eds.), *An Economic History of Ulster, 1820–1939* (Manchester, 1985), p. 19.

[30]   F. Dovring, *Land and Labor in Europe 1900–1950* (The Hague, 1956), pp. 66–7.

[31]   See also P.K. O'Brien and Leandro Prados de la Escosura, 'Agricultural productivity and European industrialization', *Economic History Review*, 45 (1992), 529–30.

[32]   Kennedy, 'The rural economy', p. 18.

actually caused the agricultural labour supply to be augmented during the Famine and immediate post-Famine years, notwithstanding the overall decline in population.'[33] Thus, with no great surprise, the tillage acreage, after all, could easily be well maintained. Crotty's point is well made even though by current consensus he probably misinterpreted the landholding statistics attached to the 1841 Census.

Thus there were rather more complicated labour supply issues in play than at first meets the eye. The simple observation that the population engaged in agriculture declined from 1.46 million to 0.781 million from 1851 to 1911 (a decline of 46.5 per cent as compared with a total population decline of 33 per cent) is only a starting point for analysis, not yet a grand conclusion of any useful substance.

A methodology

With a relatively fixed acreage of land under cultivation, but a declining labour input, and even with, generally speaking, a fairly flat long-run trend in the real volume of agricultural output from 1850 to 1914, an improvement in labour productivity is evident. The movement in the land to labour ratio seems to capture this, but this is not yet a very revealing index, and even coupled with the basic statistics of land-use change it cannot yet establish much about labour shortage or labour saving. What is needed is some measure of labour usage which can combine the labour-intensive needs of the arable with the supposed less labour-intensive needs of pasture. In a report of 1836 it was established that the labour involved in ploughing a fixed area of land was one quarter of that which was required in spade husbandry, but that the plough required five times the labour per unit area of the stockman. Thus crude labour usage ratios were formulated.[34] But farming is hardly ever such a simple catalogue of exclusive choices. For example, the total labour input into animal production is a composite of the stockman's labour, but also that of the ploughman or spade husbandman, because a measure of the labour input which goes into providing the non-pasture fodder for the animals is also required. Thus the simple notion that pasture production traditionally is less labour intensive is only part of the story and perhaps more misleading than it is revealing. As we will discover, livestock farming was more labour intensive than is popularly thought.

[33] Crotty, *Irish Agricultural Production*, pp. 49–50.
[34] J.H. Andrews, 'Limits of agricultural settlement in pre-Famine Ireland', in L.M. Cullen and F. Furet (eds.), *Ireland and France 17th to 20th Centuries: Towards a Comparative Study of Rural History* (Ann Arbor, MI, and Paris, 1980), p. 53.

The investigation into labour usage can be rationalised into two strategies. Firstly, the output of agriculture can be valued and then related to the standard estimates of the agricultural labour force, from which combination a measure of labour productivity can be derived. Secondly, labour usage can be estimated on the farm in a way which captures both total usage and the distribution of that usage between different enterprises (arable and pasture, or arable and livestock). Furthermore, those arable or tillage operations which were derived from livestock farming (fodder crops) can be separated, thereby adding another dimension to the otherwise more simplistic approach of constructing tillage to pasture ratios.

A form of the first strategy appeared in the previous chapter where a comparison was made with England and Wales. The second strategy is adopted here. The method is to reduce all farm operations into a common index based on standard man days. At the extremes the annual labour required to grow an acre of wheat, oats, barley and so on, was much different from the less labour-intensive requirements of pasture cultivation and the tending of animals, though within the animal sector the daily milking of cows had a higher labour input than did the tending of sheep on the open pastures or moors.

The literature on standard man days has mainly been applied to modern problems. Unfortunately, if those modern estimates are adopted they will be based on a different range of agricultural technologies than existed in Ireland in the nineteenth century. Therefore the attempt must be made to construct nineteenth-century estimates. In addition, although an assessment of the total labour input for a long period of agricultural adjustment from 1850 to 1914 will be made, necessarily it will use unchanging estimates of the individual labour requirements of the different products across those sixty years. This approach, therefore, will not take into account changing technology which agricultural adjustment itself may have produced. The decline in spade husbandry, even though there was only a marginal change in the tillage area, is one such technological change or adaptation. Furthermore, even if Ireland had certain homogenous characteristics of terrain, soils and climate, there were different regional emphases of land use, and perhaps different regional methods of cultivating the same or similar crops. There may have been developmental changes which make it unrealistic to make comparisons over any considerable period. For example there may have been time and space-based differences in the degree of mechanisation. None of these considerations can satisfactorily be catered for in the method adopted.

In a limited way W.E. Vaughan has pointed us in this direction already. He estimated labour inputs for the two years 1854 and 1874.[35] These were based on the published accounts of Richard Barrington, a nineteenth-century tenant farmer from Fassaroe in county Wicklow, for the period 1837 to 1885.[36] These accounts allow an assessment of the labour input devoted to a range of crops, and expressed in man days per acre. For estimates of the labour involved in livestock production Vaughan consulted the work of the contemporary Commissioner of Valuation, Sir Richard Griffiths, who instructed the valuers and surveyors associated with the country-wide valuation of lands and tenements in Ireland from the 1850s onwards.[37] Vaughan does not sufficiently explain his method of estimation to replicate it here, and therefore his novel use of the Barrington material leaves us tantalisingly close to a measure of labour inputs, but not close enough. Therefore, Fassaroe has been revisited, so to speak, in appendix 6 and estimates of the labour requirement for a narrow range of crops for the period from the 1830s to the 1880s have been made. These estimates will form the basis of the analysis which follows.

Establishing labour inputs

Appendix 6 contains details of the reconstruction of farm labour inputs employed on the Barrington family farm. Table 6.1 is a summary of the resulting estimates that are applied in this chapter to mid-nineteenth-century Irish agriculture. The second part of the table is a set of modern estimates based on the *Farm Management Survey* of 1984/5 conducted by the University of Manchester.[38]

Ideally the relationship which each crop bears with the other crops in the two estimates should be the same. Therefore if the ratio of the labour requirements for potatoes relative to cereals, for example, was the same in both of the estimates it would suggest that the *precise measure* of standard man days for that crop was not as important as long as the

---

[35] W.E. Vaughan, 'Agricultural output, rents and wages in Ireland, 1850–1880', in L.M. Cullen and F. Furet (eds.), *Ireland and France 17th to 20th Centuries*, pp. 85–97, especially pp. 87, 91–2. See also W.E. Vaughan, 'Landlord and tenant relations between the Famine and the Land War, 1850–78', Ph.D. thesis, University of Ireland, 1973, p. 37.

[36] Richard Barrington, 'The prices of some agricultural produce and the cost of farm labour for the last fifty years', *Journal of the Statistical and Social Inquiry Society of Ireland*, 9 (1887), 137–53.

[37] Sir Richard Griffith, *Instructions to the Valuators and Surveyors*, appointed under the 15th and 16th Vict, c. 63, for the Uniform Valuation of Lands and Tenements in Ireland (Dublin, 1853), especially p. 33.

[38] University of Manchester, *Farm Management Survey 1984/5* (University of Manchester, Department of Agricultural Economics, April, 1986), pp. ix–x.

Table 6.1 *Imputed labour requirements in mid-nineteenth-century Irish agriculture* (in standard man days per acre)

	Potatoes	Turnips	Wheat	Oats	Barley	Total for 5 acres
(A) Derived from the Barrington family farm, Fassaroe 1837–85						
1837–85[a]	39.2	25.3	17.9	17.3	18.4	118.0

(B) Derived from Richard Griffith *c.* 1850
    Hay (cutting and making)   4.6
(C) Derived from modern estimates, 1984–5[b]

Cereals	1.0	Grass		0.8
Peas/Beans for stock	1.0	Permanent grass		0.4
Potatoes	12.1	Rough grazing		0.4
Brassicas	16.2	Fallow		0.4
Beets	20.2			
Carrots	8.1			

Notes: [a] Estimates used in subsequent calculations are: potatoes @ 40 man days per acre plus 4.5 for labour for manure; turnips @ 25 man days per acre plus 4.5 for labour for manure; wheat @ 18, oats @ 17 and barley @ 19.
[b] Originally reported in SMDs per hectare but here converted into statute measures

Sources: See appendix table 6.4; Sir Richard Griffith, *Instructions to the Valuators and Surveyors, appointed under the 15th and 16th Victoria, c. 63, for the Uniform Valuation of Lands and Tenements in Ireland* (Dublin, 1853), p. 33; University of Manchester, Farm Management Survey 1984/5 (University of Manchester, Department of Agricultural Economics, April, 1986), pp. ix–x.

relative ratios it had with other crops was more or less constant over time. Although the use of modern estimates is inappropriate to apply to earlier periods, Barbara Solow's view was that, 'changes in techniques in the intervening years are unlikely to have been large enough to obliterate the main features that are evident'.[39] She cited, but did not use, an estimate of standard man days derived from the Small Farmers Scheme in Northern Ireland in 1958.[40] With constant cross-crop relationships the trend in labour usage would be accurate even if the precise measurements

[39] Solow, *The Land Question*, p. 106.
[40] *Ibid.*, pp. 107–8, based on L. Symons (ed.), *Land Use in Northern Ireland* (London, 1963), p. 57. Solow's intention was to show that apart from dairying, tillage was more labour intensive than livestock production. For 1871 she demonstrated that small farms with scarce land were labour intensive, and therefore arable, while large farms with relatively scarce labour were mainly livestock.

were not. A comparison of the estimated labour usage on the Barrington farm with modern estimates indicates that technological changes in arable production released labour more dramatically from cereals and pulses than from other crops, and labour saving in potato production was greater than from other root and green crops. If temporary grass is equated with hay, and permanent grass with pasture, then the modern estimates suggest that the former is twice as labour intensive as the latter. Given the improved technology employed in the hay harvest I doubt whether the same ratio between hay and pasture existed for nineteenth-century Ireland. Interestingly, in alternative modern assessments the labour requirement for hay is often similar to that required for grains. This fact might be significant later in a discussion of the livestock labour requirements because it suggests that substituting hay for grains does not necessarily save labour, and therefore the adjustment to livestock production was not necessarily labour saving, though in modern estimates the labour requirement is based on highly mechanised farm operations, hardly a description of late-nineteenth-century Ireland.

There is a large modern literature on the parameters or weights used in calculating standard man days, and predictably for the same crops or the same animals there are a range of possible weights employed by different investigators, and even by the same investigator in separate publications (as in table 6.2).[41] The second half of table 6.2 uses potatoes as the standard against which to measure the labour requirements of other crops. There is reasonable equality in the potatoes/turnips and potatoes/hay ratios when comparing Fassaroe with modern estimates, but the other potato/crop ratios vary quite widely.[42] However, since modern estimates by no means agree with one another, and since the empirical base of the Fassaroe estimates is probably as good as exists in many

[41] See for example, Symons, *Land Use*, p. 57; J.T. Coppock, 'Crop, livestock, and enterprise combinations in England and Wales', *Economic Geography*, 40 (1964), especially 68–70; Coppock, *An Agricultural Geography of Great Britain* (London, 1971), p. 150; Coppock, *An Agricultural Atlas of England and Wales* (London, 1976 ed.), pp. 226–7; W.B. Morgan and R.J.C. Munton, *Agricultural Geography* (London, 1971), p. 111; MAFF, *Farm Incomes in England and Wales, 1974–5*, Farm Incomes Series No. 28 (HMSO, February 1976), para. 51; *Farm Management Survey*, pp. ix-x; C.J.W. Edwards, 'Changes in agricultural labour efficiency in Northern Ireland 1975–84', *Irish Geography*, 19 (1986), 77. On the livestock side Coppock assessed the labour requirement of a dairy cow as 15 man days in one publication, but 12 in another. See also Dovring, *Land and Labour*, appendix 4, pp. 398–418.

[42] Modern estimates often make the distinction between root crops which are gathered and cut, and hence applied to livestock, and those which are left in the ground for the direct folding of animals. The historic folding is much less labour intensive and also does not incur a technological or machine cost, but it also implies little or no stall feeding. Ireland was/is a wet country, but in terms of temperature it was/is mild, and therefore the assumption about little stall feeding is probably near to the truth.

modern studies, they are adopted here. The measured differences over time suggest that there was more technological change in the ploughing/reaping/sowing/harvesting types of crops most commonly associated with labour-intensive farm work, that is, in the grain harvest and the mowing of hay. More correctly, greater technological change is seen here than in the traditionally heavier labour-intensive operations associated with root and green crop cultivation.

The livestock side is even more complicated given the lack of nineteenth-century Irish estimates to set against modern ones. A heavy reliance on intuition, or more frankly, guesswork, may be required to complete the analysis of labour inputs. One possible guide is the comment from Sir Richard Griffiths to the effect that 'Dairymaid, support and wages, for 6 cows' cost £8 in Ireland in the early 1850s, or 26.7 shillings per cow per annum.[43] If it is assumed that this was entirely a wage or labour charge, then at Fassaroe from 1847–54 when the wage rate for agricultural labour was 1.3 shillings per day the annual labour input was 20 man days per cow. An official estimate of an agricultural labourer's cash wage rate for the early 1850s was 5.875 shillings per week, which for a six day week suggests 27 man days per cow (or 32 for a 7-day week).[44] This estimate expressly discounts added payments for the hay and corn harvest and allowances in kind. Did Griffiths' reported costs per cow include these unidentified 'support' charges? The cow expense was set alongside other expenses, including cutting and making hay, the service of a bull, undefined 'contingencies', cooperage for collecting and storing the milk and butter derived from the cow, and so on. To this extent the labour charge may indeed have been isolated.

Another contemporary source, this time for Limerick in 1863, suggests that the labour charge for keeping 30 cows by a dairyman and girls, including wages and keep, was £43.625, or about 29 shillings per cow. At the contemporary wage rate of 6 or 7 shillings per week this suggests 25 to 29 standard man days per annum per cow on the basis of a 6-day week, or 29 to 34 standard man days for a 7-day week. At Fassaroe the contemporary wage rate was 1.5 shillings per day, suggesting about 20 standard man days per cow. This includes an unspecified charge for keep. By 1871 the same 30 cows cost £65 in terms of labour charges.[45] At the contemporary Fassaroe wage rate this suggests 23.6 standard man days per cow, and at the national average contemporary wage rate of 7.58 shillings per

[43] Griffith, *Instructions*, p. 33.

[44] A. Wilson Fox, *Second Report by Mr. Wilson Fox on the Wages, Earnings, and Conditions of Employment of Agricultural Labourers in the United Kingdom*, *BPP* [Cd. 2376], vol. xcvii, 1905, p. 135.

[45] R.O. Pringle, 'A review of Irish agriculture, chiefly with reference to the production of live stock', *Journal of the Royal Agricultural Society of England*, 2nd series, 8 (1872), 62.

Table 6.2   *Estimated farm labour requirements in nineteenth-century Ireland compared with modern estimates* (in standard man days per acre per annum)

Crop	Fassaroe[a]	Coppock 1	Coppock 2	Morgan and Munton[b]	University of Manchester	Symons
Potatoes	43.6	16	20	15	12	21
Turnips[c]	29.8	9–11.5	12	3–12	16–20	
Wheat	17.9	2	3.5	1.5	1	3.5
Oats	17.3	3	4.5	1.5	1	4.5
Barley	18.5	2	3.5	1.5	1	3.5
Hay	4.6	1	2	1.5	0.8	2

Using potatoes as the standard produces the following ratios

Potatoes/						
turnips	1.5	1.4–1.8	1.7	1.2–4.9	0.6–0.8	
pots/wheat	2.4	8	5.7	10	12.1	6
pots/oats	2.5	5.3	4.4	10	12.1	4.7
pots/barley	2.4	8	5.7	10	12.1	6
pots/grain	c. 2.4	5–8	4–6	10	12.1	5.6
pots/hay	9.5	16	10	10	15.1	10.5

Notes: [a] After adding 4.5 for manure labour costs.
[b] Originally reported in SMD per hectare.
[c] Turnip range in Coppock 1 = stock feeding and mangolds
   Turnip alternative in Coppock 2 would give 21 for mangolds and 20 for cabbage
   Turnip range in Morgan and Munton = folded and unfolded
   Turnip range in Manchester = brassicas 16 and beets 20

Sources: Fassaroe See Appendix 6 and associated text.
Coppock 1	J.T. Coppock, *An Agricultural Geography of Great Britain* (London, 1971), p. 150.
Coppock 2	J.T. Coppock, *An Agricultural Atlas of England and Wales* (London, 1964), p. 214.
Morgan & Munton	W.B. Morgan and R.J.C. Munton, *Agricultural Geography* (London,1971), p. 111.
University of Manchester	University of Manchester, *Farm Management Survey 1984/5* (University of Manchester, Department of Agricultural Economics, April 1986).
Symons	L. Symons (ed.), *Land Use in Northern Ireland* (London, 1963), p. 57.

week it suggests 34 to 40 standard man days. This includes an unspecified charge for keep. If a final choice of 20 standard man days per cow equivalent is made, it does not, on these contemporary bases, exaggerate the dairy cow labour requirement.

Table 6.3 compares modern labour estimates for livestock, but only for those animals which are reported in the nineteenth-century Irish Agricultural Returns. The calculated equivalents in dairy cow terms are also reported, as are those equivalents using the mid-nineteenth-century Irish estimate of 20 standard man days per cow.[46] The labour input estimates in the second column of the lower part of table 6.3 are now used as the inferred labour inputs in Irish livestock production in the second half of the nineteenth century. On this basis, a combination of that column with the estimates in table 6.1 produces the full product estimates in table 6.4. Some of these estimates are compromises, some are based on guesswork, and there is also the uncertainty of the effects of technological change. In effect, the hidden assumption is that there was no technological change.

Labour usage in Irish agriculture

Corn or grain crops were more labour intensive 100 years ago compared with modern times. This is self-evident given the state of technological change, but perhaps tillage cultivation was even more labour intensive in Ireland than in the rest of Britain, and the subsequent rate and degree of technological change much slower and lower. Modern estimates suggest a ratio of labour inputs between corn and root and green crops (including potatoes) of between 1:4 and 1:12. At Fassaroe the ratio was 1:2 or 1:3. In contemporary and modern estimates the labour required for cultivating hay and permanent grass was and is much less than for tillage crops. The simple tillage to pasture ratio which is so often used to measure broad land-use change, therefore, misses the emphasis of that change when it comes to labour inputs. By the present calculations, for every acre of grain which was converted to grass, there was a release of 18 standard man days. Only 5 of them were re-employed if the conversion was to hay, and fewer than 2 if the conversion was to permanent pasture (based on a labour ratio of hay to grain of 1:3.6, or pasture to grain of 1:9).

The compensating factor in reducing the impact on labour usage in the switch from tillage is to question what the pasture was used for. It looks like a change in land use which saved labour, but if it resulted in the support of more animals, then the degree of re-employment of that

[46] Preferring to use here the specifically measured Fassaroe wage rates rather than the national composite estimate, and thus err on the side of underestimation.

Table 6.3   *Estimates of standard man days labour inputs for livestock* (in standard man days per animal per annum)

Animal	Coppock 1	Coppock 2	Morgan and Munton	University of Manchester	Symons
a	b	c	d	e	
Dairy	12	15	10	7	15
Cattle > 2	2.5	3	2.5	1.5	3
Cattle 1–2	2.5	3	2	1.5	3
Cattle < 1	2	3	1	1.5	3
Sheep > 1	0.75	0.25	0.75	0.25–0.5	0.66
Sheep < 1	0.25	1	0.5	0.25	
Pigs > 1	4	4	4	3.5	
Pigs < 1	0.5	1.2	2	0.5	
Poultry	0.05–0.25	0.1–0.3	0.01–0.25	0.03–0.04	

Range of estimates in cow equivalents based on 1 and 20 dairy cow equivalents

Dairy Cow	1	20
Cattle > 2	0.2–0.25	4–5
Cattle 1–2	0.2	4
Cattle < 1	0.1–0.2	2–4
Sheep > 1	0.04–0.075	0.8–1.5
Sheep < 1	0.02–0.05	0.4–1.0
Pigs > 1	0.3–0.5	6–10
Pigs < 1	0.04–0.2	0.8–4
Poultry	0.001–0.025	0.02–0.5

Notes: a In fact sows and boars at 4 and other pigs at 0.5.
b Upland sheep 0.5, lowland sheep 1.0, other sheep 0.25.
c Ewes and rams 0.75, other sheep 0.5.
d Beef cows and heifers over 2 years, 2.5 and 2 respectively, rams and ewes 0.5, all other sheep 0.25. Pigs > 1, in fact boars and gilts.
e And separately, bulls counted for 7 man days per year.

Sources: See table 6.2.

labour depended on the type of animals raised. Milk production was labour intensive, and therefore a conversion to grass which led to an extension of dairying was not at all labour saving. An inspection of the agricultural trends in Ireland in the second half of the nineteenth century reveals that the largest increase in cattle numbers was for dry cattle up to 2 years of age. These cattle employed fewer labour units per animal than did all tillage crops per acre, though the true saving in labour is a more delicate equation involving the acres of tillage released, the re-engagement of a number of acres as hay and/or pasture, and the engagement of

Table 6.4 *Standard man days weighting factors for mid-nineteenth century Irish agriculture*

Product	SMDs	Product	SMDs
Corn/Grain[a]	18	Milk cows	20
Potatoes	44	Cattle > 2	5–4
Root & Green	29	Cattle 1–2	4
Flax[b]	25	Cattle < 1	2–4
Hay	5	Sheep > 1	0.8–1.5
Pasture[c]	0.625–1.25	Pigs > 1	6–10
		Pigs < 1	0.8–4
		Goats[d]	0.8–1.5
		Poultry	0.02–0.5

Notes: a Includes peas and beans.
b A guess based on high labour use as in root and green crops.
c Modern methods have the pasture to hay ratio at 1:4 or 1:8.
d This is assumed to be the same as sheep. Probably it should be higher, maybe even the same as milk cows bearing in mind the uses made of goats for their milk. To equate with milk cows however, that is by multiplying by a large factor, would raise the total labour usage for goats to ridiculously large heights. Modern estimates make no mention of goats.

Sources: Derived from earlier tables.

a certain number or head of 'cattle'. An acre of grain released 18 standard man days, but an acre of pasture employed fewer than 2 of them. Thus for every new acre of pasture, as long as fewer than 4 extra head of cattle were kept there would be a net labour saving (cattle having a labour input weight of 4 or 5 standard man days per head). Eight extra sheep could be engaged and still effect a labour saving, but fewer than 1 extra milk cow could be engaged for every loss of 1 acre of corn and gain of 1 acre of pasture. Combinations of different animals could be used on fewer grain and more pasture acres and still result in a net labour saving, but under certain circumstances such a combination could also result in a net labour increase.

If labour saving was the object of the exercise then a net reduction in the total labour input over time should be observed. On the estimates employed in this chapter this was the case. Table 6.5 has been calculated using the estimates of standard man days in table 6.4 (upper bounds for livestock). The total labour requirement for the cultivated area rose irregularly up to a peak in 1859 when it stood at nearly 136 million standard man days.[47] It then fell away at a moderately even rate to a low

[47] There are no grass acreages for 1849 and 1850. The figures for these years are based on a back projection of five year averages.

point of 85 million standard man days in 1909, after which it recovered slightly.[48]

Even though the area of permanent grass was always the largest component of land use by some considerable margin, the total labour required to cultivate that grass exceeded the labour requirement for only two other crops, flax and hay, and certainly not for the main crops. It equalled the labour input into root and green crops (excluding potatoes) from about 1880 onwards. The labour input into hay cultivation from about 1880 actually exceeded the labour input for grass. The labour inputs for both corn and potatoes each always exceeded that for pasture: it was always twice as large for potatoes, and twice as large for corn until the mid-1890s. Thus, although the switch to grass was labour saving, this can be a misleading interpretation of changing land use: the rise of hay cultivation was a much more impressive feature of post-Famine land-use change. The ratio of the total man days engaged in hay cultivation compared with the equivalent engaged on the pasture stood at close to 0.6:1 in the early 1850s, but it was greater than one by the Great War. The man days engaged on the cultivation of hay and grass combined exceeded the equivalent which was engaged on corn production from 1901. Still in all, in absolute terms, the major labour requirement on the farm as far as the cultivated area was concerned was for the root and green crops, and most importantly for the potatoes. In 1859 potatoes employed 39 per cent of all the labour used on the land, by 1890 it used 34 per cent, and about 30 per cent on the eve of the Great War. The labour requirements for green and root crops remained buoyant at 12 or 13 million standard man days throughout the period.

The proportion of labour employed in corn production fell away steadily from 43 per cent in 1851 to 32 per cent in 1865, settling at a fairly constant 26 per cent by the end of the century. Flax was only significant as a proportion of the total labour requirement during the cotton famine of the mid-1860s.[49] Hay and grass together employed from under 14 per cent of the total labour on the land in the 1850s, to over 25 per cent by the end of the century, the hay rising more than the grass. Using constant labour weights through time for each land use assumes incorrectly that there was little or no technological change; it means that the labour

---

[48] The unreported lower bound estimates mirror this pattern but of course with lower values, at 129.736 million standard man days in 1859 and 78.871 million in 1909. If the chronology is extended it would reveal a sizeable increase up to 100 + million smds in the last two years of the Great War when the plough-up campaign had some effect.

[49] In Northern Ireland in 1958 when potatoes were reckoned at 21 standard man days, flax was reckoned at 15. If a similar ratio is assumed in the mid-nineteenth century then the 44 standard man days for potatoes equates with 31 for flax. In adopting 25 I am erring on the side of caution.

Table 6.5 *Estimates of labour inputs into Irish agriculture, 1850–1854 to 1910–1914* (in millions of standard man days)

	On cultivation (including grass)	On animals	Total	Livestock/cultivation ratio
1850–54[a]	128	49	177	0.38
1855–59	133	58	191	0.43
1860–64	130	54	184	0.41
1865–69	124	57	181	0.46
1870–74	119	60	179	0.50
1875–79	111	60	170	0.54
1880–84	105	56	161	0.54
1885–89	102	58	160	0.57
1890–94	98	62	160	0.63
1895–99	93	63	156	0.67
1900–04	90	65	154	0.72
1905–09[b]	87	67	154	0.77
1910–14[b]	87	71	158	0.81

Notes: a The separate enumeration of milk cattle did not begin until 1854. But from 1854–9 the proportion of all cattle over 2 years which were milk cows was within the narrow limits of 66.7 and 68.4 per cent with an average of 67.5. The original calculations were made using this proportion.

b There is suspicion in the rate of upturn of the livestock figure because it is unduly influenced by the redefinition of poultry in 1906/7.

Source: See text.

inputs for individual crops followed the same *trends* as their acreages, and it also assumes constant crop yields. However, the question of whether variable yields required different inputs of labour is by no means certain. It appears that high yields in any particular harvest did not necessarily require more, or significantly more labour, though the same was not true for secondary labour functions such as threshing; Belgian evidence for example suggests that high yielding grain was easier to thresh.[50]

A similar exercise on the construction of labour inputs for the livestock

---

[50] On technological changes concerned with the harvest during the period, though for Britain rather than Ireland see, E.J.T. Collins, 'Harvest technology and labour supply in Britain, 1790–1870', *Economic History Review*, 22 (1969), 453–73; For Ireland see the remarks by C. Ó Gráda, *Ireland Before and After the Famine: Explorations in Economic History* (Manchester, 1988), passim. For Belgium see P. Solar and M. Goossens, 'Agricultural productivity in Belgium and Ireland in the early nineteenth century', chapter 14 in B.M.S. Campbell and M. Overton (eds.), *Land, Labour and Livestock: Historical Studies in European Agricultural Productivity* (Manchester, 1991), p. 381.

side of agricultural production is also reported in table 6.5.[51] The proportion of labour employed in livestock production devoted to milk fell steadily from 56 per cent in the late 1850s to 45 per cent in the half decade or so before the Great War. Cattle production employed from 14 to over 20 per cent of all labour used in the livestock sector, and cattle and milk together accounted for 70 per cent in the 1850s reducing thereafter, but only modestly, to 60–65 per cent by the Great War. There was a steady rise in the labour used in poultry production: it was always high, even as high as pigs and sheep at times.[52]

In and around 1851–61 livestock production consumed 28 to 29 per cent of all labour employed on the farm, rising to 44 per cent by 1911. According to this analysis livestock requirements never exceeded crop requirements. The ratio of the labour requirement employed in livestock production compared with crop production moved evenly from 0.38 to 0.8. The more usual way of showing the adjustment from crops to livestock is to look at the ratio of tillage acres to pasture acres. This moved from about 2 to over 5 during the period.[53] The alternative labour-usage method employed here is a slightly less dramatic way of showing the same change, and it might be added that the implied labour saving is smaller than is often suggested.

Labour profitability

While the total labour requirement declined evenly over time an important question remains: was this due to a labour supply constraint which resulted in a change out of labour-intensive crop production and into less labour-intensive livestock production, or did labour usage decline because agriculturalists responded to changing market conditions and consequently created a labour surplus which they subsequently shed? This is still an open question.

Table 6.6 summarises one method of looking at productivity changes. Whatever happened down on the farm the labour force which remained behind was more intensively used. The transition to greater livestock use may initially indicate the less seasonal, more daily rhythm of work.[54] There was a decline in standard man days from 1861 down to 1901, and

---

[51] A separate enumeration for dairy cows was not made until 1854, therefore statistics for the period before this date are based on a back-projection from the post-1854 trend.

[52] Even taking into account the new methods of enumerating poultry in 1906–7, the poultry sector was still a large labour user. At a lower bound estimation the poultry requirements diminish considerably.

[53] This includes hay as pasture. If it is included as tillage the ratio moves from 1.5 to 2.5.

[54] J. Mokyr, *Why Ireland Starved: A Quantitative and Analytical History of the Irish Economy, 1800–1850* (London, 1985 edn), p. 78 n. 7 and 214–5.

especially from 1871 to 1881. This may suggest that agricultural adjustments were labour saving, but it is still hard to say what was driving the changes. Was there a labour supply constraint with a declining workforce as the evidence for such a constraint, but also was there a less than proportional decline in labour usage (or standard man days), making those that remained behind appear to work harder as the second half of table 6.6 seems to show?

Though a decline in population for any reason, coupled with a less than proportionate decline in labour usage, will automatically produce a rise in the ratio of standard man days to population, it will not similarly automatically measure labour productivity, but it is suggestive of other considerations. By simple back-projection, the question can be asked whether there was an oversupply of labour at the time of the Famine? The question is an old one. It is found wanting by Mokyr and Ó Gráda, but it has a long history.[55] In 1921 Thomas Barrington commented on the depletion of the rural population from 1846 to 1850, suggesting that, 'it might be urged with some show of reason that, as regards agriculture, and even intensive cultivation, the labour thus lost consisted of a surplus relative to requirements that could be spared without prejudice to existing methods and practices'.[56] Diminishing returns to labour were ingrained in the Irish economy according to Barrington. T. Jones Hughes has spoken of 'The curse of rural Ireland in the [mid] nineteenth century' as 'over-population or under-employment',[57] and recently the now late T.W. Freeman posed 'The big question', namely, 'Was Ireland overpopulated?' and the answer is 'an emphatic "yes" given that overpopulation is assessed in relation to the development of resources at one chosen time'. Such views pass easily into legend through the textbooks. Thus D.B. Grigg unhesitatingly declares 'There is little doubt that Ireland in the 1840s was overpopulated.'[58] In more analytical and forceful tone, but with an echo of Barrington, Crotty referred to 'the meagre subsistence' in the 1840s of an 'economically surplus labour force and its dependants'. Furthermore, he referred to a decline in the labour force in the period 1845–51, but if anything an increase in production such that the output

---

[55] *Ibid.*, ch. 3; Mokyr, 'Malthusian models and Irish history', *Journal of Economic History*, 40 (1980), 159–66; J. Mokyr and C. Ó Gráda, 'Emigration and poverty in prefamine Ireland', *Explorations in Economic History*, 19 (1982), 362–3.

[56] Barrington, 'The yields of Irish tillage crops', 225. See also J. O'Donovan, *The Economic History of Live Stock in Ireland* (Cork, 1940), p. 205, but based, it is said, on the wisdom derived from the lectures of that immense Irish historian George O'Brien.

[57] T. Jones Hughes, 'East Leinster in the mid-nineteenth century', *Irish Geography*, 3: 5 (1958), 237–8.

[58] T.W. Freeman, 'Land and people, c. 1841', in W.E. Vaughan (ed.), *A New History of Ireland. V Ireland Under the Union I, 1801–70* (Oxford, 1989), p. 263. D.B. Grigg, *The Dynamics of Agricultural Change* (London, 1982), p. 35.

Table 6.6   *Labour productivity in Irish agriculture, 1851–1911*

	Total population (inter-censal change)		Agric population (inter-censal change)		Total SMDs (inter-censal change)	
	Millions	%	Millions	%	Millions	%
1851	6.552		1.460		176.0	
1861	5.799	(−11.5)	1.173	(−19.7)	187.0	(6.3)
1871	5.412	(−6.7)	1.046	(−10.8)	179.2	(−4.2)
1881	5.175	(−4.4)	0.986	(−5.7)	164.6	(−8.1)
1891	4.705	(−9.1)	0.937	(−5.0)	159.9	(−2.9)
1901	4.459	(−5.2)	0.876	(−6.5)	154.7	(−3.3)
1911	4.390	(−1.5)	0.781	(−10.8)	156.4	(1.1)

	SMDs/Head of total population (Inter-censal change)		SMDs/Head of agricultural population (Inter-censal change)	
	Millions	%	Millions	%
1851	26.9		120.5	
1861	32.2	(20.0)	159.4	(32.2)
1871	33.1	(2.7)	171.3	(7.5)
1881	31.8	(−3.9)	166.9	(−2.6)
1891	34.0	(6.8)	170.7	(2.2)
1901	34.7	(2.1)	176.6	(3.5)
1911	35.6	(2.7)	200.3	(13.4)

Sources: Derived from earlier tables.

of the mid-1840s could have been achieved with a reduction in the workforce of 20 per cent.[59] This sounds like a severe case of diminishing returns in the 1840s carrying with it a vision of the labour force tripping over itself at every twist and turn.

David Fitzpatrick quotes contemporary observations and concludes that at times both before and after the Famine 'the supply of labour was at once "too much and too little" ', and not too surprisingly we learn that underemployment in winter was endemic.[60] A continuation of this line of argument might be to confront the problem of whether, and to what degree, underemployment was also endemic in summer. This, however, would ignore other features of the seasonal rhythm of agriculture and the activities of the labour force in other pursuits. It should not be assumed that there were periods of enforced leisure, there were cottage industries to fall back upon.[61]

[59] Crotty, *Irish Agricultural Production*, pp. 44–5.
[60] D. Fitzpatrick, 'The disappearance of the Irish agricultural labourer, 1841–1912', *Irish Economic and Social History*, 7 (1980), 78.
[61] Mokyr, *Why Ireland Starved?*, p. 216.

Ernest Barker's characterisation of congestion in the 'congested districts' of western Ireland in the late nineteenth century was that they were not congested in the sense of overpopulated, but rather in the sense that they had a disproportionate share of uneconomic holdings, and therefore needed assistance. By analogy, at the time of the Famine Ireland had many potentially uneconomic holdings. By definition, all holdings were potentially uneconomic, but in more realistic terms Barker perhaps was alluding to the small size of peasant holdings. At the time of the Famine those at the margin of subsistence were not simply vulnerable by their small size, but, collectively, they also formed a large proportion of society.[62]

One origin for the oversupply thesis is Sir Charles Trevelyan, a central character in the Famine years. He referred to pre-Famine Ireland as overrun with the unemployed, or should we say underemployed.[63] Mary Daly does not subscribe to this view but instead points to the complaints of labour shortages, and the resultant reduction in the crop acreage in 1847 as 'proof' of (i.e. to disprove) the 'common assertion' of pre-Famine unemployment.[64] It neither proves nor disproves this because as she correctly describes the events of the Famine and its aftermath, there was a great unsettling of the labour force through death, emigration and destitution, which did result in a labour constraint. This says very little about the degree of employment before the crisis. There is however, yet another view. Peter Solar contends that Irish agricultural production on the eve of the Famine did better than much historiography has suggested, and the interpretation of why Ireland was richer than popular image supposes arose, 'not because of abundant labour, but in spite of it'.[65] An available and cheap labour supply encouraged spade husbandry and generally good practices of manuring and fertilising which demonstrably improved yields and hence output. Relative riches thus arose out of abundant labour.

Were the post-Famine adjustments so efficient in labour saving as to shake out the labour force at a rate sufficient to give the impression of a rapid advance in labour productivity? It is more likely that the land was not as well worked, and in fact husbandry practices deteriorated. Declining yields were experienced everywhere. Whatever mechanisms

---

[62] E. Barker, *Ireland in the Last Fifty Years (1866–1916)* (London, 1916), pp. 77, 83–4, although almost certainly based on a misreading of the 1841 landownership data.

[63] Daly, *The Famine*, p. 64.

[64] *Ibid.*, p. 64.

[65] P. Solar, 'Agricultural productivity and economic development in Ireland and Scotland in the early nineteenth century', in T.M. Devine and D. Dickson (eds.), *Ireland and Scotland 1600–1850. Parallels and Contrasts in Economic and Social Development* (Edinburgh, 1983), p. 78.

were in place there certainly was an incompatibility between the rate of decline in the size of the labour force (the agricultural population), and the rate of decline in the standard man days or work on the farm (see table 6.6 columns 4 and 6). When the decline of work was greatest (a fall of 8.1 per cent in 1871–81) the decline in the labour force was modest when compared with the same decline in most other inter-censals. When work on the farm appeared to increase (a rise of 1.1 per cent in 1901–11) the rate of decline of the labour force was one of the greatest. In this context the measure of standard man days per head of the agricultural labour force is *akin* to a labour efficiency index. In this example, with all but one of the inter-censal changes in positive figures, it appears that there was an increase in labour efficiency over time.[66]

Much of the tillage output was eventually intended to feed the livestock, but this has not been accounted for yet: the switch in labour usage from tillage to livestock has not been properly identified.[67] For example, all labour expended on hay and grass must be counted for livestock labour purposes, and most or all of the root and green crops were also destined for livestock, and a large proportion of potatoes went into livestock, as indeed did a proportion of corn crops. The same assumptions and procedures can be adopted as in the output chapter to estimate how much tillage was intended for livestock use. From 1850 to 1914 between 56 and 64 per cent of the corn output was produced for human consumption, in the new language adopted in this chapter it used tillage standard man days, but about 40 per cent was used in livestock production, and 100 per cent of the root and green crops was grown for livestock purposes, and also up to 45 per cent of the potatoes. An appropriate reallocation of the labour force suggests that in the period up to the mid-1860s perhaps only 30 per cent of the total standard man days was used in tillage production which was destined for human consumption, and the remaining 70 per cent was employed directly in livestock production or indirectly for the livestock by producing fodder crops. By 1900 the ratio between the two was nearer to 20:80, declining ultimately to about 18:82 by the Great War. In his study of French

---

[66] See Edwards, 'Changes in agricultural labour efficiency', 75. More correctly, the labour efficiency index is the standard man days required *100/standard man days available, when standard man days are measured in the same units.

[67] It was such considerations which prompted C.J.W. Edwards to favour standard output conversion factors based on standard cash values apportioned to each crop or livestock component, in his study of modern Londonderry. This is all very well for a static cross-sectional study where fixed factors could be applied, but in a longitudinal study the factors would constantly change with price relativities reflecting market conditions. Labour inputs per enterprise are more likely to remain unchanged over times. Besides, Edwards' study was not based on labour issues. C.J.W. Edwards, 'Farm enterprise systems in east county Londonderry', *Irish Geography*, 7 (1974), 32–3.

agriculture George Grantham also identified the problem of the *indirect* input of labour necessarily engaged in fodder production, that is engaged to all intents and purposes in primary tillage production but for secondary livestock use. His views are comforting but their application in his French study cannot really help the different approach we take here. He says that by the mid nineteenth century the labour employed in raising fodder added 35 to 40 per cent to off-peak employment.[68]

Table 6.7 is a final summary of labour usage. There was an increase in labour employed in crop production for human consumption in the first decade after the Famine, and also an increase in that side of crop production which was destined for fodder. During a decade when the rate of decline of labour on the farm was at its greatest the amount of work which that labour was required to perform increased in all areas of farm activity. The great decline in farm population was coincident with, in the first instance, an increase in work and no adjustments in agriculture to speak of. From about 1861 down to the 1890s there was a large decline in the labour requirement for crops in their own right. At the same time the labour used in producing crops for fodder also decreased, but at a much slower rate. Effectively therefore there was both an actual transfer of activity towards livestock labour use but also a substantial shift in land use for livestock usage. Far from it taking as long as the 1890s for livestock to supersede crops in terms of labour requirements, livestock all along engaged more labour. This may yet be sufficient to dispel the notion that agricultural change came about because labour supply was constrained. At the very least it seriously questions it.

Labour is a cost which incurs a wage and whose worth is reflected in the value of agricultural output. Table 6.8 compares the labour requirement or usage on the farm with the value of agricultural output. Initially there was a large fall in the value of output per unit of labour input in crop production from the 1850s to early 1860s, but thereafter it looks as though there was some fluctuation, further decline, and indeed partial recovery at the end of the period. However, that initial fall had been sufficiently large, and it was not until the decade before the Great War that the value of crop output per unit of labour input began to recover back towards the 1851 position (for which see the index of crop value per unit of standard man days input). If this is a trend of labour profitability then it indicates a decline in that profitability for the production of grain crops, or at best a depressed profitability, and it seems to warrant the agricultural changes that took place. It points as much as any other

---

[68] G.W. Grantham, 'Divisions of labour: agricultural productivity and occupational specialization in pre-industrial France', *Economic History Review*, 46 (1993), 484–5.

Table 6.7   *Standard man days labour inputs in Irish agriculture,
1851–1911* (in millions of standard man days)

	Livestock labour	Crop labour employed in livestock	Total livestock labour	%	Crop labour	%	Ratio of livestock to crops
1851	49.1	72.2	121.3		54.7		2.2
1861	55.4	75.5	130.9	(7.9)	56.1	(2.6)	2.3
1871	59.0	71.5	130.5	(−0.3)	48.7	(−13.2)	2.7
1881	58.0	66.3	124.3	(−4.8)	40.3	(−17.2)	3.1
1891	60.5	63.7	124.2	(−0.1)	35.6	(−11.7)	3.5
1901	63.9	60.8	124.7	(0.4)	30.0	(−15.7)	4.2
1911	69.6	58.8	128.4	(3.0)	28.0	(4.6)	4.6

Note: Inter-censal percentage changes in brackets.
For 1861, 1871, 1881, 1891 and 1901, annual averages of 1856–65, 1866–75, 1876–85,
1886–95 and 1896–1905. But for 1851 and 1911, annual averages of 1849–55 and 1906–14.

Source: See text.

alternative suggestions to a question of market considerations which induced the changes in Irish agriculture.

On the livestock side there was a continuous and large rise in the value of output per unit of labour input from 1851 to 1881. It then fell in the 1880s (in fact 1886–95), before recovering again. By 1901 (1896–1905) it was only a little above the preceding peak of 1881, but it was well above it by 1911 (see the index of livestock value per unit of standard man days input).

In total output terms there was a rise in the value of output per unit of work up to 1881, then a fall back in the 1880s (in fact 1886–95), before a recovery by 1901 and a considerable advance up to the Great War. If this is a trend in labour profitability then it shows that the rate of growth of that profitability in livestock production was greater than the equivalent rate of growth of profitability in that part of crop production which was undertaken for the sake of producing those crops in their own right, and not as an input into secondary uses. In addition, we notice that the value of crop output per unit of labour input in purely monetary terms was higher than the value of livestock output in the first half of the period. Livestock production used a lot of labour because it depended on labour hungry crop inputs. The labour saving arguments, therefore, take on a new dimension.

Table 6.8 *Labour input and the value of final agricultural output in Ireland, 1851–1911*

	Standard man days (millions)			Values (In £s million)		
	Livestock input	Crop input	Total input	Livestock output	Crop output	Total output
1851	121.3	54.7	176.0	12.1	15.9	28.0
1861	130.9	56.1	187.0	23.2	13.6	36.8
1871	130.5	48.7	179.2	29.5	12.6	42.1
1881	124.3	40.3	164.6	31.0	9.4	40.4
1891	124.2	35.6	159.8	28.6	7.3	35.9
1901	124.7	30.0	154.7	31.6	6.5	38.1
1911	128.4	28.0	156.4	38.7	7.6	46.3

Value of output per unit of SMD input

	Livestock		Crops		Total	
	£s	Index	£s	Index	£s	Index
1851	0.10	(100.0)	0.29	(100.0)	0.16	(100.0)
1861	0.18	(177.7)	0.24	(83.4)	0.20	(123.7)
1871	0.23	(226.6)	0.26	(89.0)	0.23	(147.7)
1881	0.25	(250.0)	0.23	(80.2)	0.25	(154.3)
1891	0.23	(230.8)	0.21	(70.5)	0.22	(141.2)
1901	0.25	(254.0)	0.22	(74.5)	0.25	(154.8)
1911	0.30	(302.1)	0.27	(93.4)	0.30	(186.1)

Notes: Livestock SMD has been adjusted to include crop labour intended for fodder crops.
The years 1861, 1871, 1881, 1891, 1901 based on average for 1856–65, 1866–75 ... etc. 1851 based on average of 1850–2. To include 1853–5 would be to influence unduly the values by the inclusion of high prices in the Crimean War period without the compensating influence of lower output values in the late 1840s, 1911 is in fact the average of 1906–14.

Source: Earlier tables in this chapter and chapter 4.

If labour supply was constrained, and if by the current estimates there was an increase in the standard man days per worker over time, then the price of labour should have increased. On the basis of ten widely scattered Irish farms the average wage per week for agricultural labourers rose from 5.88 shillings in 1850/54 to 10.67 shillings in 1903. This near doubling of wages was continuous over time (except in 1862 and 1863), with occasional plateaux of one or two years. For the shorter period 1874

Table 6.9   *Wage trends 1850–1903 (weekly wages in shillings per worker)*

	1st sample 10 farms	2nd sample 22 farms	Composite sample
1850–55	5.896		
1856–65	6.513		
1866–75	7.500	7.583[a]	7.514
1876–85	8.570	8.108	8.340
1886–95	9.446	8.758	9.102
1896–1903	10.380	9.458	9.919

Note: [a] In fact 1874–5
Source: A. Wilson Fox, *Second Report . . . on the Wages, Earnings and Conditions of Employment of Agricultural Labourers in the United Kingdom, BPP* [Cd.2376], vol. xcvii, 1905, p. 137.

to 1903 similar data are available from twenty-two farms.[69] Table 6.9 summarises the annual average weekly cash wage for the period 1850–1903. The average wage on the larger, but short-lived sample was smaller. In the following analysis the smaller sample is employed. This may exert an upwards bias on the wage bill, but it will be counteracted by a downwards bias on any measures of profitability.

The trend in the national wage bill can be related to the equivalent trend in the value of output in two separate ways. Firstly, the average wage centred on census years is combined with the supposed agricultural labour force to produce a surrogate total wage bill. Secondly, the average wage is combined with the labour requirements – the standard man days which are assumed to be the labour inputs – and produces a second surrogate total wage bill. The first calculation makes unwritten assumptions about levels of employment and continuous, non-seasonal employment: the second method assumes constant wage rates in winter and summer, but it is couched purely in terms of costs and effort and can avoid other distortions due to seasonal and other effects.[70] The results of

[69] Wilson Fox, *Second Report*, p. 137.
[70] On such issues see Pringle, 'A review of Irish agriculture', 54; J.W. Boyle, 'A marginal failure: the Irish rural laborer', in S. Clark and J.S. Donnelly (eds.), *Irish Peasants: Violence and Political Unrest 1780–1914* (Manchester, 1983), p. 322. A.L. Bowley made the distinction in Irish agricultural wages between that level which assumed uniform employment and that which allowed for want of employment at certain times of the year. At the time of the Famine the annual earnings of Irish agricultural labour could be halved if want of employment was taken into consideration. By the late 1880s his view was that the distinction had pretty well disappeared. A.L. Bowley, 'The statistics of wages in the United Kingdom during the last hundred years. (Part IV.) Agricultural Wages – Concluded. Earnings and General Averages', *Journal of the Royal Statistical Society*, 62:3 (1899), 555–70, especially 558–9, 566–7.

Table 6.10   *Labour costs and labour output*

	Average wage	Size of labour force	Standard man days	Value of output[a]	
	shillings per week	000s	mills SMD	£s millions	(Index)
1850–55	5.896	1.460	176.043	28.0	(76.1)
1856–65	6.513	1.173	186.975	36.8	(100.0)
1866–75	7.500	1.046	179.204	42.1	(114.4)
1876–85	8.570	0.986	164.582	40.4	(109.8)
1886–95	9.446	0.937	159.857	35.9	(97.6)
1896–1905	10.380	0.876	154.697	38.1	(103.5)
1906–14	n.a	0.781	156.434	46.3	(125.8)

	Surrogate wage 1[b]		Surrogate wage 2[c]	
	£s millions	Index	£s millions	Index
1850–55	22.4	112.7	8.6	85.2
1856–65	19.9	100.0	10.1	100.0
1866–75	20.4	102.7	11.2	110.4
1876–85	22.0	110.6	11.8	115.8
1886–95	23.0	115.9	12.6	124.0
1896–1903	23.6	119.0	13.4	131.9

Notes: [a]   Output for 1850–5 in fact 1850–2 – see note in table 6.8.
[b]   First wage bill is average weekly wage times 52 weeks.
[c]   The second wage bill is yearly standard man days divided by 6 to give standard man weeks, multiplied by the average weekly wage.

Sources: Tables 6.6, 6.8 and 6.9.

these crude attempts at trending are given in table 6.10, in combination with the value of output.

The first surrogate wage bill is derived from the size of the labour force multiplied by the average weekly wage, and further multiplied by fifty-two to give an annual bill. The second is the number of standard man days, which is a yearly measure, divided by six to give standard man weeks (allowing for a sabbath day of rest), multiplied by the weekly wage. There is a massive difference in the two estimates. Much of this might be accounted for by the fact that the agricultural labour force here defined was not all in receipt of weekly wages. Perhaps 70 per cent or more of the labour force was composed of the farmer or grazier and his extended family.[71] The farmers' and graziers' labour inputs were not costless, though the agricultural labourers' wage may not be the appropriate measure. In addition, much family labour was probably retained 'even when the marginal product of such labour had fallen below the level of the local agricultural wage'.[72] To this extent the true wage bill in the first estimate is greatly overestimated. In addition, on the basis of a fifty-two week year it is automatically assumed that there was a continuous level of employment. Again, this will certainly lead to an overestimate of the wage bill.

If the wage cost in the first estimate is anything like the true wage cost then the standard man days exercise is grossly underestimated. Conversely, if the standard man days wage bill is at all close to reality then the level of unemployment or seasonal unemployment must have been profound. This can be *partly* accounted for by seasonal migration to Scotland and England, though the size of this annual exodus may easily be exaggerated.[73]

W.E. Vaughan has estimated a national wage bill which amounted to £10.646 million in 1854 and £13.565 million in 1874, giving an increase over the twenty years of 27.4 per cent.[74] The trend at least compares with one version of these new estimates which shows a rise from £8.65 million to £11.2 million over an equivalent period, or an increase of 29.5 per cent (from the second estimated surrogate wages bill).

The two wage estimates can be used to construct a crude cost-benefit schedule showing the final value of agricultural output for every £1 expended on labour (table 6.11). Using the first wage estimate suggests an increase in the returns to labour up to about 1871 (which of course is really a return to the employer not the employee), and a decline thereafter to the turn of the century. The second wage estimate follows a similar trend but with a decrease from c. 1871 down below the 1851 base. In this sense the employer was not getting as good value for his money. The beginning of the diminishing returns occurred between 1871 and 1881 (more specifically on the time periods chosen it would appear to have taken place between 1866–75 and 1876–85). This change was coincident with the depression of 1879–82, and perhaps it was related to the high

---

[71] For which see Wilson Fox, *Second Report*, p. 262. See also Boyle, 'A marginal failure', pp. 312–3; Fitzpatrick, 'The disappearance', *passim* but especially p. 89 for an assessment of the distribution in 1912.

[72] L. Kennedy, 'Traders in the Irish rural economy, 1880–1914', *Economic History Review*, 32 (1979), 209.

[73] Boyle, 'A marginal failure', pp. 320–1.

[74] Vaughan, 'Agricultural output, rents and wages', p. 87; though elsewhere in Vaughan, *Landlords and Tenants*, pp. 4–5, he suggests that labourers accounted for about £9 million in the 1850s.

Table 6.11   *Value of labour output per unit value of labour input,*
1850–1903

For every £ expended on labour the return to the employer was

	Based on labour force derived wages bill		Based on SMD derived wages bill	
	£s	Index	£s	Index
1850–5	1.3	100.0	3.2	100.0
1856–65	1.8	148.1	3.6	112.0
1866–75	2.1	165.1	3.8	116.1
1876–85	1.8	147.0	3.4	106.2
1886–95	1.6	124.7	2.9	88.1
1896–1903	1.6	128.8	2.8	88.0

Sources: Derived from table 6.10.

point of agricultural output in 1876, which then gave way to the decline
into depression.

Conclusion

Observably, from all the data in this chapter there was a declining labour
force but a fairly steady labour input measured in standard man days
across the product mix of the Irish farm, and encompassing its changes
over time. This may not initially suggest a labour constraint but rather a
case of underemployment earlier in the trend. Yet the rising unit wage
bill does suggest a labour constraint. The unit wage even rose during the
classic deflationary period of the 1870s to 1890s. Thus, there was a rise in
the profitability of each £1 expended per unit of labour input to about
the 1870s, but then generally it declined. But the rate at which labour was
saved by changing agricultural processes was not at a fast pace. As much
as anything this may be due to the common view that livestock production is not, or more appropriately was not, as labour intensive as arable
farming. As the analysis in this chapter shows, general textbook arguments tend to neglect the labour input into supplying fodder, and this
was not without significance in an economy which had suffered a major
reverse in what was an abundant and cheap labour supply.

7 Conclusion – structure, output and performance, and the distribution of the spoils

Introduction

This study has endeavoured to explore the ways that the Irish agricultural economy responded to the various changes in the internal and external economic circumstances which confronted it in the second half of the nineteenth century and the first decade or so of the twentieth century. In isolation from the international world, and solely in comparison with the rest of the UK, the Irish agricultural industry seemed to respond positively to economic stimuli, and looked in an increasingly healthier position than its larger neighbour across the Irish Sea. When the comparisons were widened to embrace countries on the European mainland some of that initial enthusiasm for the Irish performance could still be maintained, though in a diluted form, but in specific comparison with some of the leading Western European economies the light which appeared to emanate from the Irish industry dimmed considerably. The models which have been used in the central chapters above were related to considerations of gross agricultural income and physical output.

In this final chapter of conclusions we pose what has become a fundamental question in nineteenth-century Irish economic history: to whose benefit did the structural changes in Irish agriculture accrue? Was it a question of a drain of benefits back to the colonial master? Of course not. Instead, after the Famine, slowly perhaps but inexorably, colonialism as a force of economic power (as distinct from a political reality) was turned on its head. Much of Ireland earned its economic independence by 1914, and soon afterwards full political independence followed. This conclusion is shaped by the need to bring together the changes in agricultural structure which have been identified with the estimated output and conjectured performance, in a bid to disentangle the division of the spoils among the broad strands of Irish agrarian society, and to explore the extent to which that division was related to the final culmination of forces which led to Irish peasant independence.

A major issue in late-nineteenth-century Irish history has been the economic origins of the Irish independence movement. In colonial situations, whether of a formal kind as applied to Imperial India, or of an informal kind as some might suggest still applied to Ireland in the nineteenth century in spite of the formal political creation of the UK, or indeed of some sort of state-sponsored variety such as the serfdom of Russia, as soon as a degree of economic power was ceded by the colonialist or the state to the colonised then political power often soon followed. In the case of India, the political power which was sought mainly by peaceful means before the Second World War was reinforced by the economic power that the country gained as a result of Britain's indebtedness in financing that war. The Emancipation of the Serfs in 1861 gave the Russian peasantry only a modicum of economic power and independence, but the follow-up legislation embodied in the Stolypin reforms of 1905 enhanced their economic self-determination, and this acted as a prelude to the social revolution of 1917. In Ireland, notwithstanding a backcloth of political unrest over a long period, the amount of economic power which the Irish peasantry had over their own land was limited.

In the early 1880s a new phase began. Tenant unrest during the Land War of the early 1880s was partly appeased by British attempts to grant a greater degree of economic self-determination but without ceding political power. The seeds of economic independence had been sown thinly before 1880, but they took firm root thereafter. This independence then grew slowly but surely until during the first decade of the new century the replacement of tenant tenures with owner-occupancy proceeded at a remarkable rate. The trend of this revolution is shown in table 7.1. The agricultural census enumerators only started collecting this kind of information in 1906. By 1910, 58 per cent of holdings were owner-farmed, and by the Great War over 60 per cent of Irish holdings were Irish owned and Irish occupied.

In this concluding chapter the economic background to the Land War of the 1880s is reviewed. It is suggested that the economic maldistribution of income derived from the land before the 1880s contributed to tenant unrest, and this led inevitably and successfully to economic independence.[1]

---

[1] The relationship between landlords and their tenants has not properly been addressed in this book until this chapter. Necessarily therefore direct reference to such features as the Ulster custom of tenant right has been omitted. On this specific issue see D.L. Armstrong, *An Economic History of Agriculture in Northern Ireland 1850–1900* (Plunkett Foundation, Oxford, 1989), especially chapter 1, pp. 3–84. See also W.E. Vaughan, *Landlords and Tenants in Mid-Victorian Ireland* (Oxford, 1994), chapter 4, pp. 67–102. Some of the broad arguments in this chapter have been rehearsed in M.E. Turner, 'Rural economies in post-Famine Ireland, c. 1850–1914', in B.J. Graham and L.J. Proudfoot (eds.), *An Historical Geography of Ireland* (London, 1993), especially pp. 324–30. That rehearsal is here revised.

Table 7.1   *The progress of owner-farming in Ireland*

Year	Total no. of holdings	Owned	Tenanted
1906	597,344	174,548	422,796
1907	599,872	230,421	369,451
1908	601,765	277,862	323,903
1909	603,827	337,295	266,532
1910	605,896	350,794	255,102
1911	607,960	389,751	218,209
1912	563,526	399,644	209,152
1913	565,015	338,694	226,321
1914	566,137	348,855	217,282
1915	569,426	360,872	208,554
1916	572,045	365,803	206,242

Sources: Derived from *Annual Agricultural Statistics*

Rising expectations and the economic origins of the Land War of the 1880s

From the new output estimates presented in chapter 4 a peak in the nominal value of final output occurred in 1876 of £48.4 million. While this was below the overall peak of 1868, it has to be recognised that 1868 was an exceptional year in the chronology. The year 1876 therefore can be viewed as the opposite pole of prosperity from the low of 1862 (measured solely by value of output).[2] This conforms rather well with the 'rising expectations' argument put forward by J.S. Donnelly. He suggested that the conflicting relative fortunes of the landlords and their tenants acted as a precursor to the Land War of the 1880s. The argument is set in terms of tenant unrest. They became discontented because they were faced with relatively fixed costs in the form of rents, but those rents rose in real terms against the backcloth of squeezed incomes after 1876. Donnelly does not suggest there was a rental problem before 1876. On the contrary, he suggests that tenants rather than landlords received the main benefits from the increases in product prices and the resulting increases in the value of production from the early 1850s to the mid-1870s. His belief is that the events of 1877–9 effectively reversed this flow of benefits from the tenants' point of view; they had fixed costs but now declining incomes.[3] The

[2]   See also B. Solow, *The Land Question and the Irish Economy, 1870–1903* (Cambridge, MA,1971), pp. 121–2. For a good summary of the agricultural distress of the late 1870s see S. Clark, *Social Origins of the Land War* (Princeton, 1979), pp. 225–45.

[3]   J.S. Donnelly, *The Land and the People of Nineteenth-Century Cork: The Rural Economy and the Land Question* (London, 1975), pp. 184–200.

coincidence of this reversal of fortunes with the period of the Land War and the land reforms which followed is appealing. Yet it leaves as rather a matter of fact that the tenants were indeed the main beneficiaries of whatever improvements there had been in the third quarter of the century. This is by no means certain.

In another study Donnelly has emphasised the extremes of the earlier depression of 1859–64, but the identification of this earlier crisis raises an important question. There was a Land War in the 1880s, but why was there not one in the 1860s?[4] By Donnelly's estimation, 'the value of the seven major crops was depressed by an average of twenty per cent for **three** years' in 1877–9 compared with 1876, 'whereas in the earlier depression their value was reduced by an average of twenty-one per cent for five years', compared with 1859.[5] His calculations are based on total output, not on what we defined in chapter 4 as final output, and therefore they cannot be compared directly with the new estimates presented in this study. Additionally, the cropped side of agriculture was so much more subservient to livestock by about 1880. From the new estimates a comparison of final output can be made paralleling Donnelly's comparison during the two depressions. Final output in each year from 1860 to 1864 compared with 1859 differed by +7, −13, −14.6, −10 and −6.2 per cent, whereas the differences from 1877 to 1879 compared with 1876 were −9, −12.4 and −23.1 per cent. At the very least this comparison shows that 1879 was the deepest depression year (also identified by Donnelly, but demonstrably also the case throughout the UK), and that there was a deepening of a depression which had built up from 1876, even though there was a more sustained period of depression in the early 1860s. It has been suggested already that the output estimates for the 1859–64 depression have probably been underestimated, but it should be noted that 1876 was an exceptionally good year, the best since 1868, and 1879 was exceptionally bad, the worst since 1867. Choosing dates for comparisons in this sort of exercise is a bit of a lottery; Donnelly would have found a worse crisis earlier if he had used 1858 as his base, yet 1876 is the best possible year he could have chosen to highlight the extremes of the later depression.

How might a farmer view rising or falling expectations? His *initial* observation might not be in terms of realised prices or values. Instead he might anticipate expected future returns by observing his standing crops

[4] For arguments against the rising expectations theory see W.E Vaughan, *Landlords and Tenants in Ireland 1848–1904* (The Economic and Social History Society of Ireland, Dublin, 1984), especially pp. 31–5.

[5] J.S Donnelly, 'The Irish agricultural depression of 1859–64', *Irish Economic and Social History*, 3 (1976), 52–3, his emphasis.

or harvested crops. Donnelly acts the farmer in a sense when he looks at crop yields, but he may not have presented the data in a consistent way. He looks at crop yields for 1859–64 as a percentage of the average of 1856–8. He then looks at 1877–9 as a percentage of the average of 1870–6. Curiously this is an inconsistent set of statistics because in value terms he plays up the good year of 1876, but in yield terms he plays it down by diluting it with six other years. A reworking of his deviation of yields from the average for 1877–9 using 1874–6 for comparison produces table 7.2, where 1879 in particular is exposed as a deep depression, deeper than Donnelly demonstrated. This was especially the case for the fodder crops, the impact upon which was vital in an economy which since the early 1860s had moved further and further towards the pastoral side. In addition, to the extent that the fall in livestock and livestock product output values in the 1860s was probably underestimated, so also was the slump in output in 1879. One area where this was likely to have been the case was in the use of the Barrington price series, because it is suspected that this series does not capture the full drop in prices for certain animals in 1879, or the correct timing of the drop.[6]

In 1879 there was a sharp dip in the value of output: it was a year of exceptionally bad weather with low tillage yields everywhere, and probably a sharper fall in livestock output than has been calculated in chapter 4. The fixed weights for carcass sizes and for milk yields employed in this study, in the reality of the circumstances of the depression, surely could not have been sustained with the inferior fodder which was available. There was some compensation for the seller for his falling physical output because prices rose for some products, but a rise in the prices of potatoes and hay meant an increase in fodder costs. Besides, prices were increasingly determined internationally in the late nineteenth century, or else they were determined in Britain, the main market for Irish produce. The year 1879 was indeed a deep trough in the recent better fortunes of Irish agriculture, and it was the year which signalled the onset of the arable depression in Britain as a whole, a depression which was to last for well over a decade.[7] So, 1879 was a particularly deep trough in agricultural incomes, but it must be viewed against the background of relatively fixed costs. The agitation associated with the Land War began that year, and it continued into the 1880s with

[6] See A.W. Orridge, 'Who supported the Land War? An aggregate-data analysis of Irish agrarian discontent, 1879–1882', *The Economic and Social Review*, 12 (1981), 218; Donnelly, *The Land and People of Cork*, pp. 253–4.

[7] See M.E. Turner, 'Output and prices in UK agriculture, 1867–1914, and the Great Agricultural Depression reconsidered', *Agricultural History Review*, 40 (1992), 38–51.

Table 7.2 *Crop yields 1877–1879*

	Oats		Wheat		Barley		Flax		Pots		Turn		Hay	
	a	*b*	*a*	*b*	*a*	*b*	*a*	*b*	*a*	*b*	*a*	*b*	*a*	*b*
1877	91	84	101	87	94	85	113	91	59	48	82	76	121	117
1878	102	94	111	96	98	88	125	100	88	71	109	104	121	117
1879	88	81	84	73	78	70	93	75	38	31	50	46	100	97

Note: (*a*) Donnelly, percentages of the mean of 1870–6   (*b*) revised, percentages of the mean of 1874–6

Sources: Derived from *Annual Agricultural Statistics*

the result that 25 per cent of the rents which were due for payment in 1879–82 were withheld by the tenants.[8]

Is there other evidence of prosperity during the prelude to the Land War? The level of savings which rose through rising bank deposits might be one example, though Philip Ollerenshaw cautions against using their size as a barometer for agricultural incomes and prosperity, especially in the 1870s when there was an extension of bank branches anyway and a transfer of funds from, metaphorically, beneath the mattress to the bank. This was a period when hoarded wealth was released and therefore the subsequent rise in bank deposits may in turn inevitably exaggerate the subsequent decline in deposits during the ensuing depression, and lead to a misleading longer-run assessment of deposits as that barometer. Nevertheless, there was a large decline in bank deposits in the late 1870s and early 1880s. Donnelly, quoting well-known sources, says they fell from a peak of £32.8 million in 1876 to £28.3 million in 1881, not attaining their 1876 levels again until 1889. Credit restrictions were in operation as well. Such restrictions applied to farmers particularly, and this was during a time when they needed credit the most. In a wider rural context, the restrictions had a knock-on effect towards the service sector, including the shopkeepers and the other suppliers on whom many of the farmers relied.[9]

Against this background, the ensuing period of the Land War can be seen as a major watershed in the political moves towards greater Irish self-determination. Agriculture was so central to the economy at the time that the competing influences of contemporary Irish political rhetoric and the more pragmatic approach of the agriculturalist in winning the early tussles in the process of self-determination, has been an ongoing debate. Barbara Solow's characterisation of the period of depression of 1879–82 and the preceding years 1877–8 is couched in terms of agricultural failure. Not without reason therefore does she head the appropriate chapter of her book with the nationalist T.M. Healy's interpretation that 'The Land League was not begotten by oratory, but by economics.'[10] In this economy which was so dependent on one sector, agriculture, the depression of incomes in that sector carried over to other groups within the economy.[11] The rent strike which occurred has been touched upon,

[8] Vaughan, *Landlords and Tenants*, p. 30, who curiously added a value judgement when he said that '*only* about 25 per cent of rents due between 1879 and 1882 were not paid', my emphasis.

[9] P. Ollerenshaw, *Banking in Nineteenth-Century Ireland: The Belfast Banks, 1825–1914* (Manchester, 1987), pp. 114–22, 198; Donnelly, *The Land and People of Cork*, p. 377; Clark, *Social Origins*, p. 231.

[10] Solow, *The Land Question*, pp. 117 ff.

[11] See in this context S. Clark, 'The social composition of the Land League', *Irish Historical Studies*, 17 (1971), 450.

and this therefore resulted in a loss of income to proprietors, but there was a more general inability – not always unwillingness – of farmers to discharge their other debts and accounts to the general service sector of shopkeepers and suppliers.[12] In this sense it was in the interests of both the farmers, and these servicing agents who supplied them, to obtain rent reductions, bringing town and country together during the period. This is what Liam Kennedy has referred to as their mutual 'coincidence of interests', though in county Mayo, for example, this alliance was short-lived.[13]

The connections between landlords and tenants were also important. As L.P. Curtis has put it 'Anything likely to cause hardship among the tenantry was bound, sooner or later, to affect the economic position of the landlords', because rental arrears filtered away from the tenants to opposite points of the socio-economic compass, in one direction to the landlords and in another to the shopkeepers.[14] For the landlords this resulted in mortgages. Curtis identifies a debt burden of 70 per cent of estate valuation for estates of 100–1000 acres, diminishing to a debt burden of 19 per cent of estate valuation for estates of over 15,000 acres, with an overall average of 27 per cent.[15] From a sample of 105 vendors under the Wyndham's Act of 1903 there were debt burdens ranging from 11 to 29.8 per cent with an average of 16.3 per cent. In general the debts were largest in Connacht and least in Ulster.[16] Moreover, not all of these

---

[12] Briefly touched upon by P. Bew and F. Wright, 'The agrarian opposition in Ulster Politics, 1848–87', in S. Clark and J.S. Donnelly (eds.), *Irish Peasants: Violence and Political Unrest 1780–1914* (Manchester, 1983), p. 209. See also S. Clark and J.S. Donnelly 'Introduction' to part III of Clark and Donnelly, *ibid.*, pp. 279–80; Donnelly, *The Land and People of Cork*, p. 254; Ollerenshaw, *Banking in Nineteenth-Century Ireland*, pp. 121–2; Clark, *Social Origins*, pp. 231–5. On the rise of the traders and suppliers of agricultural inputs, and distributors of outputs, and the rise of rural credit see L. Kennedy, 'Farmers, traders, and agricultural politics in pre-Independence Ireland', in Clark and Donnelly, *ibid.*, especially pp. 342–3. See also L. Kennedy, 'Retail markets in rural Ireland at the end of the nineteenth century', *Irish Economic and Social History*, 5 (1978), 46–61.

[13] Kennedy, 'Farmers, traders, and agricultural politics', pp. 343–4, 346. See also Clark, 'The social composition', 450–1; Clark, *Social Origins*, p. 245 and chapter 8 in general, pp. 246–304. D.E. Jordan Jnr, *Land and Politics in the West of Ireland: County Mayo, 1846–82*, Ph.D. Dissertation, University of California at Davis, 1982, in summary pp. 8–10, 164, and in detail in chapters 3–6.

[14] L.P. Curtis, 'Incumbered wealth: landed indebtedness in post-Famine Ireland', *American Historical Review*, 85 (1980), 335. Vaughan, *Landlords and Tenants in Mid-Victorian Ireland*, p. 131, suggests tempering this level of indebtedness to about 22 per cent.

[15] Vaughan, *Landlords and Tenants*, 340–9.

[16] *Ibid.*, 349–53. R.W. Kirkpatrick reminds us that although distress in Ulster was not as severe as elsewhere leading up to or during the Land War, it was nevertheless, sometimes a major problem, 'Origin and development of the Land War in mid-Ulster, 1879–85', in F.S.L. Lyons and R.A.J. Hawkins (eds.), *Ireland under the Union: Varieties of Tension: Essays in Honour of T.W. Moody* (Oxford, 1980), pp. 201–35. See also Armstrong, *An*

debts arose from the immediate effects of the late 1870s depression. Towards the late nineteenth century many landlords were already encumbered by debt. The landed economic power base of the British or Anglo-Irish was already resting on fragile ground. In the aftermath of 1879, with the legislation which followed the depression, they were faced with diminished revenues from rents. What emerged was a decline in the economic power of the landed elite.[17] Any diminution of the collective power and influence of the Anglo-Irish was eventually translated into greater power for the Irish themselves.

The depression of 1879–82 visited the island again in the middle to second half of the decade – but now from a different source – and output continued to decline. As always, Ireland was dependent on favourable weather, but she was also dependent on the wider British economy. The industrial depression which struck Britain in 1884–7 had an important effect on working-class diets, an effect which spilled over on the demand side for Irish agricultural goods. An example of this is the slump in butter exports, and the relative slump in the trade in Irish pigs to Britain.[18] In addition, there was the general rise in non-Irish imports into Britain. Technological changes in packaging and food preservation with the canning and freezing industries meant that the natural protection offered by the Contagious Diseases legislation of the 1860s was no longer as effective.[19]

Agriculture and Irish independence

The two depressions of the second half of the nineteenth century were actually or potentially watersheds in the course of post-Famine Irish agricultural history. The first in 1859–64 may have been the final move in a longer trend towards pastoralism, but the second in 1879–82 heralded the arrival of the Land War and the final drive to Irish independence. Arguably therefore it was considerably more important as a landmark in Irish history. In terms of agricultural history its importance lies in the repercussions which arose from tenant unrest. Those tenants may have

---

Economic History of Agriculture, ch. 1, for a full analysis of worsening landlord/tenant relationships before the 1880s in Ulster, as well as in Ireland as a whole.

[17] Summarised in Clark and Donnelly, 'Introduction', p. 272. See also Curtis, 'Incumbered wealth', 367.

[18] Donnelly, The Land and People of Cork, pp. 311–13. P. Solar, 'The Irish butter trade in the nineteenth century: new estimates and their implications', Studia Hibernica, 25 (1989–90), 136; and for the trade of Irish pigs and sheep to Britain see table 2.7 above.

[19] R. Perren, The Meat Trade in Britain 1840–1914 (London, 1978), especially chapters 7 and 9.

been supported politically by the Irish Land League, but their grievances came from economics, in particular from their precarious tenure.

It could be said that independence began with the granting of the famous three Fs under the 1881 Land Act: fair rent, freedom of sale of tenancy and fixity of tenure. This focused attention on two things: ownership of the soil, and distribution of the returns from the soil.[20] Occupiers had managed to improve the nominal value of their output from 1850 to the mid-1870s, and then again from the mid-1890s, but between times they took part in the move to greater independence. This movement was not necessarily rapid, and by 1906 there was still no doubt as to who owned the land: of a total of 597,000 holdings only 175,000 were recorded as 'owned', by which the enumerators meant owner-occupied, whereas 423,000 were tenanted (figures subject to rounding errors). By 1914, however, out of 566,000 holdings 349,000 were owner-occupied and 217,000 were tenanted (refer back to table 7.1).

If the maldistribution of income was crucial, if economic considerations in general are crucial in explaining subsequent changes in certain aspects of agricultural structure, then from the tenant's point of view it was his net income which counted. Therefore, the important element of the equation which is missing so far is the influence of factor costs on output, and hence on income. The literature, especially on the returns to land (rent), leaves little doubt that there is a long way to go before a full understanding of the course of factor inputs is achieved. This literature will not be reviewed, but simply noted by using the summary of it from K.T. Hoppen.[21] From a combination of this and the new output estimates, table 7.3 can be constructed, using a chronological division dictated by Hoppen.

Output rose between the early 1850s and early 1870s, which, with a less than equal rise in rents and wages suggests a rise in what was essentially the tenant net income (ignoring capital in this equation). This seems partially in line with W.E. Vaughan's ideas of the distribution of farming incomes, which sees the tenants improve by more than the landlords over this period. As two broad groups that seems to be the case, but a 22 per cent rise for the tenants against a 20 per cent rise for the landlord class shows that the difference was marginal. The labourers faired least well of the three broad classes in society, though as has been pointed out in

[20] See in general M.J. Winstanley, *Ireland and the Land Question 1800–1922* (London, 1984); Solow, *The Land Question*; Vaughan, *Landlords and Tenants*. But from Armstrong, *An Economic History of Agriculture*, pp. 54–6 there are clear enough statements that the granting of the three Fs was only a start, and that the ultimate expectation was the transfer of ownership to the tenants and the eradication of landlordism.

[21] K.T. Hoppen, *Ireland Since 1800: Conflict and Conformity* (London, 1989), p. 100.

Table 7.3   *Land, labour and tenant net income, 1852–1854 to 1905–1910 (in £s million and percentage changes)*

Dates	Output		Rent		Labour		Tenant net income	
1852–4	37.0		10.0		9.3		17.7	
		(19.5)		(20.0)		(14.0)		(22.0)
1872–4	44.2		12.0		10.6		21.6	
		(−12.2)		(−4.2)		(3.8)		(−24.5)
1882–4	38.8		11.5		11.0		16.3	
		(12.6)		(−30.4)		(3.6)		(54.0)
1905–10	43.7		8.0		10.6		25.1	

Reworked on alternative time periods[a]

Dates	Output			Tenant net income	
1856–8	37.6			18.3	
1871–3	42.3	(12.5)		19.7	(7.7)
1857–8	38.0			18.7	
1877–8	43.2	(13.7)		20.6	(10.2)
1856–60	38.5			19.2	
1876–80	42.8	(11.2)		20.2	(5.2)

[a] The rent and labour input estimates remain unchanged

Sources: See text.

other chapters the true wage bill is not captured in these figures, because the true net tenant position should make allowance for the labour cost of the farmers' extended family. Therefore the difference in the rewards accruing to the tenants and their landlords was even more marginal than these figures suggest, making it very doubtful that the tenants came off the better, an argument which we continue later in this chapter. However, as it stands, at face value, in nominal net value terms the difference between the tenant net income position between the early 1850s and early 1870s lends weight to the rising expectations arguments.

A second major point to notice is the difference between the early 1870s and early 1880s. Final output went into reverse, wiping out entirely the nominal gain of the previous period, but there was no compensating adjustment in costs (rent and wages). Indeed, the wages bill actually rose. The tenant net income took a nose dive. It eventually recovered by a substantial margin, partly by a nominal recovery in the value of final output, but mainly as a result of a substantial decrease in the rent bill in the aftermath of the Land Act of 1881, and the various pieces of legislation which were enacted subsequently and which led to greater peasant independence. These were the Ashbourne Act of 1885, the Land Purchase Acts of 1891 and 1896, Wyndham's Land Act of 1903 and its successor of 1909. So, with tenurial adjustments which enhanced the position of owner-farming, and which stretched into Edwardian times, there came a recovery of tenant net incomes. In two senses therefore economic power was transferred to the Irish.

Thus, even though labour costs remained high, the occupiers had become increasingly independent, and they faced a much reduced national rent bill. This may have had significant consequences for work effort: there may have been genuine productivity improvements because the feeling and experience of independence meant there was no longer a fear 'that improvements resulting from careful treatment of the soil would be taxed by an increased rent'. The newly independent owners 'set themselves assiduously to cultivate their holdings to the best advantage', because 'it was the land legislation which re-united industry with its reward'.[22] It has been suggested that in the pre-Famine period the differences in productivity between Ireland and Belgium might be explained precisely by the emphasis on tenant farming in Ireland, but on owner-occupancy in Belgium. In the latter, the 'owner-occupiers reaped the full benefit from improvements'. *If* there had been security of tenure in Ireland 'it may still be suggested that owner-occupiers would have put more effort into building up and maintaining soil fertility than would landlords – who may have faced high costs of labour supervision – or tenants – who feared that the landlord would be the principal beneficiary'.[23] This relationship between tenure and work effort may be a big *if*, but undoubtedly there was a substantial move to owner-occupancy in Ireland, especially after about 1890. In 1870 3 per cent of Irish holdings were owner-occupied, but by 1908 and the first official Census of Production this level had expanded to 46 per cent.[24]

The element which is missing from this analysis, apart from the existing doubts over the true size of the labour cost, is a consideration of the capital input. In the period when independence was on the march, the burden of capital improvement shifted from the landlord to the tenant, though both its initial size under the landlords and its

---

[22] I am combining two views here. The first two quotes come from Barrington in the 1920s, and the third from Connell in the 1950s; T. Barrington, 'The yields of Irish tillage crops since the year 1847', *Journal of the Department of Agriculture*, 21 (1921), 291; K.H. Connell, 'The land legislation and Irish social life', *Economic History Review*, 11 (1958), 4.

[23] P. Solar and M. Goossens, 'Agricultural productivity in Belgium and Ireland in the early nineteenth century', chapter 14 of B.M.S. Campbell and M. Overton (eds.), *Land, Labour and Livestock: Historical Studies in European Agricultural Productivity* (Manchester, 1991), pp. 381–2.

[24] Solow, *The Land Question*, p. 193.

comparable size under the new situation of increasing peasant ownership seems hardly capable of measurement at the present state of empirical research.

From an earlier rendition of this history and analysis there may be detected an apparent *volte face* in my interpretation, for example, of the decades prior to the Land War. It looks as though I now disagree less with W.E Vaughan's interpretation than was originally the case, and see the tenants faring equally prosperously as the landlords.[25] In perfect fairness to a consistent situation the dates chosen so far exactly replicate those used in that earlier version, though of course the output figures have now been revised. Essentially the level of output in the early 1850s has been reduced by a more judicious choice of prices. Inevitably this has important repercussions for the distribution of income. Yet revision can take more than one form, and in another way it would be perfectly consistent to revise other aspects of the data, in particular the choice of years. It has already been identified that there were price distortions in the 1850s as a result of the Crimean War. This had knock-on implications for the value of final output. Already in this chapter it has been demonstrated that the choice of years in an analysis of changing crop yields is very important when it comes to assessing the relative severity of depressions. The same can now be said regarding tenant net incomes. For example, if the years Vaughan had *originally* chosen are now used, 1854 and 1874 (though he did not include potatoes in his final output), then in my estimates the tenant net income declines over time. In this kind of analysis it hinges on simple arithmetic, but the almost casual choice of years for comparison is crucial. The inclusion of 1852 – when there was a final output of less than £28 million – within a calculation of annual average output for the early to mid-1850s rests uneasily alongside other years when the value of output was of the order of £30 and over £40 million. The inclusion of 1852 seriously underscores the general level of subsequent years. Furthermore, it has a greater influence than the years employed to derive similar cross-sections in the 1870s, 1880s and the early twentieth century. In his latest study Vaughan suggests that a 20 per cent increase in rents actually favoured the tenants because it must be seen in the light of a 70 per cent increase in agricultural output, but he comes to that conclusion by comparing the early 1850s with the mid-1870s. In his revised estimates of output of the early 1850s, three of those years (1850–2 for example) coincided with the three lowest years in nominal output in the broader period 1850–6, but the output of the mid-1870s embraced the three highest years (1874–6). Vaughan's comparison

---

[25] Turner, 'Rural economies', especially pp. 324–30.

of the lowest trough with the highest peak inevitably leads to his conclusion, but is it a distortion of history?[26]

We should really seek neither single years, nor groups of years where there are unusual distortions in the underlying trend. In this respect the period 1856–8 is about as close to a plateau as could be achieved early in the period. Peter Solar chose the annual average of the wider period 1856–60 to make his estimate of final agricultural output. In turn 1871–3 is another plateau, but to go beyond these years is to include 1874, and this involves the beginnings of the upturn in the value of final output identified in the arguments regarding rising expectations. The period 1874–6 is both a plateau but also a peak, and it embraces the culmination of those rising expectations. Although 1876 was the second most important year in terms of nominal final output with a value of £48.4 million, it was not so different from the two preceding years when the output was £46 and £45.4 million. The decision on the precise location of the third cross-section in the early 1880s presents a different problem since the period was in the midst of a downturn which did not finally trough until 1887. Thereafter, until the mid-1890s, final output achieved another plateau varying narrowly between £35.4 and £38.0 million (1890 and 1889 respectively). In what appears to be a rough and ready cyclical pattern it might make sense to compare peaks with peaks, or troughs with troughs, or, as the current argument suggests, relative plateau with relative plateau. It is all very well doing this with final output estimates which are annual, but it is quite another thing doing it with rents and wages. Nevertheless, the second part of table 7.3 repeats part of the exercise using the annual averages of 1856–8 and 1871–3. It shows a much less rosy period for the tenants relative to the other classes. The table also includes estimates which redefine the periods for comparisons (for example 1857–8 or 1856–60, and 1877–8 or 1876–80). The outcome of this comparison produces much the same, less optimistic position for the tenants, and indeed for some comparisons it appears to be quite a good deal less optimistic. The original years analysed (as in the upper half of table 7.3) are almost the only groups of years which enhance the position of the tenant above that of the landlord. In a sense this might be seen as the fickleness of trend but it may also expose the inability of all historians so far properly to articulate the problem at issue, and therefore properly to model the appropriate way to approach its solution.

There is no doubting however the fall in tenant net incomes into the 1880s, and what followed from the subsequent land legislation. From 22 August 1881 to 31 March 1900 there were 328,220 tenants who took

---

[26] Vaughan, *Landlords and Tenants in Mid-Victorian Ireland*, pp. 50 and 249.

advantage of the Irish Land Act of 1881 to have a fair rent set. They farmed 9,859,970 acres. Therefore, just over 60 per cent of all occupiers farming 65 per cent of the cultivated land area had fair rents set for a first time or term. On closer inspection this shows a marginal bias to the larger occupiers. At the end of fifteen years, 52,396 of them with 1,432,515 acres had a fair rent set for a second time or term. Rent was reduced on average by 20.8 per cent in the first term rising to 22.4 per cent in the second.[27] The annual average output in the half decade prior to the 1881 Act was £42.8 million, and by the last half decade of the century it was £35.5 million, a fall of 17 per cent (or 7 per cent by taking 1880–4 as the base). The adjustment to a fair rent more than matched the decline in the nominal value of output.[28]

Under the provisions of the Ashbourne Acts of 1885 and 1888 and their successors the Balfour Acts of 1891 and 1896, tenants were encouraged to purchase their land by way of a 100 per cent advance from the state as cash or stock. But that tenant was liable to an annuity of 4 per cent for a period of forty-nine years. Interesting comparisons might be made here with the emancipation of the serfs in Russia after 1861 where a buy-out scheme was introduced which was also based on forty-nine years. In the Irish case, of the 4 per cent, 3 per cent was interest and the remaining 1 per cent formed part of a sinking fund. In practice, the effect was to transfer what was formerly a rent into an interest charge, which by any other name, for forty-nine years at least, was a rent. It was terminable and not perpetual and in its operation seemed to imply a decrease in 'rent' at inception, though without any allowance for future price changes (which in our period means no allowance for deflation). Suppose a tenant redeemed a £100 rent at eighteen years' purchase. On the £1,800 borrowed at 4 per cent for forty-nine years he would become liable to an interest repayment of £72, or £28 less than his *current* rent.

27 Donnelly, *The Land and the People of Cork*, pp. 296–7 where he says that in the early 1880s the size of initial rent reductions was only about 16 per cent. Though see Armstrong, *An Economic History of Agriculture*, p. 71, for somewhat different numbers, though effectively the same impact. For the first ten years of the scheme, from 1881 to 22 August 1891 the level of rent reductions was, on average 20.7 per cent, ranging from 19.4 per cent in Ulster to 22.1 per cent in Munster. See *Report of the Irish Land Commission for the period 22 August 1890 to 22 August 1891, B(ritish) P(arliamentary) P(apers)* [C. 6233], vol. xxv, 1890–91, p. 44.

28 C.F. Bastable, 'Some features of the economic movement in Ireland, 1880–1900', *Economic Journal*, 11 (1901), 33. By the time Bonn wrote his essay in 1909, 369,483 holdings had rents fixed for a first term, 131,637 for a second term. In addition, tenants on 143,564 holdings had been converted into owners and on a further 173,343 holdings purchase agreements had been conducted, M.J. Bonn, 'The psychological aspect of land reform in Ireland', *Economic Journal*, 19 (1909), 376–8. See also Curtis, 'Incumbered wealth', 332–67 for an account of mortgage debt by landlords who chose to keep their land.

The clever trick was that British capital was used to pay for the buy-out.[29] The problem arises over whether the interest charges under this and other schemes operating in the last quarter of the nineteenth century have been included subsequently in estimates of the national rent bill. Undoubtedly they should be in any calculation of farming profits.

Huttman reckoned that Irish farmers reacted to the land-reform measures if, or rather when, they perceived a personal benefit and therefore shifted their preferences from tenancy to owner-occupancy on the basis of the competitiveness of payments.[30] The end result was to reduce the costs of land inputs from the third quarter of the nineteenth century to the Great War, a reduction which Huttman has estimated was as much as 20 per cent. It is still unclear, however, whether the annuity payments have been included in the national rent estimates. Huttman has put the annuity, in terms of a per cent of rent at the date the annuity was purchased, at 70 per cent plus or minus 2 per cent.[31]

By March 1911, £66.5 million had been advanced for land purchases for something close to 6 million acres, with agreements to purchase a further 4.5 million acres which was pending at a price of £46.5 million. By mid-1913 purchases fulfilled or pending under Wyndham's Act of 1903 and its successor the Birrell Act of 1909 totalled £96 million, plus £4 million for labourers' cottages, and another £24 million for purchases of land prior to 1903. Therefore, in rounded terms about £125 million had been expended, but there was an estimated £60 million still needed to complete the transfer from occupancy to owner-occupancy.[32] Ireland was surely gaining a large share of its economic independence, achieving it rapidly, and financing it in the main by British capital. This movement to economic independence was strongest in Ulster. By 1913, 69 per cent of Ulster farmers owned their own holdings.[33]

A dominant feature of the rural economy and of Irish independence, therefore, was the centrality of the Land War. For the arguments presented here this hinges on the vital question of the distribution of rewards from Irish land and Irish labour. The significance of the new

[29] J. Johnston, *Irish Agriculture in Transition* (Dublin and Oxford, 1951), p. 8, quoting Bonn, 'The psychological aspect'. See also E. Barker, *Ireland in the Last Fifty Years (1866–1916)* (London, 1916), pp. 100–16.

[30] J.P. Huttman, 'The impact of land reform on agricultural production in Ireland', *Agricultural History*, 46 (1972), 354–5.

[31] *Ibid.*, 355. See also Barrington, 'The yields of Irish tillage crops', 291.

[32] Barker, *Ireland in the Last Fifty Years*, pp. 53–4, 98–116. See also Barrington, 'The yields of Irish tillage crops', 290–1 for a summary table of the purchase of land under the various acts from 1870–1920.

[33] L. Kennedy, 'The rural economy, 1820–1914', in L. Kennedy and P. Ollerenshaw (eds.), *An Economic History of Ulster, 1820–1939*, (Manchester, 1985), p. 60 note 129. See also Armstrong, *An Economic History of Agriculture*, pp. 71, 84.

output estimates is related to the costs of inputs, the returns to factors of production, and the related distribution of farm income between land-lords, tenants and labourers, but in particular between landlords and tenants. Any new appreciation of this distribution opens up the whole debate about the economic circumstances of the late 1870s. A happy link may have been suggested between increased independence and improved agricultural output, but the implications of this independence movement call into question the estimation of tenant net incomes (as in table 7.3), an estimation which is based on insecure and infrequent rent calculations and random and infrequent turning points in history.

K.T. Hoppen took my earliest output estimates to revise strongly held views about the distribution of farming incomes between landlords and occupiers between the 1850s and 1870s. He suggested that once the early 1880s had passed and the country had emerged from the Land War, the collective results from land reform meant that the Great Depression in agriculture, for Irish farmers at least, was an illusion. As he put it, 'To them went the triumph and the spoils; theirs the victory alone.'[34] Hoppen's calculations, however, and the debates about productivity and the distribution of incomes, are highly speculative, and now subject to revision. New, annual output estimates have emerged, but not without encountering severe problems of construction and interpretation. The rent and wage estimates are even more problematic.

The portrait of labour is not straightforward, as already indicated. Farmers' relatives often assisted in an unpaid capacity, and occupiers also acted as labourers without a wage. Besides, payments were often made largely in kind, and there was a heavy seasonal component in labour services. These and other befuddling experiences lead to a less than certain, or even moderately certain estimate of the labour force. Thus the equation for the wages bill which is based on the number of labourers multiplied by the wage rate is on shaky ground.[35] The estimate of a national wages bill is surely no better than a guess.[36]

Solow estimated the national rent bill at £11.4 million in 1865, rising to £12.1, £12.4 and £12.8 in 1870, 1875 and 1880, or a 12 per cent rise from 1865 to 1880.[37] After rents had been renegotiated under the provisions of the 1881 Land Act, she estimated the new rent bill at £11.5 million in 1886, falling to about £9.2 million in the early 1890s after the new Land Act of 1887. The contemporary Registrar General, Dr T.W. Grimshaw,

---

34 Hoppen, *Ireland Since 1800*, p. 100.
35 D. Fitzpatrick, 'The disappearance of the Irish agricultural labourers, 1841–1912', *Irish Economic and Social History*, 7 (1980), especially 80–2.
36 See also Hoppen, *Ireland Since 1800*, pp. 91–2.
37 Solow, *The Land Question*, pp. 62–9.

suggested a national rent bill in 1885 of £11 million.[38] Effectively this is the pattern adopted so far from Hoppen's summaries. Yet there are alternative estimates available resulting in quite wide divergencies in trend. J.J. Lee and others have reviewed the literature on rent or made independent national rent estimates: in 1845 it could have been at a level of £12–13 million, rising, it is said to £15–16 million by 1875, with at least one intermediate estimate for 1869 of £11.8 million; Joel Mokyr has estimated pre-Famine rent at £13.1 million; Raymond Crotty would have had the national rent bill at £16 million in 1845 rising to about £30 million by 1880.[39] If we should adopt some of these new rent figures, say the middle-of-the-road estimates of £12–13 million in the 1850s rising to £15–16 million in the 1870s and early 1880s, the conclusions reached so far would not be seriously affected; tenant net incomes might have been squeezed even more, perhaps to two-thirds their level in the early 1880s compared with the early 1850s.

Clearly the cost of factor inputs in relation to output can produce different assessments of the rewards to tenants and landlords. W.E. Vaughan put the rise in rents between the Famine and the late 1870s at about 20 per cent – which is in line with the rents quoted in table 7.3 – though on individual estates the increase varied from no increase at all in County Cavan, to increases of 50 or 60 per cent in Leitrim and Donegal. He claimed that if the course of rents had followed the course of agricultural output, as he had estimated it, then rents should have risen by another 13 per cent.[40] Therefore, his interpretation was that the return to tenant incomes and wages, to farmers and labourers, was greater than to the landlords. This is a view pretty well followed by J.S. Donnelly in his Cork study.[41] This is what might be called the orthodox view of the distribution of farming incomes. From my new estimates, and avoiding selection problems regarding particular years, output rose from an annual average of £36.5 million in the 1850s to £43.4 million in the 1870s, or about 19 per cent. This is by some clear margin less than Vaughan's

---

[38] *Ibid.*, pp. 176, 178–9; *First Report of Her Majesty's Commissioners appointed to inquire into the Financial Relations of Great Britain and Ireland, BPP* [C. 7720 and 7721], vol. xxxvi, 1895, p. 455, 'A Paper put in by Dr Grimshaw, Registrar-General for Ireland, with regard to the income of Irish agriculturists in the year 1885'.

[39] Crotty and Lee derived from J.J. Lee, 'Irish agriculture', *Agricultural History Review*, 17 (1969), 74–5; J. Mokyr, *Why Ireland Starved: A Quantitative and Analytical History of the Irish Economy, 1800–1850* (London, 1985 edn), p. 28. See also C. Ó Gráda, 'Agricultural head rents, pre-Famine and post-Famine', *The Economic and Social Review*, 5 (1973–4), 386–7.

[40] Vaughan, *Landlords and Tenants in Mid-Victorian Ireland*, pp. 45–8, 239–40; Vaughan, *Landlords and Tenants in Ireland*, pp. 14, 21; Vaughan, 'An assessment of the economic performance of Irish landlords, 1851–81', in Lyons and Hawkins (eds.), *Ireland under the Union*, pp. 173–99, especially 177–8.

[41] Donnelly, *The Land and the People of Cork*, pp. 189, 194, 199–200.

estimate of increased output, and only just on a par with the increase in rents.

This has important implications for the economic background to, and the political developments arising from, the Land War, whose central issues were tenure and rent. This can be seen in two separate ways: in economic terms through the relationship between tenants and landlords and the distribution of income from the land they farmed and owned respectively; but also in political terms through the relationship between Irishmen and the absentee British. The revised estimates suggest that during the long-term prelude to the Land War there was at best little adjustment in the relative fortunes of those various classes which had a call on agricultural income (table 7.3 upper part), but at worst an adjustment in those relative fortunes which favoured the landlords, certainly over the tenants. So when Vaughan talks of the 'remarkable prosperity' of Irish agriculture in one study and concludes in another that, 'On the whole tenants were in a relatively privileged position, holding their land at what were in real terms falling rents', we disagree, or at least we shout not proven, and we also note that there was a degree of contemporary debate on the prosperity of recovery after the Famine.[42] The new estimates indicate that at best the status quo was maintained, but they also suggest that the arguments in favour of a short-term rising

---

[42] Vaughan, 'An assessment of the economic performance', p. 180; Vaughan, *Landlord and Tenant*, p. 23. See also his 'Landlord and tenant relations in Ireland between the Famine and the Land War, 1850–1878', in L.M. Cullen and T.C. Smout (eds.), *Comparative Aspects of Scottish and Irish Economic and Social History 1600–1900* (Edinburgh, 1977), pp. 216–17. For the ease with which such views filter into texts and other works see Winstanley, *Ireland and the Land Question*; R.F. Foster, *Modern Ireland 1600–1972* (London, 1988), pp. 375–7; and O. Macdonagh, 'Introduction: Ireland and the Union, 1801–70', in W.E. Vaughan (ed.), *A New History of Ireland V Ireland Under the Union, I 1801–70* (Oxford, 1989), lviii; Jordan Jnr, 'Land and politics: County Mayo, 1846–82', pp. 139–40, even though much of what he subsequently presents from County Mayo may in fact contradict this view. His supposed 'low or moderate rents' which he says were not 'exhorbitant', appear rather large when set against the contemporary land valuations, pp. 142–51 (quotes on pp. 142 and 151 respectively); and S. Warwick-Haller, *William O'Brien and the Irish Land War* (Dublin, 1990), pp. 20–1, who further states, p. 22, that the low level of rents may have delayed economic progress, suggesting that they encouraged complacency rather than a spirit of improvement, competition and business-like methods. In one respect Warwick-Haller is correct, the whole equation depends not on whether landlords did or did not raise rents, so much as the relative movement of rents with incomes. The tradition which the new estimates now question is whether there was a large rise in incomes, and whether the apparently modest rental increases in conjunction with this rise effectively produced a diversion of spoils to the farmers. For example see Clark, *Social Origins of the Land War*, pp. 167–8. What the new estimates do is to re-evaluate, or at least question the course of incomes in the third quarter of the century. Evidence that this was a lively contemporary issue, especially focused during the depression of the early 1860s can be found in R.D. Collison Black, *Economic Thought and the Irish Question 1817–1870* (Cambridge, 1960), pp. 47–8.

expectation for tenants in the early 1870s which reversed savagely later in the decade, increasingly make more sense.

If the new estimates are secure, in trend at least, then they suggest that Hoppen's tenants' victory arising from the Land War was preceded by a long period of relatively fixed incomes and then a short sharp shock, and not the orthodox tale of a long tide of prosperity.[43] It was a period of sudden deterioration in tenant incomes, but with relatively fixed costs of land and labour, or costs which were slow to adjust, or not adjustable without the legal intervention which followed. If C. Ó Gráda's estimate of rent of £8.5 million in about 1852 is adopted (admittedly a low post-Famine base), then his view that the landlords' share of output had increased from the early 1850s to the 1870s falls easily into place, and is reinforced by the trend of the new estimates.[44]

Part of the conflict of views must be the choice of base years. Solow was confident in her view that from the 1860s onwards rental increases lagged behind the prices of agricultural products;[45] but if *we* are correct then we follow an old traditional line which suggested that 'Between 1850 and 1870 when prices were rising, there was a marked increase of rents.'[46] This issue of the lag of prices with respect to rents, or rents with respect to prices, might be important, especially in view of Donnelly's theme of rising expectations in the 1870s, expectations which received a nasty jolt at the end of the decade. If we are correct and Vaughan is wrong, then the assertiveness with which some commentators have followed the Vaughan line has to be tempered, but more importantly, the economic and perhaps social background of tenants in agriculture in the years from the Famine to the Land War needs to be reassessed, and the impact of the Land War in changing both the political as well as economic map of Ireland must also be reassessed.[47]

---

[43] Vaughan, *Landlord and Tenant*, pp. 27–35. See also Hoppen, *Ireland Since 1800*, chapter 4, 'Agricola Victor'. And for what I would call antidotes to this view, first in a study based on Ulster, an area not associated with the excesses of distress, see Kirkpatrick, 'Origin and development of the Land War', 201–35, and second in the west of Ireland see P. Bew, *Land and the National Question in Ireland 1858–82* (Dublin, 1978), chapter 1, but especially pp. 25–33.

[44] Ó Gráda, 'Agricultural head rents', 389–90, or a 30 per cent increase from the early 1850s to the mid-1870s. Ó Gráda further states that the landlords' share of final output had increased.

[45] Solow, *The Land Question*, pp. 70–1, 77.

[46] Johnston, *Irish Agriculture in Transition*, p. 5.

[47] An example of such assertiveness can be found in Clark and Donnelly, 'Introduction', in Clark and Donnelly (eds.), *Irish Peasants*, pp. 276–7; and W.L. Feingold, 'Land League power: the Tralee poor-law election of 1881', in Clark and Donnelly (eds.), *Irish Peasants*, pp. 285–6.

Conclusion

The Famine was a landmark in Irish history, though it exposed the frailty of the agricultural economy without necessarily awakening the anger of the Irish people against British rule. The depression of 1859–64 further exposed the economy but also aroused the agriculturalists to a greater awareness of market forces. Crucially, it was the depression of 1879–82 and the associated Land War which exposed the tensions at the opposite poles of the social and economic ladder, and finally led to concerted political moves towards Irish home rule: it heralded the most long-lasting change of them all, the successful move towards Irish independence. The acquisition of land by the Irish peasantry threw into reverse the opposite, centuries-old process of the British acquiring Irish land, but the really clever trick of this last move towards economic independence was that British capital, in the main, paid for the buy-out.

By 1914 the Irish people, or more particularly the peasantry, were well on the road to economic independence, and in the last analysis they had achieved that independence rapidly. The power gained by the people in securing this independence – as distinct from the more obvious and public power of Irish politicians and political movements – as a direct result of the depression of 1879–82 and the decades which led up to it, hinges on a particular interpretation of the performance of Irish agriculture. At times that performance looks good, especially when compared with Britain, but there is still some doubt about the trend of prosperity and in particular about the distribution of rewards arising from agriculture for the whole period from the Famine to the Great War.

Appendixes: A note on the origin of the data

The annual agricultural returns, known as the June Returns, have long been used by geographers and historians as a basis for studying the long-term trends in English and Welsh agriculture since 1866.[1] It was in that year that the first British agricultural census was introduced and subsequently systematised.[2] Twenty years earlier the collection of annually aggregated agricultural data was pioneered in Ireland. It began in 1847, in the wake of the onset of the Great Famine, but before the full trauma of that event had been played out. Indeed several years earlier an agricultural census of sorts was attached to the 1841 population census. This early census recorded livestock numbers at the national, provincial and county levels, though some of those data are suspected of underestimation. This is certainly the case with the cattle numbers, and also with the pig numbers if 1841 is to be regarded as a contemporary 'average year'.[3] The 1841 census also gave an appraisal of landholding distribution, across space, and also by size categories under four main landholding size groups. Some of these data are the subject of very severe criticism; there is the suspicion that the enumerators of the census used a mixture of statute and Irish acres, where the latter is 1.62 times the size of

---

[1] From a large bibliography see J.T. Coppock, *An Agricultural Atlas of England and Wales* (London, 1964 and 1976 edns), Coppock, *An Agricultural Atlas of Scotland* (Edinburgh, 1976), Coppock, 'Mapping the agricultural returns: a neglected tool of historical geography', in M. Reed (ed.), *Discovering Past Landscapes* (London, 1984), pp. 8–55. See also MAFF, *A Century of Agricultural Statistics: Great Britain 1866–1966* (HMSO, London, 1968).

[2] The returns for England and Wales were collected, or at least brought together annually as one 'document', while the equivalent returns for Scotland were presented separately. When I say that this was the first agricultural census I am not forgetting other great landmarks such as the 1801 Crop Returns, or indeed Domesday Book. The June Returns became an annual census, and given the range of data collected they can easily be regarded as the first agricultural census.

[3] See P.M.A. Bourke, 'The agricultural statistics of the 1841 Census of Ireland. A critical review', *Economic History Review*, 2nd series, 18 (1965), 376–91. See also a local study of Down and Antrim in M.E. Turner, 'Livestock in the agrarian economy of counties Down and Antrim from 1803 to the Famine', *Irish Economic and Social History*, 11 (1984), 19–43.

the former.[4] Nevertheless, the 1841 census, in association with the 1845 Poor Law Commissioners' Report is widely used as a pre-Famine benchmark in the study of Irish agriculture. If census enumeration is a learning-by-doing exercise then the post-Famine agricultural statistics when they appeared were reputedly 'free from major error'.[5]

Whether the 1841 census was used as a model for the subsequent annual returns introduced in 1847 is not clear. From 1847 until 1852 the details of the annual census were collected in late September and early October under the direction of the Board of Works or the Census Commissioners. In 1853, 1854 and 1855 the timing was brought forward to mid-July, and from 1856 onwards to the early part of June. From 1853 the administration of the collection was in the hands of the General Register Office in Dublin until the Department of Agriculture and Technical Instruction for Ireland was created in 1900. The country was divided into districts with each district the responsibility of members of the Royal Irish Constabulary. To them fell the duty to visit and inquire of every farm in the district and return the results of the inquiry to Dublin. It was then the responsibility of the government department to organise the data in generalised ways, and also to comment year by year or over a long run of years on issues of trend and general observation. Effectively therefore the police collected particulars of areas under a variety of crops and the numbers of a wide variety of animals for every farm in Ireland.

This system of data gathering prevailed, basically, from 1847 to 1918. There were minor changes, and also some major ones. For example, the inquiry was originally aggregated, tabulated, and published at various degrees of spatially defined detail. These ranged from the small baronial level, of which there were over 300 in nineteenth-century Ireland, as well as at the national, provincial, county and poor law union levels. There were 4 provinces, 32 counties, and what with boundary changes, between 160 and 163 poor law unions. The practice of reporting at the baronial level was abbreviated in 1875 and discontinued entirely from 1881.[6] The return for 1848 is incomplete because of political disorder in the Metropolitan Police Area of Dublin and in the counties of Waterford

---

[4]   Bourke, 'The agricultural statistics', 377–8; 'The extent of the potato crop in Ireland at the time of the Famine', *Journal of the Statistical and Social Inquiry Society of Ireland*, 20:3 (1959–60).

[5]   From Bourke, 'The agricultural statistics', 376.

[6]   See L.D. Stamp, *An Agricultural Atlas of Ireland* (London, 1931), p. 5. See also Saorstát Éireann, *Agricultural Statistics 1847–1926: Reports and Tables*, Department of Industry and Commerce (Dublin, 1930), p. v. Note that between 1847 and the First World War there were some, but very few poor law union changes. These took the form of amalgamations of or sometimes the division of existing ones.

and Tipperary. Although there was a complete return in 1918 it is known that farmers understated their areas under crops, and therefore for that year a method of sampling was used to correct for the underestimation. From 1919 until the mid-1920s a similar method of sampling was employed, by which time the political map of Ireland had been redrawn.

There are obviously some doubts about the correctness of the returns, especially in the early years when there were suspicions about why they were being collected and for what purpose their findings might have been used, and there were teething troubles. Thereafter the actual method of collecting the data is said to be reasonably accurate. Cormac Ó Gráda describes the annual census as 'probably unparalleled anywhere for scope, detail, and reliability', though this comment and the statement cited earlier that the outcome was 'free from major error' stand uneasily alongside Sir Robert Alexander Ferguson's evidence presented to the House of Lords Select Committee of the 1850s inquiring into the best method of obtaining accurate agricultural statistics. On checking specific returns for Tyrone in 1853 he found it hard to believe that the returns were accurate. In turn this evidence stands in contrast to that from William Donnelly, the Registrar General of Marriages and the person responsible from 1853 for the collection of Irish agricultural statistics. He attested to the increasing reliability of the statistics – which cross-checked with other evidence – and in general stated that the earliest fears that the statistics would form the basis of new taxation had been overcome.[7]

The timing of the census in June each year produces some problems for agricultural historians. At one stage in the main account above reference was made to the probable underenumeration of the pig population, since young pigs born and disposed of between censuses were never enumerated. In similar fashion in some counties, districts or regions, where the milk and butter trades were more important than the raising of calves, there might have been underenumeration of the cattle population for those animals born and disposed of between censuses. At the national level, if those calves had been sold down the line for rearing they will have been picked up by the census, but if they had been exported between censuses they will have been lost in the annual count.[8] Liam Kennedy describes some specific examples where the

[7]   C. Ó Gráda, 'Supply responsiveness in Irish agriculture during the nineteenth century', *Economic History Review*, 2nd series, 28 (1975), 312; *Report from the Select Committee of the House of Lords Appointed to Inquire into the Best Mode of Obtaining Accurate Agricultural Statistics from all Parts of the UK, B(ritish) P(arliamentary) P(apers)*, vol. viii, 1854–5, pp. 74, 86.

[8]   R.O. Pringle, 'A review of Irish agriculture, chiefly with reference to the production of live stock', *Journal of the Royal Agricultural Society of England*, 2nd series, 8 (1872), 32–3.

annual statistics are clearly inconsistent and in error. The examples he gives are a mixture of typographical and probably genuinely undetected errors by the enumerators. Occasionally they themselves pointed out likely errors of enumeration. For example, in their assessment of the agricultural produce for 1853 they pointed out that according to the Ordnance Survey measurements Ireland should contain 20,177,446 acres of land. The summation of the acres submitted by landholders that year produced a figure of 20,189,984 acres, or 12,538 acres in excess of the 'true' area. The Registrar General did not question the OS measure, but rather suggested that the enumerated figure anyway was a very close approximation, based as it was on the combined result of information furnished voluntarily by 585,313 occupiers of holdings and collected by 3,875 enumerators. In other words it would have been a miracle if the figures had coincided. Nevertheless it was suggested that the surplus 12,000 acres could easily represent land reclaimed from Loughs Swilly and Foyle, and reclaimed from the Wexford coast since the publication of the OS figures. Kennedy rightly concludes that the enumerators should be commended rather than criticised for their efforts. It was a job well done, and internally consistent in presentation.[9]

What has survived in a readily available form is the aggregation of the individual returns. This occurs at several geographical levels: nationally, provincially, at the county level, for poor law unions, and up to the 1870s under baronies as well. This annual census, or 'Returns of Agricultural Produce' and later 'Agricultural Statistics', as it was known, usually took two forms: a general abstract of the agriculture for the year; and the detailed spatial coverage. The general appraisal invariably took the form of a series of summaries for the particular year, and a comparison with the previous year or years, indicating the actual and percentage changes that had taken place, sometimes offering explanations for the changes. By the judicious selection of years it is possible to obtain a full long-term summary of the annual returns by inspecting just a few years. For example, at the county level, and hence also at the provincial and national levels, the 1871 returns give a summary of the acreages under all crops and animals for the long span of years 1847–71. Thereafter it became usual to give summaries for the previous ten years. Thus it is possible to obtain the basic farming data, at the county level at least, not

[9]   L. Kennedy, 'Regional specialization, railway development, and Irish agriculture in the nineteenth century', in J.M. Goldstrom and L.A. Clarkson (eds.), *Irish Population, Economy, and Society: Essays in Honour of the late K.H. Connell* (Oxford, 1981), pp. 180–1. The example for 1853 is taken from *Returns of Agricultural Produce in Ireland . . . 1853, BPP* [1865], vol. xlvii, 1854/5, p. xi.

by looking at all years but by an inspection of just six years, 1871, 1881, 1891, 1901, 1911 and 1914.[10]

The general abstract gave a summary at the county level of the distribution of landholdings, and in the detailed tables also at the poor law union level. From 1847 to 1851 this distinguished between holdings of less than 1 acre, and thereafter in four other broad categories, 1–5, 5–15, 15–30 acres and finally holdings of over 30 acres. From 1851 this became more detailed, and distinguished between nine size groups, less than 1 acre, 1–5, 5–15, 15–30, 30–50, 50–100, 100–200, 200–300 and over 500 acres. Yet what was a holding? Before the land reforms of the late nineteenth and early twentieth century most land was tenanted. At this time and especially after the legislation of 1903 – Wyndham's Act – there occurred a rapidly growing trend for the purchase of freeholds.[11]

Setting aside these more recent changes in tenure for the moment, it is still true to say that the annual returns are a statement of land occupancy not of ownership. This still begs the question, what was a holding, because over time the returns gave three different measures. In 1861 for example the actual number of *separate* and *distinct* holdings in Ireland was recorded as 610,045, and these related not to the number of land-holders but to the number of *distinct farms*.[12] In addition, at the county, provincial and national level this information was tabulated for the nine different size categories employed. Alternatively, by county and province and under nine separate classes the number of holdings in Ireland, '*adjoining* farms, held by the *same person*, being regarded as one Holding', was recorded as 583,719. Finally, by a third method, the number of holdings in each county and province and under nine class categories yielded for the whole of Ireland 553,664 holdings whereby 'the *total extent* of land held *by each* person', regardless of location, was counted as a single holding.

Given the fact that farms were not necessarily great respectors of county boundaries, the reported county and provincial figures were *approximate*. In 1861 the enumerators collected details on holdings for the first time in such a way that tables referring to *occupiers* could subsequently be constructed.[13] Which definition should be chosen in

---

[10] For a full digest of landownership details it was necessary to look at all the annual returns at some stage or other.

[11] Discussed more fully in the chapters above.

[12] Using the same emphasis as exists in the actual returns. This and the remainder of the paragraph based on *Agricultural Statistics of Ireland ... 1861, BPP* [3156], vol. lxix, 1863, pp. x-xii.

[13] For example, *Agricultural Statistics of Ireland ... 1871, BPP* [C. 762], vol. lxix, 1873, p. x.

subsequent analyses of landholding through time?[14] The returns themselves give a lead here. In their subsequent use of the landholding numbers, in conjunction with other variables such as crop distributions and livestock numbers, the enumerators used the first method. This was the original method and the one in most consistent use throughout the period. Yet does it matter which method is used? If there are regional differences in the distribution of large and small holdings then this method, which confines distributions within county boundaries, would exaggerate the importance of small holdings by necessarily creating more holdings than actually existed by double-counting and reclassifying holdings which straddled county boundaries. By a similar process the examples of holdings straddling boundaries would result in an underestimation of the numbers of larger holdings by truncating their sizes. In terms of fact, and for the precision of the database this is an important problem, but in terms of generalisations – and this study is designed to deal in terms of national and broad regional changes – it does not appear to be an important problem at all.[15]

The second part of the annual returns is a detailed breakdown at the various geographical levels of the acreages under a wide variety of crops and the numbers of a wide variety of animals. For the crops there is detail for five corn crops, four root and green crops, and details for flax, hay and permanent pasture. Not only are livestock distinguished one from another, but there is also a breakdown by age within livestock groups. Thus for horses there were those over 2 years of age, those under 2, and a third category for those under 1 year. There are yet other categories for mules and asses. Cattle were divided into milch cows (from 1854 onwards), and three other categories by age. Sheep and pigs were each divided into two age groups, and there were separate entries for goats and poultry. Up to 1906 the pigs were divided into those less than and those greater than 1 year of age, but from 1906 the division was made at 6 months of age.

These then are the basic data on crop acreages and livestock numbers. In addition the average yields of crops were presented in various volumetric measures, ranging from quarters per acre to barrels per acre, and also cwts and tons per acre. Yet note one important point here, the data for yields are in fact estimates, compiled in 1853 for example from average returns of produce made by each sub-inspector of the Constabulary for the various crops in his district, 'according to the best informa-

[14] See also D. Fitzpatrick, 'The disappearance of the Irish agricultural labourer, 1841–1912', *Irish Economic and Social History*, 7 (1980), 71–2.

[15] With the county as the base, Spearman rank correlation tests using the three landholding definitions produces correlation coefficients of 0.98 or better.

tion he could procure'. This was achieved by a somewhat complicated pro-forma which took the form of a questionnaire in which the quality and the quantity of the produce had to be estimated. Only a sample of landholders was selected to take part in this exercise. Prior to 1856 the produce was estimated in this way, through the Constabulary for Sub-Inspectors' Districts, but from 1856 the areal unit for assessment became the Electoral District. The change was introduced to acquire greater accuracy: the Electoral District was a much smaller area and their use thus increased the number of the average rate returns which were obtained. Even though the pro-forma was modified over time, still in all it remained the basic method of deriving average produce, and those estimates were then used to derive estimated average yields.[16] Nevertheless, given these data it becomes a short step to calculating gross output, though in most cases the enumerators have also performed this calculation. The crop yields are not only useful in their own right but when it came to estimating the output of British agriculture scholars sometimes inferred British yields from the trends in Irish yields because for much of the contemporary time there were no equivalent British data.[17]

On a less systematic basis it is possible to distinguish ewes from rams and boars from sows; to analyse beekeeping and distinguish different types of poultry; to investigate different potato varieties, and to analyse the fortunes of the flax industry by a study of the regional distribution of scutching mills. In addition, occasionally, a number of other time-based or area-based patterns can be revealed; on prices, wages and the migration of agricultural labour.

It is not necessary to consult every volume of the Returns to acquire the database, though as it happens every volume of the Annual Agricultural Statistics has been consulted for this study, and more besides. In addition there is an indispensable statistical digest which was produced in 1930 by the Department of Industry and Commerce in Dublin. They compiled a volume of statistics for the period 1847–1926, known simply as *Agricultural Statistics 1847–1926*, but infuriatingly for historians they carefully excised the data for Northern Ireland.[18] They retained data

---

[16] *Returns of Agricultural Produce ... 1853*, p. xii and Appendix pp. xlv-xlix, *Table Showing the Estimated Average Produce of the Crops ... 1875*, BPP [C. 1407], vol. lxxviii, 1876, p. 3.

[17] For example see E.M. Ojala, *Agriculture and Economic Progress* (Oxford, 1952), pp. 190–207.

[18] Saorstát Éireann, *Agricultural Statistics*. Though on a national basis for the period down to 1888 see the summary statistics of crop acreage and average yields, and livestock numbers, in T.W. Grimshaw, 'A statistical survey of Ireland, from 1840–1888', *Journal of the Statistical and Social Inquiry Society of Ireland*, 9 (1888), 321–61 and appendixes.

from those three out of the original nine Ulster counties which formed part of the new Irish Free State in their compilation, even though for most of the period under review from 1847 the data had been collected for the country as a whole. Incongruously the compilers of the volume included very detailed data for the two years 1912 and 1917 for the six Northern Ireland counties of Ulster, as indeed they did also in what is basically a brief and not very useful summary of truncated statistics for 1918.

Without itemising everything in the 1930 volume, what is presented is the annual statistics 1847–26 across a wide range of crops, including acres and yields and hence annual gross product, and of a range of animals of different kinds and ages set at the national level. This is followed by similar material at the county level for 1925 and 1926, but excluding yields and produce, and at the rural district level for 1926. The two main tables in the volume (tables 12 and 13) indicate land use across a wide range of crops, they record the numbers of different kinds of animals in different age groups, for each of the twenty-six counties of what became the Republic for each of the seven population census years 1851–1911, and for the years 1916–18 inclusive and 1925 and 1926. Although the original returns give much detail on farms or landholdings, this information is absent from this volume except for the year 1926 (and excluding the six counties of Ulster in Northern Ireland). At times the data is summarised at the national and provincial level. The compilation is prefaced by an essay on the trends in agriculture in the Republic over the long-run period and is illustrated with some maps, graphs and comparisons with other European countries. Make no mistake therefore, this is a good data set, but necessarily the lack of completeness required an exhaustive inquiry of the original parliamentary papers to write back into the exercise the six counties of Ulster.

It is a pity that a companion volume on the six counties is not available as well.[19] There is, though, a volume on the *Agricultural Output of Northern Ireland, 1925* which in its tables II, III, IV and VII gives full national figures for 1847–1927 aggregated for the six counties of Northern Ireland, indicating crop acreages, total produce by crop and weight – from which yields can be calculated – and full animal numbers

[19] Though see L. Kennedy, 'The rural economy, 1820–1914', in L. Kennedy and P. Ollerenshaw (eds.), *An Economic History of Ulster, 1820–1939* (Manchester, 1985), p. 57, note 51, where he indicates that he has collated the Ulster statistics in an unpublished paper. See also D.L. Armstrong, *An Economic History of Agriculture in Northern Ireland 1850–1900* (Plunkett Foundation, Oxford, 1989), especially pp. 106–7, 110–12, 166–71, for details of crop acreages, product prices and livestock numbers for the six counties of Ulster for the years 1850–1900 and 1926.

and landholding distributions across the full range of landholding size groups.[20]

In spite of these handy data sets, the desire in the present study to look sometimes at the county level required the inspection of the annual returns, *in situ*, in *British Parliamentary Papers*. The returns for any one year are usually included in the *Parliamentary Papers* of the following year.[21]

For the purposes of this study the acreages under the principal land-use types, the yields of the crops, the numbers of the different livestock distributions and data on landholding distribution, across space and through time, have been extracted.

The appendixes

The data are presented in a series of appendixes which follow. Appendix 1 is a digest of the annual agricultural statistics in terms of a range of crops and a variety of animals.[22] Appendix 2 is a summary of crop yields set at the national level, though the reader might usefully note that these data are also available at more detailed regional levels. Appendix 3 is a detailed summary of landholding and occupancy statistics, for land-holding from 1847 to 1914, and for occupiers from 1861 to 1914.

In chapter 4 the value of agricultural output has been estimated. The material for this calculation is available in appendixes 1 and 2 in conjunction with the data in appendix 4. Appendix 4 is a set of agricultural product prices.[23] Appendix 5 is an explanation of procedures adopted in the calculation of output, within which some other prices which are available and which some historians have preferred to use are discussed (see especially notes 18 and 19 and accompanying text).

---

[20] *The Agricultural Output of Northern Ireland, 1925, BPP* [Cmd. 87], 1928, pp. 72–83. See also R. O'Connor and C. Guiomard, 'Agricultural output in the Irish Free State area before and after Independence', *Irish Economic and Social History*, 12 (1985), 89–97 for a summary of twentieth-century sources available for the calculation of output.

[21] There is a good, fully referenced list of these for the period 1847–92 in J.S. Donnelly, *The Land and the People of Nineteenth-Century Cork* (London, 1975), pp. 393–4.

[22] The appendix also contains details on Irish livestock exports to Britain. A fuller compilation of Irish exports of both live and dead stock which covers the pre- as well as post-Famine periods, and which has been reconstructed from official sources and a multitude of other archives, is available in P. Solar, *Growth and Distribution in Irish Agriculture before the Famine*, Ph.D. Dissertation, University of Stanford, 1987, chs 3–6, pp. 94–254.

[23] For which see T. Barrington, 'A review of Irish agricultural prices', *Journal of the Statistical and Social Inquiry Society of Ireland*, 15 (1927), 249–80. See also R.M. Barrington, 'The prices of some agricultural produce and the cost of farm labour for the last fifty years', *Journal of the Statistical and Social Inquiry Society of Ireland*, 9 (1887), 137–53.

Finally, appendix 6 is an estimate of the labour inputs in Irish agriculture in the late nineteenth century based on the accounts of Richard Barrington and his father who were tenant farmers at a place called Fassaroe in County Wicklow. These labour input estimates are used in chapter 6 in a discussion of agricultural labour supply.

Appendix 1 Annual agricultural statistics, 1847–1914

PART A: *Crops – in acres*

Year	Wheat	Oats	Corn Barley[a]	Bere[a]	Rye	Peas & Beans[b]
1847	743,871	2,200,870	283,587	49,068	12,415	23,768
1848						
1849	687,646	2,061,185	290,690	60,819	20,168	53,916
1850	604,867	2,142,596	263,350	57,811	18,342	62,590
1851	504,248	2,189,775	282,617	53,347	19,697	49,717
1852	353,566	2,283,449	249,476	40,933	12,993	36,189
1853	326,896	2,157,849	272,644	28,380	12,376	35,242
1854	411,284	2,045,298	236,293	16,920	11,366	22,575
1855	445,775	2,118,858	226,629	11,185	11,632	18,485
1856	529,050	2,037,437	182,796	6,554	13,337	16,034
1857	559,646	1,980,934	211,288	6,026	15,348	13,586
1858	546,964	1,981,241	190,768	5,002	11,470	12,935
1859	464,175	1,982,662	177,894	3,751	9,447	14,851
1860	466,415	1,966,304	181,099	3,057	9,677	12,832
1861	401,243	1,999,160	198,955	3,052	8,530	14,017
1862	356,321	1,977,528	192,302	2,910	9,218	15,202
1863	260,311	1,953,883	171,892	2,730	5,929	15,153
1864	276,483	1,814,886	172,700	2,862	6,032	16,090
1865	266,989	1,745,228	177,102	2,913	7,178	16,899
1866	299,190	1,699,695	150,293	2,227	7,794	14,834
1867	261,034	1,660,511	171,001	1,931	7,671	13,552
1868	285,150	1,701,645	186,318	2,014	7,892	9,960
1869	280,460	1,685,240	221,710	2,401	8,782	9,955
1870	259,846	1,650,039	241,285	2,475	9,322	10,689
1871	244,451	1,636,136	220,979	1,855	9,700	10,913
1872	225,294	1,624,711	219,013	1,152	8,823	11,821
1873	167,554	1,510,972	230,115	820	8,404	12,873
1874	187,978	1,480,897	211,608	867	9,034	11,391
1875	158,995	1,501,867	233,903	713	9,617	11,713
1876	119,700	1,487,166	220,814	602	8,585	11,921
1877	139,297	1,476,172	226,216	680	10,444	9,786
1878	154,041	1,412,845	243,604	585	10,866	9,580
1879	157,511	1,330,261	254,292	553	9,099	10,151

Corn *(contd)*

Year	Wheat	Oats	Barley[a]	Bere[a]	Rye	Peas & Beans[b]
1880	148,708	1,381,928	218,016	561	7,107	10,157
1881	153,794	1,393,312	210,093	474	7,588	11,914
1882	152,824	1,397,307	187,254	363	7,773	11,216
1883	94,740	1,381,904	183,291	316	7,250	11,190
1884	67,890	1,348,444	167,061	346	7,149	8,729
1885	71,017	1,328,869	179,133	344	8,399	7,141
1886	69,546	1,321,983	181,598	298	10,576	6,703
1887	67,181	1,315,055	162,169	258	10,774	7,026
1888	99,013	1,280,858	170,929	380	13,942	5,823
1889	89,745	1,238,952	185,783	471	15,786	4,356
1890	92,341	1,221,013	182,058	379	14,573	4,370
1891	80,870	1,215,396	177,966	353	13,443	4,735
1892	75,408	1,226,244	175,178	408	13,117	4,433
1893	54,998	1,248,338	168,776	195	13,461	3,605
1894	49,338	1,254,837	164,595	176	11,926	3,185
1895	36,532	1,216,401	171,650	139	11,520	2,852
1896	38,019	1,193,581	173,032	383	13,715	2,090
1897	47,235	1,175,118	170,504	148	13,058	1,817
1898	52,798	1,165,359	158,012	149	12,389	2,247
1899	51,866	1,135,536	169,469	209	12,113	2,415
1900	53,821	1,105,050	173,996	177	11,407	2,738
1901	42,934	1,099,335	161,534	150	11,001	2,620
1902	44,244	1,082,144	167,788	89	9,638	2,665
1903	37,596	1,097,538	158,712	79	10,050	2,370
1904	30,825	1,078,772	158,043	60	9,414	2,075
1905	37,860	1,066,806	154,589	56	10,155	1,724
1906	43,880	1,076,310	176,435	109	10,343	2,266
1907	38,143	1,075,390	170,246	185	8,868	2,159
1908	36,677	1,060,301	154,442	154	8,050	2,091
1909	43,606	1,035,735	163,100	[a]	7,464	1,890
1910	47,631	1,073,690	168,008	[a]	8,681	2,069
1911	45,056	1,040,185	158,180	[a]	9,026	1,984
1912	44,855	1,046,000	165,367	[a]	7,765	1,700
1913	34,004	1,048,813	172,948	[a]	6,723	1,475
1914	36,913	1,028,758	172,289	[a]	7,535	1,508

	Root and green crops					
	Potatoes	Turnips	Mangels & beets[c]	Other	Flax	Hay[d]	Fruit
1847	284,116	370,344	13,766	59,512	58,312	1,138,946	
1848							
1849	718,608	360,069	18,758	70,204	60,314	1,141,371	
1850	875,357	347,331	20,390	74,494	91,040	1,200,124	

	Root and green crops *(contd)*						
	Potatoes	Turnips	Mangels & beets[c]	Other	Flax	Hay[d]	Fruit
1851	868,501	383,548	25,847	94,710	140,536	1,246,408	
1852	876,532	356,790	30,830	90,735	137,008	1,270,713	
1853	898,733	399,377	33,283	86,850	174,579	1,270,742	
1854	989,660	329,170	21,351	77,426	151,403	1,257,864	
1855	982,301	366,953	22,339	72,797	97,075	1,314,807	
1856	1,104,704	354,451	22,071	78,015	106,311	1,302,787	
1857	1,146,647	350,047	21,449	86,533	97,721	1,369,892	
1858	1,159,707	338,202	29,547	90,075	91,646	1,424,495	
1859	1,200,347	322,137	26,906	87,042	136,282	1,437,111	
1860	1,172,079	318,540	31,986	85,037	128,595	1,594,518	
1861	1,133,504	334,104	22,833	80,975	147,957	1,546,206	
1862	1,018,112	376,715	23,114	79,194	150,070	1,552,924	
1863	1,023,414	351,436	16,320	86,682	214,099	1,560,638	
1864	1,039,724	337,355	14,073	84,854	301,693	1,609,569	
1865	1,066,620	334,212	14,389	86,947	251,433	1,678,493	
1866	1,050,353	317,198	20,082	93,972	263,507	1,601,423	
1867	1,001,781	335,728	18,739	76,162	253,257	1,658,335	
1868	1,034,681	320,094	19,031	82,568	206,483	1,692,135	
1869	1,041,902	322,072	21,080	84,210	229,252	1,670,716	
1870	1,043,583	339,059	25,320	90,761	194,910	1,773,851	
1871	1,058,434	327,035	31,810	94,410	156,670	1,829,044	
1872	991,871	346,711	34,832	100,820	121,992	1,800,273	
1873	903,262	347,848	38,231	83,122	129,297	1,838,248	
1874	892,425	333,588	38,327	89,303	106,907	1,906,679	
1875	900,586	332,538	43,224	93,807	101,174	1,944,676	
1876	880,716	344,637	48,624	89,715	132,938	1,861,128	
1877	873,291	334,379	48,948	98,391	123,380	1,924,917	
1878	846,712	330,243	45,219	95,689	111,817	1,942,804	
1879	842,671	314,697	51,155	86,167	128,021	1,937,255	
1880	820,651	302,695	41,515	82,381	157,540	1,909,825	
1881	855,293	295,212	44,838	74,683	147,145	2,001,029	
1882	837,918	294,070	36,316	80,659	113,484	1,962,152	
1883	806,467	306,799	37,945	79,072	95,943	1,931,784	
1884	798,952	304,031	34,541	83,889	89,225	1,962,487	
1885	797,292	296,984	37,179	87,854	108,147	2,034,768	
1886	799,847	299,323	37,413	84,729	127,890	2,094,209	
1887	796,939	300,123	41,733	90,297	130,284	2,143,818	
1888	804,566	294,237	45,749	89,593	113,613	2,221,980	
1889	787,234	297,913	44,021	90,581	113,652	2,187,522	
1890	780,801	295,386	46,457	91,818	96,896	2,093,634	
1891	753,332	300,326	51,757	86,009	74,665	2,059,529	
1892	740,025	300,447	51,554	82,837	70,647	2,142,810	
1893	723,735	302,774	47,034	80,165	67,487	2,167,473	
1894	717,090	311,310	52,039	82,836	101,081	2,182,598	
1895	710,486	313,281	53,027	74,960	95,203	2,194,476	

	Root and green crops *(contd)*						
	Potatoes	Turnips	Mangels & beets[c]	Other	Flax	Hay[d]	Fruit
1896	705,665	308,471	54,301	79,286	72,253	2,202,424	
1897	677,216	308,966	54,649	74,616	45,537	2,176,142	
1898	664,864	306,929	55,955	77,219	34,469	2,174,470	
1899	662,914	301,449	62,714	74,964	34,989	2,118,907	
1900	654,079	297,859	68,803	77,636	47,451	2,165,715	
1901	635,321	289,759	77,457	76,906	55,442	2,178,592	
1902	629,304	288,506	77,144	75,439	49,742	2,168,464	
1903	620,393	287,548	76,074	75,768	44,685	2,224,165	
1904	618,540	285,831	75,828	70,284	44,293	2,260,160	
1905	616,755	282,105	72,697	72,816	46,158	2,294,506	
1906	616,107	278,367	67,323	71,899	55,189	2,330,016	10,602
1907	590,998	275,092	67,132	69,758	59,659	2,281,318	11,449
1908	587,144	279,044	72,134	70,575	46,916	2,298,793	12,145
1909	579,799	276,944	73,437	71,491	38,110	2,278,538	12,583
1910	592,985	275,296	75,267	68,539	45,974	2,421,587	12,994
1911	591,259	270,805	77,857	73,806	66,618	2,512,403	14,045
1912	595,184	271,771	81,700	73,434	55,062	2,487,349	15,218
1913	582,303	276,596	78,914	71,668	59,305	2,481,822	15,734
1914	583,069	276,872	81,570	73,895	49,253	2,487,513	16,090

			Summary		
	Total corn		Total root & green	Hay	Total crops[e]
				(Meadow & Clover)	
1847	3,313,579		727,738	1,138,946	5,238,575
1848					
1849	3,174,424		1,167,639	1,141,371	5,543,748
1850	3,149,556		1,317,572	1,200,124	5,758,292
1851	3,099,401		1,372,606	1,246,408	5,858,951
1852	2,976,606		1,354,887	1,270,713	5,739,214
1853	2,833,387		1,418,243	1,270,742	5,696,951
1854	2,743,736		1,417,607	1,257,864	5,570,610
1855	2,832,564		1,444,390	1,314,807	5,688,836
1856	2,785,208		1,559,241	1,302,787	5,753,547
1857	2,786,828		1,604,676	1,369,892	5,859,117
1858	2,748,380		1,617,531	1,424,495	5,882,052
1859	2,652,780		1,636,432	1,437,111	5,862,605
1860	2,639,384		1,607,642	1,594,518	5,970,139
1861	2,624,957		1,571,416	1,546,206	5,890,536
1862	2,553,481		1,497,135	1,552,924	5,753,610
1863	2,409,898		1,477,852	1,560,638	5,662,487
1864	2,289,053		1,476,006	1,609,569	5,676,321
1865	2,216,309		1,502,168	1,678,493	5,648,403
1866	2,174,033		1,481,605	1,601,423	5,520,568

Summary *(contd)*

	Total corn	Total root & green	Hay	Total crops[d]
			(Meadow & Clover)	
1867	2,115,700	1,432,410	1,658,335	5,459,702
1868	2,192,979	1,456,374	1,692,135	5,547,971
1869	2,208,548	1,469,264	1,670,716	5,577,780
1870	2,173,656	1,498,723	1,773,851	5,641,140
1871	2,124,034	1,511,689	1,829,044	5,621,437
1872	2,090,814	1,474,234	1,800,273	5,487,313
1873	1,930,738	1,372,463	1,838,248	5,270,746
1874	1,901,775	1,353,643	1,906,679	5,269,004
1875	1,916,808	1,370,155	1,944,676	5,332,813
1876	1,848,788	1,363,692	1,861,128	5,206,546
1877	1,862,595	1,355,009	1,924,917	5,265,901
1878	1,831,521	1,317,863	1,942,804	5,204,005
1879	1,761,867	1,294,690	1,937,255	5,121,833
1880	1,766,477	1,247,242	1,909,825	5,081,084
1881	1,777,175	1,270,026	2,001,029	5,195,375
1882	1,756,737	1,248,963	1,962,152	5,081,336
1883	1,678,691	1,230,283	1,931,784	4,936,701
1884	1,599,619	1,221,413	1,962,487	4,872,744
1885	1,594,903	1,219,309	2,034,768	4,957,127
1886	1,590,704	1,221,312	2,094,209	5,034,115
1887	1,562,463	1,229,092	2,143,818	5,065,657
1888	1,570,945	1,234,145	2,221,980	5,140,683
1889	1,535,093	1,219,749	2,187,522	5,056,016
1890	1,514,734	1,214,462	2,093,634	4,919,726
1891	1,492,763	1,191,424	2,059,529	4,818,381
1892	1,494,788	1,174,863	2,142,810	4,883,108
1893	1,489,373	1,153,708	2,167,473	4,878,041
1894	1,484,057	1,163,275	2,182,598	4,931,011
1895	1,439,094	1,151,754	2,194,476	4,880,527
1896	1,420,820	1,147,723	2,202,424	4,843,220
1897	1,407,880	1,115,447	2,176,142	4,745,006
1898	1,390,954	1,104,967	2,174,470	4,704,860
1899	1,371,608	1,102,041	2,118,907	4,627,545
1900	1,347,189	1,098,377	2,165,715	4,658,732
1901	1,317,574	1,079,443	2,178,592	4,631,051
1902	1,306,568	1,070,393	2,168,464	4,595,167
1903	1,306,345	1,059,783	2,224,165	4,634,978
1904	1,279,189	1,050,483	2,260,160	4,634,125
1905	1,271,190	1,044,373	2,294,506	4,656,227
1906	1,309,343	1,033,696	2,330,016	4,738,846
1907	1,294,991	1,002,980	2,281,318	4,650,397
1908	1,261,715	1,008,897	2,298,793	4,628,466
1909	1,251,795	1,001,671	2,278,538	4,582,697
1910	1,300,079	1,012,087	2,421,587	4,792,721
1911	1,254,431	1,013,727	2,512,403	4,861,224

Summary *(contd)*

	Total corn	Total root & green	Hay	Total crops[e]
			(Meadow & Clover)	
1912	1,265,687	1,022,089	2,487,349	4,845,405
1913	1,263,963	1,009,481	2,481,822	4,830,305
1914	1,247,003	1,015,406	2,487,513	4,815,265

Notes:
[a]   Barley and Bere counted together from 1909.
[b]   Peas and Beans counted together in National Summary
[c]   Mangels by title on their own to 1871, with beets thereafter.
[d]   Hay refers to meadow and clover.
[e]   Total includes flax throughout and fruit from 1906.

Source: See sources at end of appendix 3, p. 254.

PART B: *Livestock numbers*

Year	Milch cows[a]	Cattle > 2 years[a]	Cattle 1–2 years	Cattle < 1 year	Sheep > 1 year	Sheep < 1 year
1847		(2,060,582)		530,833	1,676,547	509,630
1848						
1849		1,749,490	543,549	478,090	1,402,624	374,487
1850		1,821,369	593,801	502,779	1,392,782	483,364
1851		1,847,717	588,344	531,400	1,544,220	577,908
1852		1,873,017	621,287	600,763	1,928,538	685,405
1853		1,961,678	699,084	722,547	2,191,285	951,371
1854	1,517,672	2,217,699	597,661	682,541	2,503,930	1,218,289
1855	1,561,296	2,314,715	570,339	679,346	2,496,171	1,106,171
1856	1,579,529	2,359,767	621,107	606,984	2,491,259	1,203,035
1857	1,605,350	2,407,363	597,440	616,151	2,332,434	1,119,818
1858	1,635,409	2,424,041	602,525	640,738	2,333,339	1,161,655
1859	1,690,389	2,476,278	636,684	702,636	2,376,388	1,216,416
1860	1,626,453	2,401,294	624,206	580,874	2,358,541	1,183,539
1861	1,545,168	2,391,760	558,503	521,425	2,387,930	1,168,120
1862	1,486,835	2,225,528	485,680	543,682	2,308,546	1,147,586
1863	1,396,924	2,079,978	500,677	563,576	2,223,451	1,084,753
1864	1,348,886	2,028,295	551,136	682,863	2,234,879	1,132,062
1865	1,387,448	2,102,372	655,382	739,794	2,385,250	1,309,106
1866	1,482,616	2,298,869	717,662	729,626	2,685,714	1,588,568
1867	1,521,053	2,811,237	684,010	712,556	3,090,849	1,744,670
1868	1,476,339	2,290,111	665,423	691,262	3,174,131	1,727,365
1869	1,506,038	2,329,709	676,932	727,034	3,031,862	1,619,333
1870	1,529,024	2,324,263	703,710	771,939	2,843,801	1,493,083

Year	Milch cows[a]	Cattle > 2 years[a]	Cattle 1–2 years	Cattle < 1 year	Sheep > 1 year	Sheep < 1 year
1871	1,545,662	2,387,999	746,596	841,777	2,757,453	1,475,982
1872	1,551,784	2,371,706	767,878	919,813	2,726,679	1,536,575
1873	1,528,136	2,372,679	822,990	951,433	2,883,369	1,601,151
1874	1,491,375	2,405,218	846,207	873,331	2,861,964	1,579,734
1875	1,530,366	2,484,240	794,898	836,150	2,754,047	1,499,980
1876	1,532,974	2,461,886	778,465	877,089	2,579,630	1,429,527
1877	1,522,811	2,401,659	785,108	810,831	2,532,510	1,454,999
1878	1,484,315	2,360,754	738,422	885,944	2,589,279	1,505,855
1879	1,464,818	2,305,447	813,215	949,116	2,571,921	1,445,982
1880	1,398,047	2,261,668	818,804	841,045	2,295,744	1,266,719
1881	1,392,102	2,292,601	786,738	877,256	2,098,910	1,157,275
1882	1,399,005	2,274,913	793,003	919,295	1,934,103	1,137,652
1883	1,402,324	2,263,518	852,891	980,544	1,985,082	1,234,229
1884	1,356,585	2,277,747	880,764	954,278	2,028,324	1,216,888
1885	1,417,423	2,366,199	881,738	980,914	2,138,568	1,339,488
1886	1,418,644	2,370,985	880,995	931,944	2,076,153	1,289,890
1887	1,394,135	2,355,958	865,413	936,033	2,032,821	1,345,005
1888	1,384,771	2,303,529	874,045	921,621	2,161,116	1,465,553
1889	1,363,781	2,274,038	866,835	953,301	2,252,846	1,536,341
1890	1,400,527	2,317,726	899,586	1,023,004	2,538,386	1,785,009
1891	1,442,268	2,405,796	981,172	1,061,543	2,784,985	1,937,628
1892	1,451,059	2,498,212	1,015,109	1,017,804	2,881,752	1,946,025
1893	1,441,329	2,538,747	969,302	956,008	2,681,180	1,740,275
1894	1,447,441	2,517,425	914,229	960,185	2,492,635	1,612,545
1895	1,433,988	2,429,156	911,705	1,017,171	2,349,567	1,563,882
1896	1,429,795	2,430,180	957,866	1,020,087	2,432,368	1,648,343
1897	1,434,925	2,448,169	964,934	1,051,771	2,466,593	1,691,313
1898	1,431,192	2,458,784	982,284	1,045,881	2,518,027	1,769,524
1899	1,443,855	2,455,390	993,384	1,058,683	2,554,913	1,809,594
1900	1,458,074	2,489,083	1,033,941	1,085,526	2,586,046	1,800,830
1901	1,482,483	2,523,989	1,046,203	1,103,131	2,586,451	1,792,299
1902	1,510,737	2,588,362	1,067,705	1,126,154	2,499,109	1,716,756
1903	1,495,179	2,527,357	1,036,253	1,100,502	2,305,680	1,638,924
1904	1,497,647	2,524,256	1,035,435	1,117,027	2,247,387	1,580,532
1905	1,487,064	2,528,370	1,024,638	1,092,207	2,194,651	1,554,701
1906	1,496,284	2,554,965	1,000,343	1,083,616	2,160,598	1,554,234
1907	1,561,463	2,601,725	999,665	1,075,103	2,219,313	1,597,296
1908	1,586,425	2,627,901	1,050,436	1,114,121	2,394,923	1,731,183
1909	1,548,936	2,573,392	1,035,116	1,091,056	2,437,520	1,695,838
1910	1,557,584	2,563,841	1,014,061	1,110,986	2,354,693	1,624,823
1911	1,565,418	2,580,545	1,020,280	1,110,895	2,317,573	1,589,863
1912	1,598,986	2,626,359	1,071,485	1,150,654	2,274,182	1,554,647
1913	1,605,220	2,661,187	1,109,681	1,161,757	2,126,138	1,494,586
1914	1,638,929	2,771,112	1,141,461	1,139,072	2,127,639	1,472,942

	Pigs > 1 year	Pigs < 1 year	Goats	Poultry
1847	324,088	298,371	164,043	5,691,055
1848				
1849	343,749	451,714	182,988	6,328,001
1850	332,355	595,147	201,112	6,945,146
1851	391,698	693,159	235,313	7,470,694
1852	430,209	642,449	278,444	8,175,904
1853	393,522	751,423	296,182	8,660,738
1854	336,401	1,006,148	311,492	8,630,488
1855	339,231	838,374	283,970	8,366,629
1856	252,140	666,385	269,746	8,908,226
1857	286,769	968,417	243,046	9,491,463
1858	330,385	1,079,498	228,351	9,563,185
1859	332,982	942,769	219,346	10,251,749
1860	273,953	997,119	194,465	10,060,776
1861	230,049	871,993	189,842	10,371,175
1862	228,240	926,084	175,328	9,916,630
1863	214,071	853,387	166,044	9,649,118
1864	202,877	855,603	171,307	10,424,085
1865	224,486	1,081,467	171,207	10,681,955
1866	241,017	1,256,257	186,880	10,889,747
1867	196,899	1,038,292	190,429	10,334,903
1868	144,861	724,717	199,060	10,602,782
1869	171,638	910,586	205,861	10,801,687
1870	222,042	1,238,813	211,891	11,159,002
1871	232,226	1,389,197	231,373	11,717,182
1872	201,011	1,187,560	238,961	11,737,529
1873	153,656	890,798	242,689	11,863,155
1874	153,078	946,108	256,753	12,068,375
1875	172,432	1,079,624	270,691	12,139,138
1876	195,971	1,229,071	264,009	13,618,500
1877	199,169	1,269,543	267,297	13,566,083
1878	168,361	1,101,038	278,974	13,711,714
1879	143,945	928,240	278,843	13,782,835
1880	115,311	734,958	265,789	13,430,182
1881	149,721	946,109	266,078	13,972,426
1882	189,027	1,241,101	263,272	13,999,096
1883	180,893	1,167,471	263,146	13,382,430
1884	167,682	1,138,868	254,411	12,747,460
1885	160,653	1,108,439	264,437	13,850,532
1886	160,832	1,102,310	266,176	13,909,822
1887	177,611	1,230,845	271,729	14,460,643
1888	171,091	1,226,734	295,678	14,486,400
1889	168,646	1,212,024	303,933	14,856,517
1890	189,343	1,381,023	327,144	15,408,428
1891	160,642	1,207,070	336,337	15,276,128
1892	133,923	979,549	332,726	15,335,749
1893	139,056	1,013,361	323,173	16,097,461

	Pigs > 1 year	Pigs < 1 year	Goats	Poultry
1894	163,208	1,226,116	318,907	16,180,601
1895	156,527	1,181,937	304,820	16,369,525
1896	163,912	1,240,674	306,445	17,537,570
1897	150,005	1,177,445	299,086	17,777,248
1898	137,723	1,116,189	296,437	17,687,430
1899	148,193	1,215,117	303,509	18,233,520
1900	135,511	1,133,010	306,078	18,547,307
1901	166,656	1,052,479	312,409	18,810,717
1902	175,019	1,152,591	303,654	18,504,324
1903	184,690	1,198,826	299,120	18,153,714
1904	164,538	1,150,588	290,222	18,256,959
1905	148,251	1,016,065	284,069	18,549,051
1906	239,077	1,005,116	267,342	18,976,798
1907	273,475	1,043,593	247,347	24,326,995
1908	275,236	942,604	246,286	24,031,095
1909	247,595	901,584	252,041	24,104,934
1910	267,845	932,160	242,614	24,339,015
1911	328,793	1,086,326	258,474	25,447,801
1912	309,594	1,014,363	252,722	25,525,724
1913	240,889	819,471	246,348	25,701,342
1914	308,942	996,696	242,243	26,918,749

	Horses > 2 years	Horses 1–2 years	Horses < 1 year	Mules[b]	Asses
1847	(	515,277	) 42,640		126,355
1848					
1849	462,463	28,486	34,975	22,364	117,939
1850	460,356	31,113	35,288	21,962	123,412
1851	451,789	30,322	39,595	21,606	136,981
1852	446,756	34,142	44,190	20,812	144,120
1853	447,543	39,097	53,145	21,315	148,720
1854	434,363	49,536	62,030	18,601	150,576
1855	437,108	53,600	65,579	19,857	151,742
1856	453,196	58,321	61,891	19,460	155,224
1857	473,644	60,257	63,881	19,512	160,987
1858	483,608	63,421	64,292	19,290	163,323
1859	497,856	65,547	65,672	19,304	169,354
1860	495,112	64,585	60,114	19,689	167,845
1861	501,977	62,564	49,691	20,146	173,711
1862	506,817	52,153	43,924	19,695	170,887
1863	497,301	47,641	35,036	19,993	167,244
1864	490,044	40,404	31,710	20,276	169,221
1865	478,546	35,150	34,643	19,802	168,009
1866	466,392	36,278	33,129	19,848	173,175
1867	454,980	35,006	34,194	19,506	167,233

	Horses > 2 years	Horses 1–2 years	Horses < 1 year	Mules[b]	Asses
1868	451,546	37,985	35,172	19,669	169,100
1869	448,767	39,950	39,484	19,556	171,664
1870	446,742	43,646	42,269	19,559	173,717
1871	444,980	46,944	46,171	19,817	180,373
1872	444,608	47,624	48,742	19,830	181,351
1873	434,376	48,993	49,078	19,891	177,779
1874	424,564	50,346	51,677	20,785	180,430
1875	415,916	54,355	56,081	21,767	180,355
1876	416,536	57,317	61,231	21,867	182,210
1877	422,727	63,519	66,460	22,792	185,842
1878	430,089	68,382	63,978	23,966	188,464
1879	437,750	69,493	65,077	24,570	188,839
1880	435,777	67,471	53,981	24,901	186,327
1881	434,279	62,289	51,786	26,392	187,143
1882	429,186	56,166	54,113	26,460	187,782
1883	423,116	56,019	55,097	27,195	189,760
1884	416,970	59,444	58,383	27,642	191,339
1885	418,785	64,962	63,397	29,286	197,170
1886	419,234	67,485	62,485	29,095	196,245
1887	422,522	69,595	65,288	29,829	199,512
1888	420,341	74,263	70,493	30,271	203,152
1889	427,307	77,062	69,895	29,838	206,236
1890	428,532	80,417	75,923	30,012	213,018
1891	427,721	86,091	79,007	28,660	216,268
1892	430,551	93,821	81,538	29,303	217,600
1893	436,996	97,848	79,083	29,202	218,720
1894	446,466	99,294	77,422	29,348	224,513
1895	459,856	95,381	75,020	29,860	224,408
1896	464,262	91,385	73,450	30,078	230,721
1897	461,558	83,816	64,841	29,270	230,253
1898	458,651	75,367	56,750	29,622	231,659
1899	457,375	66,759	56,152	30,129	237,177
1900	445,507	64,448	57,023	30,678	242,247
1901	428,236	73,691	62,989	28,882	238,980
1902	428,795	81,897	69,073	29,397	242,862
1903	436,563	89,327	69,856	29,795	243,241
1904	442,820	93,132	68,978	29,931	244,145
1905	447,303	94,566	67,125	29,684	244,606
1906	443,106	99,477	61,830	30,967	243,669
1907	438,106	99,790	57,463	29,791	237,540
1908	446,083	97,458	61,088	30,351	241,133
1909	443,090	98,089	58,006	30,479	243,666
1910	450,317	101,678	61,249	31,460	240,677
1911	455,191	100,536	60,604	31,740	246,353
1912	455,363	101,502	60,667	30,911	243,437
1913	460,962	98,843	54,677	30,338	243,339
1914	466,622	96,790	55,933	30,942	244,487

	Cattle	Sheep	Pigs	Summary totals Horses	Goats	Poultry	Mules & asses
1847	2,591,415	2,186,177	622,459	557,917	164,043	5,691,055	126,355
1848							
1849	2,771,129	1,777,111	795,463	525,924	182,988	6,328,001	140,303
1850	2,917,949	1,876,146	927,502	526,757	201,112	6,945,146	145,374
1851	2,967,461	2,122,128	1,084,857	521,706	235,313	7,470,694	158,587
1852	3,095,067	2,613,943	1,072,658	525,088	278,444	8,175,904	164,932
1853	3,383,309	3,142,656	1,144,945	539,785	296,182	8,660,738	170,035
1854	3,497,901	3,722,219	1,342,549	545,929	311,492	8,630,488	169,177
1855	3,564,400	3,602,342	1,177,605	556,287	283,970	8,366,629	171,599
1856	3,587,858	3,694,294	918,525	573,408	269,746	8,908,226	174,684
1857	3,620,954	3,452,252	1,255,186	597,782	243,046	9,491,463	180,499
1858	3,667,304	3,494,994	1,409,883	611,321	228,351	9,563,185	182,613
1859	3,815,598	3,592,804	1,275,751	629,075	219,346	10,251,749	188,658
1860	3,606,374	3,542,080	1,271,072	619,811	194,465	10,060,776	187,534
1861	3,471,688	3,556,050	1,102,042	614,232	189,842	10,371,175	193,857
1862	3,254,890	3,456,132	1,154,324	602,894	175,328	9,916,630	190,582
1863	3,144,231	3,308,204	1,067,458	579,978	166,044	9,649,118	187,237
1864	3,262,294	3,366,941	1,058,480	562,158	171,307	10,424,085	189,497
1865	3,497,548	3,694,356	1,305,953	548,339	171,207	10,681,955	187,811
1866	3,746,157	4,274,282	1,497,274	535,799	186,880	10,889,747	193,023
1867	4,207,803	4,835,519	1,235,191	524,180	190,429	10,334,903	186,739
1868	3,646,796	4,901,496	869,578	524,703	199,060	10,602,782	188,769
1869	3,733,675	4,651,195	1,082,224	528,201	205,861	10,801,687	191,220
1870	3,799,912	4,336,884	1,460,855	532,657	211,891	11,159,002	193,276
1871	3,976,372	4,233,435	1,621,423	538,095	231,373	11,717,182	200,190
1872	4,059,397	4,263,254	1,388,571	540,974	238,961	11,737,529	201,181
1873	4,147,102	4,484,520	1,044,454	532,447	242,689	11,863,155	197,670
1874	4,124,756	4,441,698	1,099,186	526,587	256,753	12,068,375	201,215
1875	4,115,288	4,254,027	1,252,056	526,352	270,691	12,139,138	202,122
1876	4,117,440	4,009,157	1,425,042	535,084	264,009	13,618,500	204,077
1877	3,997,598	3,987,509	1,468,712	552,706	267,297	13,566,083	208,634
1878	3,985,120	4,095,134	1,269,399	562,449	278,974	13,711,714	212,430
1879	4,067,778	4,017,903	1,072,185	572,320	278,843	13,782,835	213,409
1880	3,921,517	3,562,463	850,269	557,229	265,789	13,430,182	211,228
1881	3,956,595	3,256,185	1,095,830	548,354	266,078	13,972,426	213,535
1882	3,987,211	3,071,755	1,430,128	539,465	263,272	13,999,096	214,242
1883	4,096,953	3,219,311	1,348,364	534,232	263,146	13,382,430	216,955
1884	4,112,789	3,245,212	1,306,550	534,797	254,411	12,747,460	218,981
1885	4,228,851	3,478,056	1,269,092	547,144	264,437	13,850,532	226,456
1886	4,183,924	3,366,043	1,263,142	549,204	266,176	13,909,822	225,340
1887	4,157,404	3,377,826	1,408,456	557,405	271,729	14,460,643	229,341
1888	4,099,195	3,626,669	1,397,825	565,097	295,678	14,486,400	233,423
1889	4,094,174	3,789,187	1,380,670	574,264	303,933	14,856,517	236,074
1890	4,240,316	4,323,395	1,570,366	584,872	327,144	15,408,428	243,030
1891	4,448,511	4,722,613	1,367,712	592,819	336,337	15,276,128	244,928
1892	4,531,125	4,827,777	1,113,472	605,910	332,726	15,335,749	246,903

Summary totals *(contd)*

	Cattle	Sheep	Pigs	Horses	Goats	Poultry	Mules & asses
1893	4,464,057	4,421,455	1,152,417	613,927	323,173	16,097,461	247,922
1894	4,391,839	4,105,180	1,389,324	623,182	318,907	16,180,601	253,861
1895	4,358,032	3,913,449	1,338,464	630,257	304,820	16,369,525	254,268
1896	4,408,133	4,080,711	1,404,586	629,097	306,445	17,537,570	260,799
1897	4,464,874	4,157,906	1,327,450	610,215	299,086	17,777,248	259,523
1898	4,486,949	4,287,551	1,253,912	590,768	296,437	17,687,430	261,281
1899	4,507,457	4,364,507	1,363,310	580,286	303,509	18,233,520	267,306
1900	4,608,550	4,386,876	1,268,521	566,978	306,078	18,547,307	272,925
1901	4,673,323	4,378,750	1,219,135	564,916	312,409	18,810,717	267,862
1902	4,782,221	4,215,865	1,327,610	579,765	303,654	18,504,324	272,259
1903	4,664,112	3,944,604	1,383,516	595,746	299,120	18,153,714	273,036
1904	4,676,718	3,827,919	1,315,126	604,930	290,222	18,256,959	274,076
1905	4,645,215	3,749,352	1,164,316	608,994	284,069	18,549,051	274,290
1906	4,638,924	3,714,832	1,244,193	604,413	267,342	18,976,798	274,636
1907	4,676,493	3,816,609	1,317,068	595,359	247,347	24,326,995	267,331
1908	4,792,458	4,126,106	1,217,840	604,629	246,286	24,031,095	271,484
1909	4,699,564	4,133,358	1,149,179	599,185	252,041	24,104,934	274,145
1910	4,688,888	3,979,516	1,200,005	613,244	242,614	24,339,015	272,137
1911	4,711,720	3,907,436	1,415,119	616,331	258,474	25,447,801	278,093
1912	4,848,498	3,828,829	1,323,957	617,532	252,722	25,525,724	274,348
1913	4,932,625	3,620,724	1,060,360	614,482	246,348	25,701,342	273,677
1914	5,051,645	3,600,581	1,305,638	619,345	242,243	26,918,749	275,429

Notes: [a] Milk cows not separately distinguished until 1854. Before this date *and throughout* the table above, they are also included with cattle > 2 years.

[b] In 1847 the mules were returned with horses. By 1911 the mules are designated as mules and jennets.

Source: See sources at end of appendix 3, p. 254.

PART C   *Livestock exports to Britain, 1854-1914*

Date	Cattle			Milk[a]	Springers[a]	Calves	Total
	Fat	Store	Other				
Ave 1854–6							242,280
1861							358,664
1862							429,019
1863							441,651
1864							348,286
1865		(246,734)					246,734
1866		(399,231)					399,231
1867		(473,871)					473,871
1868							

Date	Cattle			Milk[a]	Springers[a]	Calves	Total
	Fat	Store	Other				
1869		456,035				53,071	509,106
1870		415,673				38,296	453,969
1871		423,396				60,529	483,925
1872		481,878				134,202	616,080
1873		(684,618)					684,618
1874		509,330				41,879	551,209
1875	254,681	293,176	11,757			35,704	595,318
1876	279,134	328,512	15,735			42,947	666,328
1877	246,698	356,249	7,706			38,788	649,441
1878	245,944	416,759	4,954			61,564	729,221
1879	247,897	320,244	6,845			66,384	641,370
1880	232,905	417,203	2,812			68,471	721,391
1881	279,125	250,899	3,701			37,832	571,557
1882	291,777	427,798	3,006			59,693	782,274
1883	229,603	278,518	1,819			46,927	556,867
1884	255,026	387,352	2,220			71,245	715,843
1885	243,348	342,938	1,884			52,300	640,470
1886	285,156	388,917	1,247			42,069	717,389
1887	331,119	302,878	2,283			32,973	669,253
1888	282,537	405,540	2,941			47,698	738,716
1889	248,362	372,682	1,432			47,367	669,843
1890	216,339	360,758	1,152			53,449	631,698
1891	240,183	323,075	3,985			63,559	630,802
1892	256,538	305,373	6,278			56,268	624,457
1893	316,344	318,545	8,473			45,307	688,669
1894	330,748	422,534	7,805			65,867	826,954
1895	302,555	414,859	5,622			68,571	791,607
1896	274,472	349,800	3,837			53,451	681,560
1897	259,173	419,302	5,043			62,494	746,012
1898	278,770	460,903	4,101			59,588	803,362
1899	278,064	442,921	6,219			45,068	772,272
1900	275,450	427,891	7,442			34,736	745,519
1901	261,690	344,954	6,269			29,725	642,638
1902	306,892	556,554	10,634			85,161	959,241
1903	246,887	556,506	6,724			87,528	897,645
1904	232,186	470,361	6,896			62,920	772,363
1905	224,943	455,667	6,205			62,316	749,131
1906	240,566	473,425	5,897			55,486	775,374
1907	257,073	471,299	5,375			49,939	783,686
1908	284,999	442,968	6,550	40,965	30,097	57,949	863,528
1909	271,783	453,887	11,011	42,317	27,509	58,915	865,422
1910	262,526	450,122	13,701	40,717	26,949	52,952	846,967
1911	253,302	455,130	10,851	46,451	28,418	48,300	842,452
1912	286,705	318,705	9,276	41,165	20,997	26,204	703,052
1913	366,358	363,094	8,682	16,278	11,157	25,726	791,295
1914	309,860	362,502	6,260	32,266	22,520	34,430	767,838

Date	Sheep			Swine		
	Sheep	Lambs	Total	Fat	Store	Total
Ave 1854–6			482,830			241,293
1861			407,426			358,187
1862			538,631			364,634
1863			517,232			357,938
1864			370,781			338,543
1865			332,831			383,452
1866			398,946			504,224
1867			588,906			248,319
1868						
1869			1,015,694			264,620
1870			620,834			422,976
1871			684,708			528,244
1872			518,605			443,644
1873			604,695			364,371
1874			744,234			344,335
1875	641,307	276,672	917,979	389,179	74,439	463,618
1876	474,871	211,937	686,808	436,044	77,272	513,316
1877	431,129	199,645	630,774	508,912	76,515	585,427
1878	446,628	196,371	642,999	401,167	69,380	470,547
1879	506,621	166,750	673,371	371,079	58,584	429,663
1880	502,806	211,957	714,763	333,653	39,237	372,890
1881	415,703	161,924	577,627	349,532	33,463	382,995
1882	393,848	164,556	558,404	453,443	49,463	502,906
1883	312,108	148,621	460,729	433,793	27,224	461,017
1884	355,466	177,819	533,285	437,227	19,451	456,678
1885	430,410	198,680	629,090	370,639	27,925	398,564
1886	493,983	240,230	734,213	391,509	29,776	421,285
1887	321,644	226,924	548,568	438,155	42,765	480,920
1888	400,836	236,748	637,584	495,680	49,292	544,972
1889	373,313	240,374	613,687	428,103	45,448	473,551
1890	387,220	249,761	636,981	543,417	59,745	603,162
1891	569,698	323,477	893,175	459,596	43,988	503,584
1892	713,528	366,674	1,080,202	457,977	42,974	500,951
1893	705,299	402,661	1,107,960	405,242	51,329	456,571
1894	574,471	382,630	957,101	515,647	69,320	584,967
1895	351,975	300,603	652,578	500,700	46,520	547,220
1896	397,164	340,142	737,306	574,677	35,912	610,589
1897	435,709	368,806	804,515	653,459	41,848	695,307
1898	449,558	383,900	833,458	556,723	32,062	588,785
1899	452,070	419,883	871,953	650,850	37,703	688,553
1900	478,081	384,182	862,263	673,847	41,355	715,202
1901	484,516	358,809	843,325	559,232	36,897	596,129
1902	599,319	456,483	1,055,802	603,108	34,864	637,972
1903	444,762	380,917	825,679	541,601	28,319	569,920
1904	372,159	367,107	739,266	478,922	29,158	508,080
1905	350,953	349,673	700,626	362,791	1,032	363,823
1906	293,174	364,239	657,413	409,510	19,920	429,430

Date	Sheep			Swine		
	Sheep	Lambs	Total	Fat	Store	Total
1907	284,922	358,857	643,779	447,808	30,786	478,594
1908	332,384	354,278	686,662	444,613	25,896	470,509
1909	384,116	377,431	761,547	314,737	9,946	324,683
1910	385,446	438,252	823,698	299,254	17,033	316,287
1911	329,737	379,829	709,566	305,256	22,829	328,085
1912	316,940	365,335	682,275	351,230	15,677	366,907
1913	334,329	283,654	617,983	171,401	10,189	181,590
1914	225,379	328,133	553,512	148,243	4,859	153,102

	Horses			
	Stallions	Mares	Geldings	Total
Ave 1854–6				
1861				
1862				
1863				
1864				
1865				
1866				
1867				
1868				
1869				
1870				
1871				
1872				
1873				
1874				
1875				
1876				
1877				
1878				
1879				
1880				
1881				
1882				
1883				
1884				
1885				
1886				
1887	68	11,801	15,769	27,638
1888	67	12,388	17,373	29,828
1889	80	13,647	18,097	31,824
1890	105	14,625	19,422	34,152
1891	125	14,055	19,216	33,396
1892	113	14,273	18,095	32,481
1893	151	13,356	16,883	30,390

	Horses			
	Stallions	Mares	Geldings	Total
1894	163	14,484	18,942	33,589
1895	188	15,370	19,002	34,560
1896	191	18,046	21,619	39,856
1897	153	17,590	20,679	38,422
1898	150	18,200	20,454	38,804
1899	122	19,471	22,494	42,087
1900	103	16,320	19,183	35,606
1901	194	11,467	13,946	25,607
1902	222	11,143	13,895	25,260
1903	265	12,867	14,587	27,719
1904	235	12,909	14,356	27,500
1905	202	14,192	16,329	30,723
1906	257	15,316	18,243	33,816
1907	235	15,303	18,088	33,626
1908	187	13,785	16,497	30,469
1909	256	13,941	15,742	29,939
1910	255	14,699	15,783	30,737
1911	262	15,742	16,453	32,457
1912	233	15,753	16,072	32,058
1913	283	16,467	17,725	34,475
1914	269	18,685	17,892	36,846

Exports as percentages of annual enumeration

	Cattle	Sheep	Swine	Horses
Ave 1854–6	6.8	13.1	21.1	
1861	10.3	11.5	32.5	
1862	13.2	15.6	31.6	
1863	14.0	15.6	33.5	
1864	10.7	11.0	32.0	
1865	7.1	9.0	29.3	
1866	10.7	9.3	33.7	
1867	11.3	12.2	20.1	
1868				
1869	13.6	21.8	24.5	
1870	11.9	14.3	29.0	
1871	12.2	16.2	32.6	
1872	15.2	12.2	31.9	
1873	16.5	13.5	34.9	
1874	13.4	16.8	31.3	
1875	14.5	21.6	37.0	
1876	16.2	17.1	36.0	
1877	16.2	15.8	39.8	
1878	18.3	15.7	37.1	
1879	15.8	16.8	33.8	

	Cattle	Sheep	Swine	Horses
1880	18.4	20.1	43.9	
1881	14.4	17.7	35.0	
1882	19.6	18.2	35.2	
1883	13.6	14.3	34.2	
1884	17.4	16.4	34.0	
1885	15.1	18.1	31.4	
1886	17.1	21.8	33.4	
1887	16.1	16.2	34.1	5.0
1888	18.0	17.6	39.0	5.3
1889	16.4	16.2	34.2	5.5
1890	14.9	14.7	38.4	5.8
1891	14.2	18.9	36.8	5.6
1892	13.8	22.4	45.0	5.4
1893	15.4	25.1	39.6	5.0
1894	18.8	23.3	42.1	4.6
1895	18.2	16.7	40.9	5.5
1896	15.5	18.1	43.5	6.3
1897	16.7	19.3	52.4	6.3
1898	17.9	19.4	47.0	6.6
1899	17.1	20.0	50.5	7.3
1900	16.2	19.7	56.4	6.3
1901	13.8	19.3	48.9	4.5
1902	14.1	20.5	46.7	4.5
1903	22.3	26.6	45.3	4.4
1904	17.3	20.7	39.8	4.6
1905	17.1	19.6	36.2	4.8
1906	15.8	18.5	30.6	5.5
1907	16.7	16.9	36.3	5.6
1908	18.0	16.7	38.6	5.0
1909	18.4	18.4	28.2	5.0
1910	18.1	21.2	26.3	5.0
1911	17.9	18.5	23.2	5.3
1912	14.5	18.0	27.7	5.2
1913	16.0	17.3	17.1	5.6
1914	15.2	15.9	11.7	5.9

Notes: [a] In 1907 the exports of milk and springers were counted with store cattle.

The distribution within animals is complete from 1875, but before that date we may have an incomplete picture. Judging by trend it looks as though the totals are correct, signalling recovery from the 1859-64 depression. We may note that the Agricultural Statistics of 1891 in presenting the figures from 1875 indicated that 'some of the younger included in the Statistics of Exports may of necessity escape enumeration in June of each year when the returns of livestock are collected'.

It looks as though the 'accounting year' changed over time. Up to 1906 it seems to have been the calendar year, but from 1907 the year turned on 31 May. In addition to the sources cited below see also, P. Solar, *Growth and Distribution in Irish Agriculture before the Famine*, Ph.D. Dissertation, University of Stanford, 1987, pp. 119–20 for a table of Irish livestock exports which takes the calves, sheep and pigs series back to the Famine.

Appendix 2 Crop yields in Ireland, 1847–1914 – weight per acre (cwts)

Year	Wheat	Oats	Barley[a]	Bere[a]	Rye	Beans[b]	Pease[b]
1847	16.5	14.7	17.4	17.2	20.5	(14.2)	
1848	11.3	13.3	16.8	15.4	19.8	(13.6)	
1849	13.3	13.3	16.8	16.4	20.5	(13.3)	
1850	11.0	13.5	17.8	16.6	19.5	(12.7)	
1851	12.5	13.8	17.6	16.6	20.0	13.9	11.6
1852	13.8	14.4	18.2	17.4	19.8	14.1	12.0
1853	14.5	13.8	18.6	16.8	20.5	13.4	12.8
1854	14.8	15.4	18.6	16.2	19.8	13.7	12.6
1855	14.3	13.7	17.4	15.6	20.0	15.9	13.2
1856	13.0	12.8	15.0	15.4	13.5	14.3	10.8
1857	12.5	12.6	14.6	14.6	13.0	13.9	9.7
1858	13.5	12.6	15.2	15.4	12.8	16.3	10.3
1859	13.3	11.6	14.0	15.0	12.0	12.9	8.6
1860	11.5	12.6	15.0	14.2	11.8	15.9	8.0
1861	9.0	11.2	12.4	12.8	10.8	13.0	9.4
1862	8.1	10.3	12.4	12.3	9.3	11.8	9.7
1863	13.5	12.8	15.9	15.8	10.6	15.4	10.0
1864	13.3	12.1	15.9	16.4	8.5	15.6	12.0
1865	13.0	12.3	14.9	14.8	10.4	16.8	10.8
1866	11.3	12.0	15.7	15.3	10.3	18.3	12.3
1867	11.7	12.5	15.7	15.5	10.2	14.0	11.4
1868	13.9	12.5	17.1	16.2	10.9	18.0	11.6
1869	11.9	11.6	15.5	15.9	10.6	16.2	12.0
1870	12.2	12.8	15.5	15.2	11.3	18.3	12.0
1871	12.1	12.7	15.2	14.5	10.8	18.8	14.1
1872	11.4	11.4	14.3	14.6	9.7	19.5	12.4
1873	11.7	12.8	15.9	14.8	10.4	15.4	12.6
1874	15.4	13.5	19.2	17.6	11.3	19.0	13.4
1875	14.6	15.3	18.1	17.4	12.6	21.6	11.3
1876	17.0	14.3	17.6	18.6	15.0	17.0	8.7
1877	13.6	12.1	15.5	14.9	12.4	14.7	8.4
1878	15.0	13.5	16.1	16.9	14.1	17.4	9.8
1879	11.4	11.7	12.8	12.2	8.7	14.9	8.6
1880	15.0	14.2	15.8	14.7	10.3	20.3	13.5
1881	14.9	14.1	15.8	14.5	10.5	16.9	12.6
1882	13.6	13.1	14.7	14.5	10.9	18.3	12.0

Year	Wheat	Oats	Barley[a]	Bere[a]	Rye	Beans[b]	Pease[b]
1883	13.7	13.6	15.4	14.4	11.4	16.4	12.1
1884	14.6	13.4	16.1	13.7	11.1	15.2	12.6
1885	15.4	13.6	16.1	13.4	12.7	18.0	13.4
1886	14.5	13.9	15.3	13.8	11.6	16.3	14.0
1887	15.2	11.5	12.8	12.5	10.8	11.3	12.4
1888	13.8	13.8	15.8	13.3	13.4	12.6	11.8
1889	16.0	14.2	17.5	13.0	12.9	18.2	12.2
1890	15.3	14.6	16.8	13.5	11.6	23.4	15.7
1891	17.3	15.5	18.6	13.4	13.6	23.3	16.3
1892	15.7	14.7	16.4	14.6	12.5	20.1	13.4
1893	16.2	15.5	16.4	13.4	13.2	19.2	10.5
1894	16.6	15.4	17.1	13.1	12.7	21.6	13.7
1895	16.3	15.0	16.6	13.9	13.4	16.3	13.1
1896	16.8	14.2	18.2	12.3	12.7	16.3	13.6
1897	15.4	13.8	15.1	12.7	10.8	18.5	12.7
1898	18.8	16.0	18.9	13.6	12.9	20.8	11.7
1899	17.9	15.8	17.9	14.8	12.9	22.8	12.8
1900	16.7	15.8	16.0	14.1	12.8	19.0	13.5
1901	18.3	16.2	18.0	14.4	13.7	20.8	14.8
1902	19.4	17.3	21.1	14.4	14.0	23.7	15.9
1903	16.8	15.3	16.4	15.8	13.4	21.9	15.9
1904	18.1	15.9	14.9	14.5	13.0	20.7	13.6
1905	20.2	16.3	19.9	14.1	13.5	21.8	13.9
1906	18.6	16.7	17.5	14.7	13.8	26.6	15.5
1907	18.6	16.0	17.7	16.5	13.5	24.1	15.1
1908	20.4	17.2	19.7	17.0	14.6	21.5	13.3
1909	21.6	18.7	21.9		15.5	23.9	16.4
1910	19.3	17.5	17.5		15.2	22.5	15.0
1911	19.7	16.3	19.2		14.5	19.2	14.9
1912	18.7	18.3	18.8		16.0	22.4	15.0
1913	20.4	18.0	19.8		15.0	26.1	17.4
1914	20.5	17.6	20.1		15.2	24.1	16.8

	Potatoes	Turnips	Mangels[c] & Beets	Cabbage	Flax	Hay
	tons	tons	tons	tons	stones	tons
1847	7.2	15.5	18.0		48.0	1.9
1848	3.9	14.3	17.5	13.5	38.4	2.0
1849	5.6	16.1	18.5	14.4	39.6	2.0
1850	4.6	15.7	17.9	13.9	39.4	2.0
1851	5.1	15.9	18.0	13.9	38.6	2.0
1852	4.8	15.9	18.1	14.2	41.4	2.1
1853	6.4	16.4	17.7	14.1	40.2	2.0
1854	5.1	15.8	17.2	13.3	37.6	1.9
1855	6.4	16.6	18.0	13.0	38.6	2.0
1856	4.0	12.9	13.0	11.9	28.3	1.9
1857	3.1	12.5	13.9	10.9	23.7	1.9

	Potatoes	Turnips	Mangels^c & Beets	Cabbage	Flax	Hay
	tons	tons	tons	tons	stones	tons
1858	4.2	12.9	13.7	11.5	30.7	1.9
1859	3.6	10.7	11.4	10.4	25.3	1.6
1860	2.3	8.3	9.1	8.3	29.6	2.0
1861	1.6	10.2	10.3	10.4	24.4	1.8
1862	2.1	10.1	9.6	8.4	25.9	1.8
1863	3.4	11.9	11.3	10.0	31.9	1.8
1864	4.1	10.3	10.5	9.3	34.2	1.6
1865	3.6	9.9	13.3	10.4	25.2	1.8
1866	2.9	11.9	12.5	10.5	24.9	1.8
1867	3.1	11.6	12.7	9.4	22.4	1.9
1868	3.9	11.0	12.9	9.7	19.4	1.7
1869	3.2	12.3	12.5	9.6	20.6	1.8
1870	4.0	11.6	12.7	9.8	25.3	1.9
1871	2.6	13.0	13.5	10.0	13.2	1.8
1872	1.8	11.4	12.4	9.9	22.4	1.9
1873	3.0	12.7	13.5	9.9	24.6	1.8
1874	4.0	13.2	14.1	10.9	27.0	1.8
1875	3.9	15.9	16.6	12.4	35.5	2.2
1876	4.7	13.2	14.4	10.2	32.7	1.9
1877	2.0	10.7	12.3	9.3	28.8	2.3
1878	3.0	14.2	15.2	10.2	31.7	2.3
1879	1.3	6.5	8.0	6.4	23.9	1.9
1880	3.6	14.3	14.6	9.9	25.9	2.0
1881	4.0	12.9	13.4	9.8	30.8	2.0
1882	2.4	11.5	11.9	9.3	29.5	2.1
1883	4.3	14.0	13.9	9.5	30.5	2.0
1884	3.8	11.5	12.7	9.2	28.1	1.9
1885	4.0	11.9	13.4	9.4	30.4	2.0
1886	3.3	13.3	13.5	9.7	29.2	2.1
1887	4.5	9.1	10.9	7.4	18.6	1.7
1888	3.1	11.3	12.9	9.1	29.0	2.3
1889	3.6	13.1	14.1	10.2	27.3	2.2
1890	2.3	14.4	14.3	9.4	33.4	2.2
1891	4.0	14.5	15.6	10.1	29.5	2.1
1892	3.5	13.5	14.5	9.3	21.9	2.1
1893	4.2	16.0	16.3	10.9	36.5	2.1
1894	2.6	13.7	14.6	9.5	34.0	2.4
1895	4.9	14.3	15.6	10.2	20.5	2.1
1896	3.8	15.5	14.4	10.3	22.8	2.1
1897	2.2	13.4	13.7	8.7	25.6	2.3
1898	4.4	16.8	18.0	10.5	31.2	2.4
1899	4.2	14.3	17.0	9.4	32.7	2.3
1900	2.8	14.9	17.2	10.4	34.3	2.4
1901	5.3	16.9	18.8 14.4	10.7	38.3	2.2
1902	4.3	17.1	19.0 13.6	10.5	36.2	2.4
1903	3.8	12.5	13.5 14.8	9.7	30.9	2.3

	Potatoes	Turnips	Mangels[c] & Beets	Cabbage	Flax	Hay
	tons	tons	tons	tons	stones	tons
1904	4.3	17.5	17.6 17.4	10.3	33.7	2.4
1905	5.6	16.7	17.6 13.5	10.2	37.7	2.3
1906	4.3	17.8	20.0 15.1	10.4	34.9	2.1
1907	3.8	14.9	17.6 12.4	9.9	31.2	2.3
1908	5.5	19.4	21.9 15.9	10.3	27.0	2.1
1909	5.5	17.9	19.6	11.3	30.1	2.0
1910	4.8	16.8	19.5	11.3	30.9	2.4
1911	6.2	19.5	22.3	13.2	27.0	1.8
1912	4.3	13.9	15.9	11.9	37.6	2.0
1913	6.4	18.8	20.6	13.2	34.1	2.2
1914	5.9	16.0	19.2	12.6	26.4	1.7

Notes:

[a] Barley and bere counted together from 1909.

[b] Peas and beans counted together 1847–50.

[c] Mangels on their own to 1871, with beets until 1900, separate until 1908 and combined again from 1909.

Sources: see sources at the end of appendix 3, p. 254.

Appendix 3 Landholding distribution 1847–1914

					Numbers of landholders in these acreage		
	<1	1-5	5–15	15–30	30–50	50–100	100–200
1847	73,016	139,041	269,534	164,337	(>30 to	>500	157,097...
1849	31,989	98,179	213,897	150,120	(>30 to	>500	156,960...
1850	35,326	91,618	203,331	145,380	(>30 to	>500	152,567...
1851	37,728	88,083	191,854	141,311	70,093	49,940	19,753
1852	35,058	81,561	182,308	139,136	70,079	51,389	20,436
1853	35,795	79,418	178,701	138,804	70,297	51,764	20,774
1854	38,165	80,976	179,140	137,640	70,821	52,512	21,026
1855	36,642	81,281	178,794	138,183	70,849	52,720	21,116
1856	36,474	82,035	179,931	138,424	71,156	53,279	21,292
1857	37,499	81,585	179,733	139,192	71,580	53,492	21,463
1858	38,198	83,219	181,267	139,618	71,791	53,544	21,566
1859	37,506	82,647	180,993	139,659	72,333	53,678	21,603
1860	38,416	82,844	181,358	140,873	72,413	54,018	21,671
1861	41,561	85,469	183,931	141,251	72,449	53,933	21,531
1862	43,716	84,463	183,031	140,218	72,342	54,147	21,698
1863	46,096	82,451	180,145	138,540	72,050	54,240	21,934
1864	48,653	82,037	176,368	136,578	71,961	54,247	22,065
1865	48,769	80,972	175,723	136,773	71,761	54,504	21,856
1866	48,236	79,742	174,357	136,499	72,154	54,825	21,900
1867	50,669	78,064	173,475	136,503	71,785	54,752	21,991
1868	49,709	77,108	172,040	136,580	72,205	54,840	22,106
1869	49,394	75,895	171,699	137,086	72,511	55,070	21,876
1870	49,164	74,295	170,986	137,987	72,383	55,320	21,726
1871	48,448	74,809	171,383	138,647	72,787	55,062	21,696
1872	52,885	73,574	170,335	138,734	72,889	55,216	21,839
1873	51,977	72,088	168,044	138,163	72,984	55,257	21,881
1874	50,267	70,025	167,450	137,956	73,043	55,385	21,839
1875	51,459	69,098	166,959	137,669	73,045	55,618	21,909
1876	52,433	67,524	164,810	137,114	72,761	55,365	22,060
1877	51,910	66,637	164,917	137,791	73,024	55,867	22,141
1878	51,221	66,359	163,062	137,493	73,227	56,138	22,156
1879	50,140	65,269	162,233	136,649	73,171	56,513	22,223
1880	50,613	64,292	161,335	136,518	72,923	56,229	22,413
1881	50,996	67,071	164,045	135,793	72,385	55,601	22,214

size groups				Percentage distribution			
	200-500	> 500	TOTAL	< 1	1–30	30–100	> 100
1847		....)	803,025	9.1	71.3		19.6
1849		....)	651,145	4.9	71.0		24.1
1850		....)	628,222	5.6	70.1		24.3
1851	7,847	1457	608,066	6.2	69.3	19.7	4.8
1852	8,044	1460	589,471	5.9	68.4	20.6	5.1
1853	8,177	1619	585,349	6.1	67.8	20.9	5.2
1854	8,167	1640	590,087	6.5	67.4	20.9	5.2
1855	8,243	1642	589,470	6.2	67.6	21.0	5.3
1856	8,243	1655	592,489	6.2	67.6	21.0	5.3
1857	8,231	1617	594,392	6.3	67.4	21.0	5.3
1858	8,383	1592	599,178	6.4	67.4	20.9	5.3
1859	8,409	1585	598,413	6.3	67.4	21.1	5.3
1860	8,427	1625	601,645	6.4	67.3	21.0	5.3
1861	8,329	1591	610,045	6.8	67.3	20.7	5.2
1862	8,214	1556	609,385	7.2	66.9	20.8	5.2
1863	8,247	1573	605,276	7.6	66.3	20.9	5.2
1864	8,303	1559	601,771	8.1	65.6	21.0	5.3
1865	8,373	1554	600,285	8.1	65.5	21.0	5.3
1866	8,339	1576	597,628	8.1	65.4	21.2	5.3
1867	8,309	1569	597,117	8.5	65.0	21.2	5.3
1868	8,181	1572	594,341	8.4	64.9	21.4	5.4
1869	8,210	1541	593,282	8.3	64.8	21.5	5.3
1870	8,188	1570	591,619	8.3	64.8	21.6	5.3
1871	8,190	1568	592,590	8.2	64.9	21.6	5.3
1872	8,159	1582	595,213	8.9	64.3	21.5	5.3
1873	8,250	1528	590,172	8.8	64.1	21.7	5.4
1874	8,225	1497	585,687	8.6	64.1	21.9	5.4
1875	8,197	1529	585,483	8.8	63.8	22.0	5.4
1876	8,176	1510	581,753	9.0	63.5	22.0	5.5
1877	8,158	1518	581,963	8.9	63.5	22.1	5.5
1878	8,191	1552	579,399	8.8	63.3	22.3	5.5
1879	8,296	1546	576,040	8.7	63.2	22.5	5.6
1880	8,340	1559	574,222	8.8	63.1	22.5	5.6
1881	8,204	1430	577,739	8.8	63.5	22.2	5.5

Numbers of landholders in these acreage

	<1	1-5	5–15	15–30	30–50	50–100	100–200
1882	51,395	65,426	161,878	135,361	72,565	55,659	22,230
1883	49,041	63,942	159,256	134,682	73,060	55,743	22,242
1884	49,808	62,231	157,775	134,447	72,710	56,050	22,451
1885	49,744	61,876	156,738	134,898	73,477	56,172	22,602
1886	47,853	61,187	157,696	134,815	73,929	56,294	22,690
1887	48,122	60,824	156,562	134,879	73,845	56,485	22,871
1888	47,951	60,266	156,146	135,311	73,763	56,476	22,796
1889	49,929	61,590	156,561	135,096	73,402	56,520	22,925
1890	50,809	60,767	155,763	134,215	73,686	56,571	23,025
1891	55,628	63,464	156,661	133,947	73,921	56,361	22,811
1892	54,201	62,825	156,025	133,614	73,532	56,673	22,926
1893	55,897	62,882	155,925	133,442	73,843	56,629	22,969
1894	57,809	62,781	156,223	133,656	73,493	56,958	22,992
1895	59,508	62,275	155,681	133,513	73,862	57,082	23,045
1896	60,807	62,221	155,333	133,211	74,005	57,243	22,997
1897	62,189	62,030	155,064	133,308	74,081	57,454	23,041
1898	63,711	61,816	154,441	133,749	74,071	57,405	23,025
1899	69,220	62,275	154,840	133,683	73,994	57,527	23,023
1900	71,848	62,154	154,751	133,530	74,049	57,576	23,051
1901	74,328	62,855	154,418	134,091	74,255	57,407	23,107
1902	74,607	62,864	154,437	133,984	74,240	57,568	23,041
1903	74,890	62,292	154,631	134,308	74,366	57,446	23,058
1904	75,701	62,185	154,290	134,480	74,329	57,683	22,933
1905	78,001	62,126	154,560	134,370	74,611	57,707	22,857
1906	80,693	62,256	154,228	135,133	74,753	57,827	22,837
1907	83,574	62,068	153,728	135,233	74,799	58,029	22,905
1908	84,869	61,730	153,299	136,058	75,192	58,241	22,863
1909	85,644	61,936	153,565	136,216	75,658	58,510	22,835
1910	86,131	62,148	154,088	136,681	75,918	58,710	22,817
1911	86,906	62,354	154,354	136,839	76,384	58,979	22,789
1912	98,059	50,355	129,706	123,489	70,897	57,276	23,094
1913	101,064	49,674	128,823	123,292	71,329	57,076	23,223
1914	102,076	49,167	128,690	123,626	71,551	57,331	23,202

Occupiers

Date	<1	1-5	5–15	15–30	30–50	50–100	100–200
1861	39,210	75,141	164,006	127,899	65,896	49,654	20,375
1862	41,661	74,808	163,560	127,397	65,806	50,014	20,651
1863	44,063	72,632	159,206	123,960	64,663	49,712	20,583
1864	46,724	72,121	155,021	121,755	64,389	49,391	20,621
1865	46,942	71,614	155,105	121,997	64,176	49,637	20,480
1866	46,563	70,706	154,139	121,663	64,438	49,712	20,583
1867	49,110	69,410	153,508	121,991	64,033	49,749	20,539
1868	48,165	68,838	152,948	121,999	64,592	49,747	20,664
1869	47,879	67,468	152,685	122,511	65,185	50,090	20,396

size groups				Percentage distribution			
	200-500	> 500	TOTAL	<1	1–30	30–100	>100
1882	8,210	1483	574,207	9.0	63.2	22.3	5.6
1883	8,294	1465	567,725	8.6	63.0	22.7	5.6
1884	8,283	1499	565,254	8.8	62.7	22.8	5.7
1885	8,258	1548	565,313	8.8	62.5	22.9	5.7
1886	8,328	1560	564,352	8.5	62.7	23.1	5.8
1887	8,317	1570	563,475	8.5	62.5	23.1	5.8
1888	8,372	1561	562,642	8.5	62.5	23.1	5.8
1889	8,367	1585	565,975	8.8	62.4	23.0	5.8
1890	8,373	1594	564,803	9.0	62.1	23.1	5.8
1891	8,280	1567	572,640	9.7	61.8	22.8	5.7
1892	8,293	1565	569,654	9.5	61.9	22.9	5.8
1893	8,270	1585	571,442	9.8	61.6	22.8	5.7
1894	8,232	1574	573,718	10.1	61.5	22.7	5.7
1895	8,263	1557	574,786	10.4	61.1	22.8	5.7
1896	8,297	1550	575,664	10.6	60.9	22.8	5.7
1897	8,245	1563	576,975	10.8	60.7	22.8	5.7
1898	8,188	1556	577,962	11.0	60.6	22.7	5.7
1899	8,189	1534	584,285	11.8	60.0	22.5	5.6
1900	8,219	1539	586,717	12.2	59.7	22.4	5.6
1901	8,186	1528	590,175	12.6	59.5	22.3	5.6
1902	8,147	1521	590,409	12.6	59.5	22.3	5.5
1903	8,141	1516	590,648	12.7	59.5	22.3	5.5
1904	8,096	1527	591,224	12.8	59.4	22.3	5.5
1905	8,046	1526	593,804	13.1	59.1	22.3	5.5
1906	8,056	1561	597,344	13.5	58.9	22.2	5.4
1907	7,952	1584	599,872	13.9	58.5	22.1	5.4
1908	7,969	1544	601,765	14.1	58.3	22.2	5.4
1909	7,893	1570	603,827	14.2	58.2	22.2	5.3
1910	7,821	1582	605,896	14.2	58.2	22.2	5.3
1911	7,745	1610	607,960	14.3	58.2	22.3	5.3
1912	8,537	2112	563,525	17.4	53.9	22.7	6.0
1913	8,447	2087	565,015	17.9	53.4	22.7	6.0
1914	8,409	2085	566,137	18.0	53.3	22.8	6.0

				Percentage distribution			
	200-500	> 500	TOTAL	<1	1–30	30–100	>100
1861	9,046	2437	553,664	7.1	66.3	20.9	5.8
1862	8,797	2372	555,066	7.5	65.9	20.9	5.7
1863	8,821	2545	546,185	8.1	65.1	20.9	5.8
1864	8,897	2542	541,461	8.6	64.4	21.0	5.9
1865	8,933	2533	541,417	8.7	64.4	21.0	5.9
1866	9,006	2545	539,355	8.6	64.2	21.2	6.0
1867	8,957	2540	539,837	9.1	63.9	21.1	5.9
1868	8,830	2560	538,343	8.9	63.9	21.2	6.0
1869	8,730	2489	537,433	8.9	63.8	21.4	5.9

Date	Occupiers						
	< 1	1-5	5–15	15–30	30–50	50–100	100–200
1870	47,751	66,344	152,477	123,545	64,893	50,382	20,328
1871	47,030	67,054	152,987	124,457	65,427	50,286	20,421
1872	51,544	65,776	152,200	124,691	65,502	50,438	20,564
1873	50,758	65,051	150,778	124,471	65,991	50,565	20,764
1874	49,165	62,732	149,716	123,563	65,805	50,531	20,648
1875	50,322	62,104	149,723	123,579	65,722	50,569	20,696
1876	51,498	61,168	148,252	123,642	65,841	50,778	20,882
1877	51,097	60,436	149,054	124,065	66,029	51,044	20,782
1878	50,449	60,221	147,094	124,019	66,399	51,110	21,007
1879	49,347	59,414	146,349	123,136	66,103	51,752	20,855
1880	49,821	58,679	145,770	123,395	65,763	51,513	21,363
1881	49,955	60,268	147,796	122,624	65,617	51,483	20,914
1882	50,500	59,269	146,020	122,452	65,692	51,236	21,332
1883	48,301	58,295	144,065	122,055	66,307	51,451	21,227
1884	49,091	56,779	142,949	121,940	65,816	51,484	21,378
1885	49,027	56,573	142,421	122,452	66,667	51,737	21,515
1886	47,195	56,249	143,857	122,833	67,195	52,026	21,666
1887	47,488	56,006	143,027	122,852	67,493	52,309	21,829
1888	47,350	55,422	142,618	123,314	67,223	52,406	21,870
1889	49,335	56,731	143,154	123,057	66,970	52,633	22,100
1890	50,201	55,944	142,281	122,381	67,382	52,595	22,199
1891	54,737	56,915	140,277	121,150	67,446	52,650	22,326
1892	53,455	56,805	140,435	121,121	66,942	52,857	22,373
1893	55,146	56,983	140,689	120,590	67,420	52,787	22,387
1894	57,033	57,006	140,979	121,074	66,983	53,317	22,546
1895	58,807	56,595	140,710	121,326	67,182	53,561	22,507
1896	60,123	56,672	140,312	121,072	67,494	53,680	22,486
1897	61,437	56,320	139,710	120,783	67,748	53,770	22,549
1898	62,994	56,261	139,348	121,474	67,669	53,824	22,617
1899	68,500	56,794	140,071	123,236	67,817	54,033	22,644
1900	71,103	56,652	139,834	121,355	67,785	54,082	22,618
1901	73,234	56,098	137,258	120,642	67,878	54,168	22,841
1902	73,352	56,233	137,365	120,740	67,722	54,314	22,867
1903	73,926	55,771	137,832	121,273	67,987	54,108	22,763
1904	74,775	55,770	137,410	121,515	67,882	54,335	22,668
1905	77,149	55,994	138,151	121,483	68,308	54,492	22,593
1906	79,834	56,090	137,762	122,429	68,895	54,880	22,650
1907	82,598	55,181	136,231	121,938	68,461	54,929	22,698
1908	83,798	54,268	133,096	120,322	67,641	54,262	22,443
1909	84,633	54,793	134,525	121,828	69,011	55,431	22,735
1910	84,799	54,899	134,784	122,064	69,144	55,538	22,779
1911	85,634	55,424	136,213	123,570	70,514	56,707	23,071
1912	86,021	54,400	133,424	121,715	68,951	55,270	22,285
1913	86,672	53,834	132,544	121,480	69,204	55,426	22,354
1914	101,863	47,695	124,197	119,821	69,424	55,751	22,794

				Percentage distribution			
	200-500	> 500	TOTAL	< 1	1–30	30–100	> 100
1870	8,764	2528	537,012	8.9	63.8	21.5	5.9
1871	8,672	2499	538,833	8.7	63.9	21.5	5.9
1872	8,627	2496	541,838	9.5	63.2	21.4	5.8
1873	8,799	2368	539,545	9.4	63.1	21.6	5.9
1874	8,778	2406	533,344	9.2	63.0	21.8	6.0
1875	8,697	2413	533,825	9.4	62.8	21.8	6.0
1876	8,737	2363	533,161	9.7	62.5	21.9	6.0
1877	8,837	2382	533,726	9.6	62.5	21.9	6.0
1878	8,723	2420	531,442	9.5	62.3	22.1	6.0
1879	8,842	2477	528,275	9.3	62.3	22.3	6.1
1880	8,679	2461	527,444	9.4	62.2	22.2	6.2
1881	8,698	2329	529,684	9.4	62.4	22.1	6.0
1882	8,825	2350	527,676	9.6	62.1	22.2	6.2
1883	8,956	2295	522,952	9.2	62.0	22.5	6.2
1884	8,958	2329	520,724	9.4	61.8	22.5	6.3
1885	8,825	2339	521,556	9.4	61.6	22.7	6.3
1886	8,898	2358	522,277	9.0	61.8	22.8	6.3
1887	8,925	2252	522,181	9.1	61.6	22.9	6.3
1888	8,977	2285	521,465	9.1	61.6	22.9	6.4
1889	8,868	2304	525,152	9.4	61.5	22.8	6.3
1890	8,847	2380	524,210	9.6	61.2	22.9	6.4
1891	8,897	2272	526,670	10.4	60.4	22.8	6.4
1892	8,977	2310	525,275	10.2	60.6	22.8	6.4
1893	9,053	2309	527,364	10.5	60.3	22.8	6.4
1894	8,921	2277	530,136	10.8	60.2	22.7	6.4
1895	8,951	2234	531,873	11.1	59.9	22.7	6.3
1896	8,981	2223	533,043	11.3	59.7	22.7	6.3
1897	8,937	2260	533,514	11.5	59.4	22.8	6.3
1898	8,915	2256	535,358	11.8	59.2	22.7	6.3
1899	8,866	2193	544,154	12.6	58.8	22.4	6.2
1900	8,939	2191	544,559	13.1	58.4	22.4	6.2
1901	8,917	2202	543,238	13.5	57.8	22.5	6.3
1902	8,856	2200	543,649	13.5	57.8	22.4	6.2
1903	8,856	2199	544,715	13.6	57.8	22.4	6.2
1904	8,865	2182	545,402	13.7	57.7	22.4	6.2
1905	8,811	2136	549,117	14.0	57.5	22.4	6.1
1906	8,720	2148	553,408	14.4	57.2	22.4	6.1
1907	8,674	2287	552,997	14.9	56.7	22.3	6.1
1908	8,801	2300	546,931	15.3	56.3	22.3	6.1
1909	8,732	2372	554,060	15.3	56.2	22.5	6.1
1910	8,745	2373	555,125	15.3	56.2	22.5	6.1
1911	8,676	2445	562,254	15.2	56.1	22.6	6.1
1912	8,462	2366	552,894	15.6	56.0	22.5	6.0
1913	8,403	2328	552,245	15.7	55.7	22.6	6.0
1914	8,588	2367	552,500	18.4	52.8	22.7	6.1

Sources for appendixes 1, 2, and 3

For crop acreages, crop yields, livestock numbers and livestock exports

In the introduction to these appendixes it was explained that there are easy ways to extract a full run of the basic data without necessarily consulting every volume of the *Annual Agricultural Returns*. The *British Parliamentary Papers* most convenient for this purpose are:

For 1847–71 *Agricultural Statistics of Ireland for the Year 1871* [C. 762], vol. lxix, 1873 pp. l, lxiii, lxxxi.
For 1872–81 *Agricultural Statistics of Ireland for the Year 1881* [C. 3332], vol. lxxiv, 1882, pp. 38, 43, 53.
For 1882–91 *Agricultural Statistics of Ireland for the Year 1891* [C. 6777], vol. lxxxviii, 1892, pp. 23, 56, 61, 71.
For 1892–1901 *Agricultural Statistics of Ireland for the Year 1901* [Cd. 1170], vol.cxvi, 1902, pp. xvii, 66–7, 83, 110–11.
For 1902–11 *Agricultural Statistics of Ireland for the Year 1911* [Cd. 6377], vol. cvi, 1912–13, pp. xx-xxi, 68–69, 85, 132–3.
For 1912–14 *Agricultural Statistics of Ireland for the Year 1914* [Cd. 8266], vol. xxxii, 1916, pp. xv, 66–7, 83, 128–9.

For livestock exports see C. 6777, Cd 1170, Cd 6377, and Cd 8266 above, and also:

Agricultural Statistics of Ireland for the Year 1916 [Cmd. 112], vol. li,1919, p. xiv.
Report from the Committee ... Transit of Animals by Sea and Land [C. 116], vol. lxi, 1870, appendix XXV, p. 110.
Minutes of Evidence ... Upon the Inland Transit of Cattle [C. 8929], vol. xxxiv, 1898, appendix III.
Agricultural Statistics, 1907 ... Acreage and Livestock Returns of Great Britain [Cd. 3870], 1908, where pp. 272-5 is 'Trade in livestock with Ireland'.

See also R. Perren, *The Meat Trade in Britain 1840-1914* (London, 1978), p. 96.

For landholding distribution and land occupancy

See the *Annual Agricultural Statistics volumes.*

Appendix 4 Irish agricultural prices

Chapter 4 alluded to problems regarding the use of product prices in the estimation of output, problems which were not addressed in the calculation of my first published estimates.[1] The main problem is the choice of prices, or the choice of base year prices from which to recalculate other years with the help of Thomas Barrington's product price indexes (reproduced as Table A4.6).[2] Barrington's purpose was to produce a set of indexes which could be used in comparative price terms. The actual prices, in nominal £-s-d, were not fundamentally important so long as the *trend* of each product price correctly captured the true course of actual prices. The Registrar General of Ireland in the late nineteenth century, Dr Thomas Wrigley Grimshaw, presented evidence to a Royal Commission in the 1890s concerned with financial relations between Great Britain and Ireland. Amongst other things he was concerned with the estimation of the value of Irish agricultural output. He was fully aware of some of the problems of estimation which modern historians also face. His concern was with the identification of what he called 'standard prices'. He said:

I am fully aware of the difficulties and fallacies attending attempts to estimate the value of agricultural produce from the available data as to prices. I take the prices which have been carefully collected by the Farmers Gazette and published in Purdom's Almanack as the most extensive and reliable set of standard prices available.[3]

---

[1] For example M.E. Turner, 'Output and productivity in Irish agriculture from the Famine to the Great War', *Irish Economic and Social History*, 17 (1990), pp. 62–78. W.E. Vaughan in *ibid.*, 'Potatoes and agricultural output', 79–92 offered some valuable criticisms of my procedures as they applied to the potato prices employed. I am grateful to Peter Solar for advice in addressing some of those criticisms and for guiding me to a better policy in selecting a base year price which to employ for estimating the output of other years.

[2] T. Barrington, 'A review of Irish agricultural prices', *Journal of the Statistical and Social Inquiry Society of Ireland*, 15 (1927), esp. pp. 251–3.

[3] *Royal Commission on the Financial Relations between Great Britain and Ireland, B(ritish) P(arliamentary) P(apers)* [C. 7720 and C. 7721], vol. xxxvi, 1895, appendix VII 'Statement by Dr Grimshaw with regard to the values of output of Irish agriculture for certain periods', p. 451, re 'standard prices'.

In my original estimates of agricultural output I used precisely those prices which Grimshaw advocated, in the form which he presented them in his evidence to the Cowper Commission published in 1887.[4] These were repeated in full subsequently in some parliamentary papers, specifically those which gave a digest of Dublin prices on a regular basis from the 1880s onwards.[5] It is clear that Barrington was guided both by Cowper and the subsequent publication of Dublin prices in parliamentary sources. However, as long as there is confidence in the trend of prices from these sources, though not necessarily in the precise nominal values, especially those earlier in the trend, then Barrington remains a valuable set of indexes. The procedure I originally adopted was to use an existing set of product indexes straight off the shelf (Barrington) and then apply a set of prices to one year in those indexes to generate the prices over the course of the remaining series. My original procedure adopted a base of 1840 for most products, with 1845 for a small minority. It was rather crucial to make sure that the prices in that base year were reflective of the price trend. This is where the mistake was made. With hindsight it cannot be established that they were reflective because of the way in which those early prices were treated. If this procedure was at fault, then it had a knock-on effect throughout the series such that the newly generated prices for subsequent years diverged from the more reliable printed estimates which are available for later in the trend.

This begs the question of why the procedure involving the base prices was faulty in the first place. The Cowper Commission's reproduction of published prices quoted that they were, 'The Mean of the Minimal and of Maximal Prices of Irish Agricultural Produce in the year 1840, and in each of the 40 years, 1846–85: From Purdom's Irish Farmers' and Gardeners' Almanac for 1886.'[6] In other words there were usually two sets of prices per product. These were mainly Dublin prices, with additionally country fair prices for some animals and for the leading butter markets like Cork (as well as Dublin for butter). Lo and behold,

---

4   Cowper Commission, *Report of the Royal Commission on Land Law (Ireland) Act 1881 and the Purchase of Land (Ireland) Act 1885*, in three volumes, BPP [C. 4969, 4969 I, 4969 II], vol. xxvi, 1887. The prices are contained in [C. 4969], in Grimshaw's evidence in appendix C, pp. 960–67.

5   For example, *Agricultural Statistics, Ireland, 1907–8. Return of Prices of Crops, Live Stock and other Irish Agricultural Products*, BPP [Cd. 4437], vol. cxxi, 1908, pp. 86–9. For all his criticism of Cowper it appears that Vaughan also used Cowper either in the same way or certainly as the basis of his own original estimates. See W.E. Vaughan, 'Landlord and tenant relations between the Famine and the Land War, 1850–78', Ph.D. Thesis, University of Ireland, 1973, p. 337, and Vaughan, 'Agricultural output, rents and wages in Ireland, 1850–1880', in L.M. Cullen and F. Furet (eds.), *Ireland and France 17th to 20th Centuries: Towards a Comparative Study of Rural History* (Ann Arbor, MI, and Paris, 1980), p. 95.

6   Cowper Commission, *Report*, at the head of each page pp. 960–67.

there is a congruence here with Grimshaw's recommendation as to 'standard prices'. In general, it turns out that the prices *reconstructed* from indexes but based on the 1840s price base were higher later in the nineteenth and early twentieth centuries than the *actual* prices quoted in official sources. This suggests that the base prices are too high, and this arises from simply taking an arithmetic mean of quoted mean maxima and mean minima.[7] The procedure was akin to proceeding from the least known or least secure of evidence to the most known or most secure of evidence. There is, therefore, a case to be said for writing history backwards, so to speak, and proceeding from a base of prices quoted in twentieth-century official sources and thus deriving prices earlier in the trend. In either procedure it is important to have faith in the *trend* index.

Appendix 4, table A4.1 is an indication of the problem of proceeding from the 1840s base and projecting nominal prices forward in time. It compares the nominal prices for 1908 and 1912, as derived from the 1840s base, with the quoted 'official' prices for those same years. It also shows the prices used in the *official* estimation of agricultural output employed under the auspices of the *Census of Production* (for 1908) and the *Food Production Survey* (of 1912).[8] The table shows a marked compatibility between the quoted prices and the prices employed in the contemporary estimates of output, but a marked divergence between those prices and the ones derived from the Barrington index with its base in the 1840s. The Barrington derived prices were all higher, except flax, which was not an important product by this time, and butter, which was a very important product. The differences between the derived prices and the officially quoted prices for 1912 ranged from the trivial plus or minus 4 or 5 per cent for barley and flax, to the reasonably worrying plus 7 per cent for wheat and minus 10 per cent for butter, 12 per cent for wool, and 14 per cent for potatoes, to the positively alarming 23, 27, and 35 per cent for oats, eggs and hay. The official prices and output prices were not

---

[7] A method which Ó Gráda also used when necessary for his estimates for 1854, C. Ó Gráda, 'Irish agricultural output before and after the Famine', *Journal of European Economic History*, 13 (1984), p. 159. For the pre-Famine period, on the basis of the newspaper evidence which he employed, Solar at times also resorted to taking the averages of quoted high and low prices 'to represent the price of the good' in question, P. Solar, *Growth and Distribution in Irish Agriculture before the Famine*, Ph.D. Dissertation, University of Stanford, 1987, pp. 29, 60n, and 62n, and for a discussion of the regional divergence in prices for the same products and other issues regarding prices see in particular pp. 31–42.

[8] *The Agricultural Output of Ireland 1908* (Department of Agriculture and Technical Instruction for Ireland, Dublin, 1912), pp. 20, 23; *Food Production in Ireland. Minutes of Evidence*, BPP [Cd. 8158], vol. v, 1914–16, p. 70; *Agricultural Statistics, Ireland, 1915. Return of Prices, of Crops, Live Stock and other Irish Agricultural Products*, BPP [Cd. 8452], vol. xxxvi, 1917–18, pp. 14–15.

A4.1:   *Comparison of prices in 1908 and 1912* (in shillings per cwt)

Product	1908 prices derived from 1840s base	1908 'Official Prices'	1908 prices derived from agricultural output
Wheat	8.6	7.9	7.1
Oats	7.2	5.7	5.3
Barley	7.8	7.4	6.8
Potatoes	4.6	3.5	1.9
Flax	49.6	51.0	40.8
Hay	4.8	2.9	2.1
Butter	96.1	107.1	104.3
Wool	78.4	71.2	63.0
Eggs[a]	11.3	8.4	8.5

	1912 prices derived from 1840s, base	1912 'Official Prices'	1912/13 prices derived from agricultural output
Wheat	8.3	7.8	7.7
Oats	8.6	6.8	6.7
Barley	9.0	8.6	8.6
Potatoes	4.8	3.6	4.1
Flax	63.0	65.0	66.0
Hay	5.3	3.1	3.4
Butter	95.2	106.8	104.3
Wool	114.2	98.0	100.4
Eggs[a]	12.6	9.3	9.2

Note: [a] Shillings per 120 eggs.

Sources: The Agricultural Output of Ireland 1908 (Department of Agriculture and Technical Instruction for Ireland, Dublin, 1912), pp. 20, 23; *Food Production in Ireland. Minutes of Evidence* [Cd. 8158], vol. v, 1914–16, p. 70; *Agricultural Statistics, Ireland, 1915. Return of Prices, of Crops, Live Stock and other Irish Agricultural Products* [Cd. 8452], vol. xxxvi, 1917–18, pp. 14–15.

as close in 1908 as the similar prices for 1912, they were almost invariably lower.[9]

[9]   In this context see C. Ó Gráda, 'Irish agriculture north and south since 1900', ch. 17 of B.M.S. Campbell and M. Overton (eds.), *Land, Labour and Livestock: Historical Studies in European Agricultural Productivity* (Manchester, 1991), p. 443n, who says there are inconsistencies between the 1908 and 1912 output estimates, inconsistencies which cannot be explained by the sometimes small differences in the basic agricultural statistics (small changes in livestock numbers for example). Ó Gráda preferred to use the 1912 estimates. My own comparison of the 1908 output prices with the general price trend suggests the same. Vaughan specifically pointed out that the 'presenter' of the 1908 estimates had used a much smaller price for potatoes than the official printed price, Vaughan, 'Potatoes', pp. 85–6.

These differences in product prices whilst intrinsically important in their own right, and also in deriving as near definitive values for agricultural output as is possible, need not be important if the relationship they hold with one another remains constant over time. In other words, the proportional contributions of the different products to final output may be the same across all the different price sources. In this case the relative prices would be little affected or not affected at all. Demonstrably this is not precisely the case because the *derived* butter and flax prices listed in Appendix 4, table A4.1 for 1908 and 1912, are not relatively synchronised. For both of the years they are too low relative to the official prices. Nevertheless, the relationship between most derived product prices and their counterpart official prices is certainly in the correct direction if not actually close.

The 1840s base was certainly overpriced, but it was almost certainly accurate in price relative terms. Taking the mean of maxima and minima was the source of overpricing, with the repercussive effects later in the series. Furthermore, in terms of value as distinct from nominal price, I readily agree with Vaughan that the errors so produced had a greater effect on the contribution of potatoes than on any other product, though such a mistake is not uncommon amongst historians.[10] The new estimates in chapter 4 above reverse the original procedure: they are derived from the base price pertaining in 1912. Appendix 4, table A4.2 compares the prices so derived, averaged for the period 1856–60 and for the individual year 1854, with the prices employed by Solar and Ó Gráda in their own estimates for those years.[11]

Both Solar and Ó Gráda chose their prices from a variety of sources, many of them local. For example Ó Gráda employed the quoted cattle prices from the Ballinasloe market in county Galway. I indicate in appendix 5 in the section regarding sheep and in notes 18 to 21 that there may be problems associated with using such prices. In my own case, in pursuing an annual assessment, I have continued to use the Barrington index but with the new basing policy. The most consistent pattern which now emerges is that on the whole my prices are lower than those employed by Solar and Ó Gráda. Yet there are significant exceptions.

[10] Vaughan, 'Potatoes', pp. 83–4, though he made the self-same error in his own original work by not differentiating potatoes from other products, Vaughan, 'Landlord and tenant relations', pp. 337–8, and specifically for potatoes, see the potato values which have been attributed to Vaughan, in which it is clear that a simple arithmetic mean of the mean maximal and mean minimal was used, in, T.W. Moody, *Davitt and Irish Revolution 1846–82* (Oxford, 1981), p. 570. It seems also that K.H. O'Rourke may have made the same mistake, 'Rural depopulation in a small open economy: Ireland 1856–1876', *Explorations in Economic History*, 28 (1991), 409–32, where the script and note 29 on page 420 lead to this conclusion.

[11] Ó Gráda, 'Irish agricultural output', 165; Solar, *Growth and Distribution*, p. 364.

Table A4.2   *Comparison of price data in the 1850s (in shillings per cwt)*

Product	Ó Gráda 1854	Turner 1854	Solar 1856/60	Turner 1856/60
Wheat	15.0	14.0	11.3	10.1
Oats	9.0	7.2	7.6	5.7
Barley	8.9	8.5	8.4	8.3
Potatoes[a]	2.7	4.9	2.3	3.2
Flax	66.0	65.0	55.0	75.3
Hay	5.0	2.9	2.6	2.2
Butter	91.0	102.1	104.9	111.8
Wool	112.0	93.6	158.7	132.9
Eggs[b]	4.2	3.7	6.7	4.1
Pigmeat	60.5	46.2	52.2	48.2
Mutton		57.1	63.2	56.4

Note: [a] Ó Gráda reports as 4 shillings per stone. I have assumed this is a misprint for 4 pence per stone.
[b] Shillings per 120 eggs. Ó Gráda reports as 60 shillings per 12 dozen. I have assumed this must be a mistake for 60 pence per 12 dozen. The standard method is to report in terms of Great Hundreds, which is equal to ten dozen eggs. If Ó Gráda meant 60 pence per Great Hundred then this is 5 shillings equivalent.

Sources: This Appendix; C. Ó Gráda, 'Irish agricultural output before and after the Famine', *Journal of European Economic History*, 13 (1984), 165; Peter Solar, *Growth and Distribution in Irish Agriculture before the Famine*, Ph.D. Dissertation, University of Stanford, 1987, p. 364.

The Turner flax price is higher, but this is never a significant crop. More importantly the Turner potato prices are higher and so are the butter prices, both of which are important products. Countervailing this is the very much higher cattle prices used by both Solar and Ó Gráda.

It was the potato prices and hence values which most disturbed Vaughan in his criticism of the earlier estimates. He compared the average of the Cowper Commission mean maxima and mean minima prices, and market prices from St Sepulchre's market and country markets like Enniskillen and Westport during the last quarter of the years 1852–5.[12] These are reproduced in Appendix 4, table A4.3, along with the equivalent potato prices used in the revised estimates in chapter 4. Clearly my revisions are moving in the right direction, if not nearly the whole way in the right direction toward meeting the criticism of overvaluation.

An important point which Vaughan makes and substantiates with

[12] Vaughan, 'Potatoes', p. 84.

Table A4.3   *Comparison of potato prices, 1852–1855 (in pence per cwt)*

	Cowper Prices[a]	Market prices from Vaughan	Turner prices
1852	60	41	44.9
1853	58	42	43.3
1854	78	44	58.3
1855	56	42	41.8
Annual Average	63	42	47.1

Note: [a] This is the average of the average maxima and average minima.
Sources: From the procedures described in this appendix; W.E. Vaughan, 'Potatoes and agricultural output', *Irish Economic and Social History*, 17 (1990), p. 84.

reference to official prices for 1906 is that the average price over an extended period is more likely to occur nearer the minimum range of prices than the maximum, and therefore taking the average of the mean maximum and mean minimum certainly leads to overvaluation.[13] This can be demonstrated in some detail in appendix 4, table A4.4. For the period 1894–1904 the average highest potato price and the average lowest potato price in each month for each of the four provinces of Ireland have been tabulated, along with the actual average price. There is a certain consistency in the relationship between the actual average and the average of the highest and lowest. In Leinster for example the various measures of central tendency suggest that the true average was about 80 per cent of the highest average prices and less than 120 per cent of the lowest. Within narrow limits this was almost the same for Connacht (81 per cent of the highest and a little over 110 per cent of the lowest). In Munster and Ulster the actual average was close to 65 per cent of the average highest potato prices, but for both provinces it fluctuated rather more as a percentage of the average lowest prices (about 140 to 125 per cent).[14]

A similar exercise can be carried out employing the mean maximal and mean minimal prices from the Cowper Commission. It shows that the Ó Gráda potato price for 1854 at 32 pence per cwt. for example, is much lower than both the mean maximal and mean minimal for that year of 96 and 60 pence respectively, as is my own price of 58 pence. Solar's annual average potato price of 1856–60 at 2.32 shillings or 27.8 pence per cwt. compares similarly with the mean maximal of 62 pence and the mean

[13] *Ibid.*, p. 84.
[14] *Agricultural Statistics, Ireland, 1907–8. Return of Prices of Crops, Live Stock and other Irish Agricultural Products, BPP* [Cd. 4437], vol. cxxi, 1908, p. 35.

Table A4.4   *Comparison of mean high, mean low, and average potato prices, 1894–1904, by Province (in pence per cwt)*

	High	Low	Ave	Ave as % of high %	Ave as % of low %	High	Low	Ave	Ave as % of high %	Ave as % of low %
Leinster						Connacht				
1894	60	35	47	78.3	134.3	46	32	38	82.6	118.8
1895	57	35	43	75.4	122.9	43	31	35	81.4	112.9
1896	55	24	36	65.5	150.0	28	20	23	82.1	115.0
1897	48	30	39	81.3	130.0	44	30	34	77.3	113.3
1898	64	37	46	71.9	124.3	46	36	32	69.6	88.9
1899	42	30	35	83.3	116.7	32	24	26	81.3	108.3
1900	54	39	44	81.5	112.8	38	29	34	89.5	117.2
1901	46	36	39	84.8	108.3	36	29	30	83.3	103.4
1902	41	32	34	82.9	106.3	37	26	30	81.1	115.4
1903	48	40	44	91.7	110.0	44	33	37	84.1	112.1
1904	50	39	38	76.0	97.4	57	42	46	80.7	109.5
			Mean	79.3	119.4			Mean	81.2	110.4
			Median	81.3	116.7			Median	81.4	112.9
Munster						Ulster				
1894	60	29	43	71.7	148.3	50	29	35	70.0	120.7
1895	50	30	39	78.0	139.3	51	28	37	72.5	132.1
1896	46	20	29	63.0	170.6	40	17	24	60.0	141.2
1897	63	32	43	68.3	179.2	54	24	33	61.1	137.5
1898	71	37	39	54.9	121.9	61	32	40	65.6	125.0
1899	50	26	33	66.0	132.0	51	25	32	62.7	128.0
1900	54	29	37	68.5	108.8	58	34	40	69.0	117.6
1901	62	26	38	61.3	126.7	52	30	36	69.2	120.0
1902	66	29	38	57.6	158.3	49	24	31	63.3	129.2
1903	69	33	44	63.8	137.5	60	32	39	65.0	121.9
1904	64	33	43	67.2	138.7	53	31	39	73.6	125.8
			Mean	65.5	141.9			Mean	66.5	127.2
			Median	66.0	138.7			Median	65.6	125.0

Source: Agricultural Statistics, Ireland, 1907–8. Return of Prices of Crops, Live Stock and other Irish Agricultural Products, British Parliamentary Papers [Cd. 4437], vol. cxxi, 1908, p. 35.

minimal of 38.8 pence, and my own average for that period of 38.6 pence. In other words, both Solar and Ó Gráda are a long way below the Cowper Commission prices, but then so also are my own new prices. In fact over the period of the Cowper Commission prices which coincide with my estimates, 1850–84, my potato price is mostly between 60 and 65 per cent of the mean maximal (in other words not too far adrift of the Ulster and Munster figures quoted above for the later period), and on

Table A4.5   *Agricultural product prices in 1912*

	Shillings	Pence	
Wheat	7	9.5	per cwt of 112 lbs
Oats	6	10.25	per cwt of 112 lbs
Barley	8	7	per cwt of 112 lbs
Hay	3	1.75	per cwt of 112 lbs
Potatoes	3	6.75	per cwt of 112 lbs
Flax	8	1.5	per stone of 14 lbs
Eggs	9	3.25	per 120 eggs
Butter	106	10	per cwt of 112 lbs
Pork	55	1	per cwt of 112 lbs
Wool	0	10.5	per lb
Beef	60	1	per cwt of 112 lbs
Mutton	62	10	per cwt of 112 lbs
Cattle – 2–3 years old	221	4	per head
Cattle – 3 Years old	254	6	per head

Source: Agricultural Statistics, Ireland, 1915. Return of Prices, of Crops, Live Stock and other Irish Agricultural Products [Cd. 8452], vol. xxxvi, 1917-18, pp. 14-15.

average still not yet 100 per cent of the mean minimal Cowper price (in fact 98 per cent on average).

The final output estimates in chapter 4 therefore still employ the Barrington indexes, but now based on quoted prices for the year 1912 (for which see Appendix 4, table A4.5). In so doing I am encouraged by what R.C. Geary said on 31 May 1951. In relation to potatoes, he stressed that although the market prices available probably exaggerated the value of potatoes in general, they nevertheless mirrored the trend of potato production.[15]

[15] R.C. Geary, commenting on H. Staehle, 'Statistical notes on the economic history of Irish agriculture, 1847–1913, *Journal of the Statistical and Social Inquiry Society of Ireland*, 18 (1950–1), p. 465.

Table A4.6 *Thomas Barrington's index of Irish agricultural prices*

	Wheat	Oats	Barley	Hay	Potatoes	Flax	Eggs
	(Base 100 = 1840, or nearest year with comparable prices)						
1840	100	100	100	100	100		100
1845	98	118	118	92	88	100	96
1846	131	187	175	69	323	95	100
1847	98	107	116	81	254	90	124
1848	87	96	100	56	292	94	106
1849	66	83	85	48	215	118	100
1850	73	101	87	60	219	124	102
1851	66	89	85	50	175	112	94
1852	61	77	80	46	231	110	95
1853	124	121	135	106	223	124	111
1854	122	144	129	125	300	142	111
1855	150	165	179	98	215	167	109
1856	102	112	149	75	150	171	122
1857	86	108	127	85	254	153	129
1858	69	100	102	85	146	207	120
1859	86	116	120	127	177	142	122
1860	99	139	135	90	265	149	124
1861	95	112	116	83	227	121	137
1862	73	100	109	87	181	157	132
1863	69	100	104	83	127	166	135
1864	64	92	102	92	131	144	139
1865	101	120	115	79	146	238	144
1866	107	140	156	90	173	211	131
1867	125	164	149	110	200	175	137
1868	117	139	155	117	177	207	138
1869	98	136	148	92	154	171	138
1870	84	127	118	110	185	135	150
1871	102	129	129	100	196	198	144
1872	103	123	133	87	312	146	150
1873	98	109	80	150	223	150	161
1874	76	144	132	133	142	148	178
1875	77	131	126	144	162	166	181
1876	79	119	122	144	163	142	193
1877	87	129	131	106	277	166	148
1878	72	136	122	87	208	139	164
1879	90	137	118	112	285	153	183
1880	76	120	114	94	142	146	176
1881	83	127	113	108	185	121	217
1882	76	124	115	108	185	112	215
1883	70	125	111	108	191	106	215
1884	62	116	106	117	170	112	211
1885	58	113	100	100	156	106	203
1886	53	99	82	81	151	115	193
1887	55	98	91	105	139	102	205
1888	61	107	100	85	156	106	202

	Wheat	Oats	Barley	Hay	Potatoes	Flax	Eggs
1889	57	114	100	75	166	104	201
1890	59	123	99	82	189	93	197
1891	65	135	112	139	222	104	199
1892	59	135	107	170	152	117	197
1893	53	131	103	149	160	142	229
1894	50	115	107	123	191	123	195
1895	50	97	103	102	191	96	209
1896	56	101	97	109	133	83	193
1897	64	106	98	111	189	90	183
1898	56	104	108	91	214	102	193
1899	54	104	101	98	173	112	199
1900	56	109	105	129	208	123	199
1901	56	116	108	123	191	117	199
1902	54	120	110	131	177	112	197
1903	55	109	107	117	216	107	207
1904	63	112	110	135	208	119	213
1905	61	114	108	131	177	119	231
1906	56	119	108	127	177	123	241
1907	71	125	109	116	243	123	249
1908	70	115	113	120	214	111	251
1909	68	120	111	147	177	125	269
1910	62	111	106	159	202	152	265
1911	66	125	109	134	216	149	271
1912	68	138	131	132	220	142	279
1913	66	120	112	125	237	126	283
1914	77	143	111	132	210	162	297

	Butter	Pork	Wool	Beef	Mutton	Store cattle 2 yrs	Store cattle 3 yrs
1840	100	100	100	100	100		
1845	92	92	106	103	120	100	100
1846	105	117	94	104	120	129	170
1847	99	149	82	105	125	157	190
1848	87	115	65	110	120	143	180
1849	74	88	71	79	105	97	150
1850	74	92	92	81	100	89	130
1851	83	94	89	86	105	100	135
1852	77	82	102	74	88	114	125
1853	98	122	123	110	128	100	110
1854	107	124	94	120	130	107	160
1855	106	129	98	112	125	157	170
1856	119	138	118	115	131	171	180
1857	117	141	141	128	133	100	190
1858	119	108	118	121	124	171	180
1859	127	123	137	117	119	150	200
1860	104	138	153	112	135	157	200

	Butter	Pork	Wool	Beef	Mutton	Store cattle 2 yrs	Store cattle 3 yrs
1861	125	132	141	126	137	150	170
1862	104	118	153	119	140	186	200
1863	113	118	165	121	130	154	210
1864	123	126	206	130	147	179	240
1865	140	145	165	136	160	186	200
1866	136	145	157	149	157	179	190
1867	109	108	125	141	120	121	150
1868	147	144	133	138	145	143	200
1869	130	158	118	151	145	150	190
1870	141	144	110	148	162	157	200
1871	136	105	184	156	170	214	235
1872	133	126	180	161	175	220	275
1873	147	141	161	172	174	229	230
1874	153	147	143	161	175	214	280
1875	150	151	145	164	160	229	220
1876	155	133	119	161	186	243	260
1877	126	142	115	150	190	214	250
1878	121	136	106	155	185	243	250
1879	113	132	83	135	160	214	230
1880	129	146	122	135	160	239	245
1881	114	141	100	133	160	221	240
1882	114	137	93	137	166	233	247
1883	110	133	86	140	173	240	240
1884	105	123	86	132	163	214	228
1885	91	119	82	118	146	185	205
1886	88	116	86	107	136	165	188
1887	96	108	91	104	127	251	174
1888	95	114	91	112	140	226	238
1889	101	118	86	116	152	253	252
1890	91	107	84	119	152	255	249
1891	109	106	86	114	130	223	233
1892	113	129	86	109	130	180	202
1893	102	137	82	110	132	175	198
1894	93	113	82	115	137	189	200
1895	92	105	82	119	149	214	212
1896	95	96	82	109	130	209	212
1897	93	113	74	110	139	219	221
1898	93	112	65	106	132	228	214
1899	101	101	61	115	135	238	221
1900	101	121	62	119	142	243	235
1901	105	131	49	115	137	243	235
1902	102	133	48	122	136	233	231
1903	99	123	67	116	147	255	245
1904	95	112	88	114	149	250	238
1905	105	133	112	110	147	243	233
1906	107	137	114	110	152	240	228

	Butter	Pork	Wool	Beef	Mutton	Store cattle 2 yrs	Store cattle 3 yrs
1907	103	135	103	115	157	255	240
1908	113	129	69	119	146	260	247
1909	106	149	92	121	122	267	252
1910	108	161	98	125	146	284	266
1911	112	137	94	121	136	299	271
1912	112	148	96	125	143	294	271
1913	109	171	109	129	159	308	282
1914	114	160	115	132	159	313	289

Note: Because of missing entries in the data it has been necessary to interpolate on a few occasions, for flax for 1853–4, mutton for 1906, and for eggs for 1887–92.

Source: T. Barrington, 'Review of Irish Agricultural Prices', *Journal of the Statistical and Social Inquiry Society of Ireland*, 15, 1927, esp. 251–3.

Appendix 5 Weighting procedures adopted in calculating agricultural output

This appendix contains the detailed procedures adopted in the calculation of agricultural output presented in chapter 4. The proportion of the total output of each product which is assigned to final output is referred to by the term weight, a term which refers to proportionalities in the main rather than to physical size. See in particular table 4.1 for a digest of some of the assumptions or weights used by other historians in their own estimation of output.

Wheat and rye

Most commentators agree that something like 90–92 per cent of total wheat output actually went to market or was consumed on the farm, with little or no variation over time, leaving 8 per cent or so for seed for the following year. The *Agricultural Output of Ireland* for the year 1908 suggested that 92 per cent was sold or consumed on the farm, and the *Food Production* report for 1912–13 suggested 94 per cent. The enumerating authorities of the *The Agricultural Output of Northern Ireland* in 1925 (for Northern Ireland alone) assigned 91.4 per cent of wheat output in this way.[1] I have therefore adopted a weight of 92 per cent, in line with practically all other estimations.

Although there are data on acreages and yields for rye there are not prices equivalent to the other grains. The acreage was never very high, and the crop was never a large proportion of tillage output. I have assumed the same weighting and price as for wheat, though I note that both the 1908 and 1912–13 estimates of output counted rye as wholly an animal fodder. Either way, its value is so small that its influence on final output is very small.

---

[1]   *The Agricultural Output of Ireland 1908* (Department of Agriculture and Technical Instruction for Ireland, Dublin, 1912), p. 21; *Food Production in Ireland* (1912–13). *Minutes of Evidence, B(ritish) P(arliamentary) P(apers)* [Cd. 8158], vol. v, 1914–1916, pp. 70–71; *The Agricultural Output of Northern Ireland, 1925, BPP* [Cmd. 87], 1928, p. 71.

Barley and oats

The disposable proportions of output of these crops are thought to have changed significantly over time. More was consumed by animals later in the period, the value of which output was realised in other ways. In the case of both crops there were larger acreages earlier on and it is in the earlier period that the larger weights are applied. The tendency therefore might be to exaggerate the value of their outputs earlier on. This is less a problem with barley because like wheat it was considered to be more of a cash crop than a fodder crop. I use upper and lower bounds of 92 and 70 per cent for barley and 60 and 28 per cent for oats. The upper bound barley estimate is based partly on Ó Gráda and Solar, and partly on the assumption that early on most of the crop was marketed as a cash crop and enjoyed the same weight as wheat. In 1908 it was suggested that 70 per cent of the barley crop was sent to market or consumed on the farm, in 1912–13 the suggested proportion was 84 per cent, and in Northern Ireland in 1925 the suggested proportion was 73.7 per cent.[2] Therefore my use of 70 per cent is certainly not an overestimate.

The upper bound oats estimate is based on Ó Gráda, and the lower bound is based on the 1908 assessment.[3] A tenant farmer in Wicklow in the 1880s believed that two-thirds of the oats crop was consumed on the farm.[4] In 1912–13 it was suggested that 27 per cent was consumed by the farmer and his family, but in Northern Ireland in 1925 as little as 7 per cent of the oats output was regarded in this way.[5]

In the cases of these two crops and those other items listed below which are based on upper and lower bound estimates (for example variable potato weights, pig carcass sizes and milk yields), I have assumed that the move to more or fewer oats marketed, or smaller to larger carcass sizes or milk yields was linear over time. This is in the absence of any over-

---

[2]   C. Ó Gráda, 'Irish agricultural output before and after the Famine', *Journal of European Economic History*, 13 (1984), 160. P. Solar, *Growth and Distribution in Irish Agriculture before the Famine*, Ph.D. Dissertation, University of Stanford, 1987, p. 364, reckons the annual average for 1856–60 was 91 per cent; *Agricultural Output of Ireland 1908*, p. 21; *Food Production*, 1912–13, pp. 70–1; *Agricultural Output of Northern Ireland, 1925*, p. 71.

[3]   Ó Gráda, 'Irish agricultural output', 160. As I understand it there is a printing error in this article to the effect that it states that two-fifths of the oats crop constituted output. It should correctly read three-fifths. And calculated from *Agricultural Output of Ireland 1908*, p. 21. Solar for 1856–60 reckoned 57 per cent, *Growth and distribution* p. 365.

[4]   R.M. Barrington, 'The prices of some agricultural produce and the cost of farm labour for the last fifty years', *Journal of the Statistical and Social Inquiry Society of Ireland*, 9 (1887), 141.

[5]   *Food Production, 1912–13*, pp. 70–1; *Agricultural Output of Northern Ireland, 1925*, p. 71.

whelming evidence which points to precise technical changes which might have determined more appropriate, more positive, turning points.

Flax

Official and unofficial modern estimators agree that the whole of the flax crop should count towards output.[6]

Potatoes

There are some variations over the putative size of final potato output. In his main tabular estimates of final output for the 1850s and 1870s Vaughan does not include the output from potatoes, though he does include them in parenthetical appendixes. Ó Gráda suggests that because so few potatoes came to market the prices we use are likely to be on the high side, since the best potatoes were saved for the market. Nevertheless, that part of the crop which was consumed on the farm by humans must count towards output. The weights I have adopted vary from a high of 45 per cent in the aftermath of the Famine, reducing to 32 per cent by the Great War. These are essentially compromises based on the work of others. For example, on the basis of pre-Famine usage Bourke has suggested that 47 per cent of the crop was devoted to human consumption, while Solar puts it at 53 per cent before the Famine. Solar further suggests 41 per cent for the period 1856 to 1860, and Ó Gráda for 1854 says 50 per cent. The lower bound weight follows the 1908 estimate, though the 1912–13 estimate suggested as much as 59 per cent. In Northern Ireland by 1925 still as much as 41 per cent of the potato crop was counted as marketed output.[7]

Hay and straw

In 1908 it was suggested that 5.9 per cent of the hay and straw crop counted directly towards final output, in 1912–13 the equivalent proportion was 3.5 per cent, and in Northern Ireland in 1925 it was 3 per cent.[8]

---

[6] For example, *Agricultural Output of Ireland 1908*, p. 21; *Food Production, 1912–13*, pp. 70–1.

[7] P.M.A. Bourke, 'The use of the potato crop in pre-Famine Ireland, *Journal of the Statistical and Social Inquiry Society of Ireland*, 21 (1967–8), 93; Solar, *Growth and Distribution*, p. 365; Ó Gráda, 'Irish agricultural output', p. 162; *Agricultural Output of Ireland 1908*, p. 21; *Food Production, 1912–13*, pp. 70–1; *Agricultural Output of Northern Ireland, 1925*, p. 71.

[8] *Agricultural Output of Ireland 1908*, p. 21; *Agricultural Output of Northern Ireland, 1925*, p. 71.

At these levels hay will never be a significant proportion of total output. A different method of assessment is to assume a certain level of annual consumption by off-farm horses. Horse numbers did not fluctuate wildly over the course of the period. It could be assumed therefore that the hay requirement of off-farm horses was fairly constant and adopt the 1908, or some other, standard throughout. However, the hay acreage rose dramatically over the course of the period, in tandem with the general shift towards the pastoral economy. A constant weighting factor therefore based on a much smaller early hay acreage will likely underestimate the value of hay output early on.

The beast can be attacked, so to speak, from the other end, through the horse population. Horses for agricultural purposes represented a fairly constant 85 to 88 per cent of all horses over 2 years of age. Therefore off-farm consumption of hay was required for about 13 per cent of horses over 2 years of age. All other horses, below 2 years and in an unbroken state, can be counted as consumers of hay with no value added in the year of enumeration. They realised their values in the years ahead. Unbroken horses were generally the young ones. It could be assumed that they were less active than those horses over 2 years of age – less active in the sense that they were not yet working. In consequence they probably consumed less hay than the 2 year olds. Similarly horses less than 1 year of age consumed less hay than horses in the other 2 categories. The weights I have adopted therefore are, 1.5 tons of hay for all non-farm horses over 2 years of age (that is for 13 per cent of all horses over that age), 0.75 tons per head for unbroken horses between 1 and 2 years of age, and 0.5 tons per head for horses less than 1 year of age.[9]

Straw prices were on or about one-half the hay price per cwt., and the Northern Ireland estimates of 1925 show that 1.4 cwt. of wheat straw was acquired from each cwt. of wheat grain, 1.5 cwt. of oats straw from each cwt. of oats grain, and that of the final tonnage of straw from these two crops 3.8 per cent was counted towards final output. In my estimate of straw output a compromise of these and other estimates has been used: 4 per cent of 1.4 cwt. of wheat, barley, oats and rye straw per cwt. of each grain produced, multiplied by half the price of hay.[10]

Other tillage

The weighting of the tillage sector is thus complete because all root and green crops are counted as fodder, realising their value in the output of

[9] Ó Gráda's hay estimate allowed 2.4 tons consumption for every off-farm horse, 'Irish agricultural output', p. 162.

[10] *Agricultural Output of Northern Ireland, 1925*, p. 71.

livestock and livestock products, as was the estimated 1.5 tons of hay required to winter a cow.[11] Some proportion of the root and green crops was marketed and consumed by humans, and other authorities have assumed a mark-up of about 5 per cent to complete the estimate of tillage output. The 1908 estimates concluded final root and green crop output accounted for 1.6 per cent of all Irish final crop output. The sales of grass seed, fruit and timber were reckoned more valuable. Together it was estimated that these constituted another 14 per cent of final output. A true mark-up might be worth at least 15 per cent of final output. In 1912–13 the equivalent figures were less than 1 per cent for root and green crops, and 7 per cent for the other crops, suggesting a final mark up of about 8 per cent.[12]

Cattle

The agricultural statistics each year separated cattle into three age groups: those greater than 2 years of age, those between 1 and 2, and those less than 1 year of age. By adding together all those cattle aged over 2 and between 1 and 2 in year 't' and subtracting this total from all those cattle aged over two in year '$t+1$' gives the 'disappearance' of cattle of 2 years and over between the two years. Similarly, by subtracting the number of cattle aged between 1 and 2 in year '$t+1$' from the number of cattle less than one year in year 't' gives the 'disappearance' of cattle aged 1 year between the two years.[13] There was a certain mortality loss: Ó Gráda assumed such loss at 5 per cent and Solow adopted the 1908 estimate of mortality loss. I adopt the same estimate.[14]

---

[11] See also J. Wilson, 'Tillage versus grazing', *Journal of the Department of Agriculture and Technical Instruction for Ireland*, 5 (1905), 220, 222, 232, where the ratio of oat grain to oat straw was estimated at 1:1.5, but barley at 1:1.1 or 1.2. In 1908 the ratios were, for wheat 1.75:1, for oats 1.5:1, for barley 1.25:1 and for rye 1.5:1, *Agricultural Output of Ireland 1908*, pp. 8, 20–1. In 1912–13 the ratio of total grain production to total straw production was 1:1.56, *Food Production, 1912–13*, pp. 70–1.

[12] R.O. Pringle, 'A review of Irish agriculture, chiefly with reference to the production of live stock', *Journal of the Royal Agricultural Society of England*, 2nd series, 8 (1872), 59; *Agricultural Output of Ireland 1908*, p. 21; *Food Production, 1912–13*, p. 70.

[13] Same method as used by B. Solow, *The Land Question and the Irish Economy, 1870–1903* (Cambridge, MA, 1971), pp. 214–15, and Ó Gráda, 'Irish agricultural output', 164.

[14] In Northern Ireland in 1925/6 there were mortality losses of only 16 head of all classes of cattle per 1,000 head, and similarly in 1923/4 and 1926/7, and 37 per 1,000 in 1924/5. Mortality loss was always greatest in stock under 1 year of age; it was as high as 60 per 1,000 in 1924/5. Apparently the mortality loss in Northern Ireland was 'extremely moderate' compared with reported losses by the Ministry of Agriculture and Fishery for England and Wales, which estimated an average for the mid-1920s of 45 per 1,000. The difference was accounted for because Irish farms on average were smaller and the farmer was 'able to devote personal attention to the rearing of calves born' an 'advantage enjoyed by small farmers in this respect *is a well known fact*'. *Agricultural Output of Northern Ireland, 1925*, pp. 26–7, my emphasis.

Thus through the annual change in stock 'disappearances' at the two cattle ages, in conjunction with cattle prices at the same ages, the annual value of final cattle output is derived. In notation, allowing for mortality:

$$[(Q>2+Q1-2)-Q>2\})\text{Price}>2]+(Q<1-Q1-2)\text{Price}<2$$
$$t \qquad t \quad t+1 \quad t+1 \qquad t\ t+1 \qquad t+1$$

An alternative assessment based on the limited export data which is available is possible. A fairly constant proportion of total cattle enumerated each year was exported to Britain, from at least 1875, and a lower, but again fairly constant proportion stretching back to the 1860s. In addition, the proportional distribution between fat and store cattle remained constant.[15] The subtraction of this export trade from the total of Irish cattle disappearances gives a figure which after deduction for mortality could be assumed was the retained disappearance which was slaughtered and consumed in Ireland. Details on slaughtering are not generally available, but by 1925, in Northern Ireland at least, the agricultural census enumerators collected details from public and private slaughterhouses. There are beef prices, and there are enough estimates of carcass weights to allow a calculation of the beef which entered the market.[16] However, unless those cattle were slaughtered on the farm, the value of cattle output should be the market price of cattle on the hoof before further value added in the slaughtering and retailing process. In the case of cattle therefore, in the absence of details of slaughter, I am ignoring beef prices. In so doing inevitably I am ignoring also the value of hides and other cattle by-products.[17] By treating them as live cattle these by-products are necessarily included in the cattle price, but without value added in subsequent processing.

[15] Between 1875 and 1891 from 13.6 to 19.6 per cent of all cattle enumerated were exported to Great Britain from Ireland with a mean of 16 per cent. About 40 per cent were fat cattle and 53 per cent were store cattle, with a residual 7 per cent as calves. Export figures are reported in chapter 2 above with full references to sources. See also the relevant section in chapter 4.

[16] See P.G. Craigie, 'On the production and consumption of meat in the United Kingdom', *Report of the British Association for the Advancement of Science* (1884), 841–7; See also R. Perren, *The Meat Trade in Britain 1840–1914* (London, 1978), pp. 2–3, and *Agricultural Output of Northern Ireland, 1925*, pp. 24–5.

[17] It should be noted that in the period 1847–53 the annual disappearance of cattle less than 2 years old was positive. That is, in this post-Famine period of rebuilding stocks the value of cattle was zero or even negative in the sense that there were costs in the acquisition of stock and mathematically at least no marketed output. In addition, the method ignores the importation of animals, of which there were usually some.

Sheep

The calculation of sheep output follows the methodology for cattle. The annual disappearance of sheep is calculable and there are lamb prices available. After a deduction of 12 per cent for mortality it is possible to value the net 'disappearance' of sheep. Using lamb prices undoubtedly underestimates this value since some of the sheep which reached market were larger than lambs. Other estimators have used the mean of variable prices for variable ages of animals based on the October/November Ballinasloe fair.[18] This was a fair held in Galway. It may have been that its specialist nature encouraged the sale of better than average animals at, therefore, better than average prices. This last point seems to be borne out when Ballinasloe prices are compared with those quoted for Dublin.[19] But Ballinasloe constituted a small proportion of the sheep and cattle trade. Sheep presented for sale at Ballinasloe as a proportion of all sheep which disappeared from the annual enumeration was less than 7 per cent in the mid-nineteenth century reducing to a little over 1 per cent at the end, and cattle were only 3 per cent in the mid-century reducing to 1 per cent at the end.[20] In addition, in some years a good proportion of those presented for sale remained unsold.[21] For these reasons it is likely that historians who employ Ballinasloe market prices overestimate the value of Irish livestock output. For this reason I use the Barrington price indexes coupled with the *official* prices throughout.

[18]   W.E. Vaughan, 'Agricultural output, rents, and wages in Ireland, 1850–1880', in L.M. Cullen and F. Furet (eds.), *Ireland and France 17th to 20th Centuries: Towards a Comparative Study of Rural History* (Ann Arbor, Michigan and Paris, 1980), p. 95. Ó Gráda also employed Ballinasloe prices for his cattle output estimates, 'Irish agricultural output', 164.

[19]   See *Agricultural Statistics, Ireland, 1907–8. Return of Prices of Crops, Livestock and Other Agricultural Products, BPP* [Cd. 4437], vol. cxxi, 1908, pp. 90–3, the prices of cattle and sheep at Ballinasloe 1841–1907 and compare with the official annual prices data from the 1880s on pp. 24–5 for 1888–1907 and *Agricultural Statistics, Ireland, 1915. Return of Prices of Crops, Livestock, and Other Irish Agricultural Products, BPP* [Cd. 8452], vol. xxxvi, 1917–18, pp. 14–15 for the overlapping period 1896–1915.

[20]   Comparing *Agricultural Statistics, 1907–8*, pp. 90–3 with annual animal 'disappearances'. The tone that J. O'Donovan uses in his *The Economic History of Live Stock in Ireland* (Cork, 1940), p. 260 also exaggerates the size of the Ballinasloe fair. He talks of the 'enormous figure of 99,658' sheep in 1856 and the average of 70–80,000 sheep up to the depression of the late 1870s. They sound like large numbers but they constituted always less than 10 per cent of the sheep which 'disappeared' annually.

[21]   In the period 1847–1907 an average of 20 per cent of all cattle brought to the market remained unsold, and the comparable figure for sheep was 12 per cent. There were wide fluctuations from year to year from highs of 46 per cent of cattle in 1892 and 39 per cent of sheep in 1883, to lows of 4 per cent of cattle in 1870 and less than 1 per cent of sheep in 1870. In every decade the proportion of cattle left unsold exceeded 15 per cent and the unsold sheep exceeded 10 per cent in every decade except the 1880s (9 per cent) *Agricultural Statistics, 1907–8*, pp. 90–3.

There could be great variation in prices across Ireland with some local price determination related to purely local factors, such as the relative competition or monopoly of traders.[22] This is essentially a retail prices issue, not ostensibly farm gate prices, even if those farm gate prices we purport to use are in fact akin to international commodity prices.

This still leaves the problem of no index for those sheep which were sold which were not lambs. Official average prices for lambs as distinguished from sheep over 1 year and those over 2 years are available from 1890. The ratio of those lamb prices to the average of those other sheep varied from 1890 to 1915 from 1:1.3 to 1:1.6, but with a marked central tendency towards 1:1.4.[23] Thus I have used the following weights: lamb prices for both lambs and sheep with the sheep weighted by a factor of 1.4, in association with calculated sheep disappearances from the annual agricultural statistics in the ratio of sheep to lambs equal to the ratio of sheep exports to lamb exports to Britain, that is to say, two-thirds sheep to one-third lambs.[24] As a check on these assumptions I have calculated a new value based on the (untrue) assumption that all sheep were slaughtered and sold at Dublin mutton prices with an average carcass weight of 60 lbs. for sheep and 40 lbs. for lambs with a 12 per cent mortality.[25] Such alternative estimates do not greatly affect the value of sheep output.

Pigs

The value of pig output is as problematic as for other animals. To begin with, 'official' prices are reported as bacon prices. This need not be such a large problem since the greater proportion of pigs were destined for pigmeat in the form of bacon (about 68 per cent).[26] However, the annual average number of live pigs exported from 1875–91 was of the order of

[22] L. Kennedy, 'Traders in the Irish rural economy, 1880–1914', *Economic History Review*, 32 (1979), 203.

[23] *Agricultural Statistics 1907–8*, pp. 90–3; *Agricultural Statistics 1915*, pp. 14–15.

[24] Between about 1870 and 1890 17–18 per cent of the annual sheep enumerated were exported to Britain in the proportion 2:1, sheep to lambs. Before 1870, as far as the official statistics reveal, the export trade was about 13 per cent of all sheep enumerated. See *The Agricultural Statistics of Ireland, 1891, BPP* [C. 6777], vol. lxxxviii, 1892, p. 23; Perren, *The Meat Trade*, pp. 96–7; and O'Donovan, *Economic History of Live Stock*, pp. 214–20, 439.

[25] For Britain Craigie used a carcass weight of 70 lbs and for Ireland Ó Gráda used 60 lbs, as did Solar. Craigie, 'On the production and consumption of meat', 842; Ó Gráda, 'Irish agricultural output', 163; Solar, *Growth and Distribution*, p. 365. Mine is essentially a compromise based partly on a quoted average carcass weight for adult sheep presented to the Belfast abattoir in 1925 of 56 lbs for adult sheep and 4 1/2 lbs for lambs. *Agricultural Output of Northern Ireland, 1925*, p. 28.

[26] In 1908 the total output of pigs (in thousands) was estimated at 1731 of which 1178 were cured in Ireland, 387 were exported and 166 were slaughtered in Ireland for consumption as fresh pork, *Agricultural Output of Ireland 1908*, p. 14.

36 per cent of total enumeration and this had risen from 30 per cent in the mid-1860s.[27] In addition, it is generally agreed that at least 100 per cent of the annual enumeration of pigs was eventually sold. That is, there was underenumeration because of multiple littering in any calendar year. For example, in 1908 it is said that 142 per cent of the annual enumeration entered the market, and in 1912–13 it was 125 per cent. I have used the following weights: upper and lower bounds based on pigs of 1.2 to 1.5 cwt. per carcass, multiplied by 130 per cent of the annual enumeration, at the current bacon prices.[28]

Wool

Irish historians have usually used a physical weight of 5–6 lbs. of fleece per head of sheep over 1 year of age.[29] Ojala's UK estimates were based on 3.6 lbs per head of *all* sheep enumerated. This was derived from the 1908 census returns and supported by contemporary nineteenth-century estimates by Craigie which showed a fairly constant wool output per head over the period 1845–82.[30] Solar adopted a fleece weight of 5 lbs for two-thirds of the sheep population.[31] I have adopted a variant of this, 5 lbs per head of sheep over one.

Milk and milk products

This is another problematic category, not least because the 'Official' prices are not milk prices. The customary practice among estimators has been to convert all milk production into equivalent butter production for which there are annual butter prices. Problems still remain, however: how much did each cow in milk produce, and how much milk went into producing 1 lb. of butter? Furthermore, what proportion of milk production was consumed by calves and pigs and should be deducted from milk output to arrive at final output? In 1908 it was taken that calves consumed 35 gallons of milk per head per annum, and pigs a further 5

---

[27] *Agricultural Statistics, 1891*, p. 23; Perren, *The Meat Trade*, pp. 96–7; O'Donovan, *Economic History of Live Stock*, pp. 214–20, 439.

[28] J. Johnston, *Irish Agriculture in Transition* (Dublin and Oxford, 1951), p. 8, quoting M.J. Bonn, 'The psychological aspect of land reform in Ireland'. See also E. Barker, *Ireland in the Last Fifty Years (1866–1916)* (London, 1916), pp. 100–16.

[29] C. Ó Gráda, *Post-Famine adjustment, Essays in Nineteenth Century Irish Economic History*, Ph.D. Dissertation, University of Columbia, 1973, p. 136; Vaughan, 'Agricultural output', p. 94.

[30] E.M. Ojala, *Agriculture and Economic Progress* (Oxford, 1952), p. 203; P.G. Craigie, 'Statistics of agricultural production', *Journal of the Royal Statistical Society*, 46 (1883), 28–30.

[31] Solar, *Growth and Distribution*, p. 365.

gallons, which with a human consumption of 20 gallons per head meant that 24 per cent of all milk produced was consumed as milk in some shape or form, thus allowing 76 per cent to be made into butter, cream and cheese, most of which (in the ratio 1:481) went into butter production. In Northern Ireland in 1925, 8.4 per cent of milk production was fed to calves, and 3.9 per cent to pigs. Of the remainder, 42.6 per cent was made into butter on the farms and 14.5 per cent into butter through creameries. Finally, 30.6 per cent was consumed by humans. Thus 87.7 per cent was counted as final output.[32] Donnelly's Cork study does not allow a proper comparison because there is no mention of pigs, but in the third quarter of the nineteenth century perhaps 30–31 per cent of milk production was consumed as liquid milk, 3–4 per cent was fed to calves and 65 per cent was made into butter.[33] Before 1914 another estimate suggests that 7.5 per cent of milk production was fed to animals, 11–14 per cent was fed to humans, 75 per cent went into butter, and small amounts were manufactured into cheese, condensed milk and cream. Of the butter, 39 per cent was said to be manufactured in creameries, 16 per cent in butter factories and 45 per cent was manufactured on the farm. There were, therefore many ways to use the output of milk.[34]

In 1908 it took 3 gallons of milk to produce each 1 lb of butter, and in Northern Ireland in 1925 it took between 2.45 and 2.9 gallons of milk to produce each 1 lb of butter, depending on whether it was made in a creamery (2.45) or on the farm (2.9)[35] Pringle estimated that from 12 to 14 quarts of milk was required to produce 1 lb of butter (that is, 3 to 3.5 gallons per lb).[36]

And what was the size of the milk yield? For the period 1856–60 Solar adopted a yield of 304 gallons per cow per annum. In 1908 it was estimated that each cow produced 400 gallons per annum, and in Northern Ireland in 1925 the yield was estimated at 407 gallons per cow per annum, the same level as in the Free State at the same date.[37] Solow

[32] *Agricultural Output of Ireland 1908*, p. 14; *Agricultural Output of Northern Ireland, 1925*, pp. 32–3; O'Donovan reported that 89.5 per cent of milk production in the Free State in 1926–7 was considered as output, based on the Census of Agricultural Production, *Economic History of Live Stock*, p. 351.

[33] J.S. Donnelly, *The Land and the People of Nineteenth-Century Cork: The Rural Economy and the Land Question* (London, 1975), p. 247.

[34] O'Donovan, *Economic History of Live Stock*, pp. 326–7.

[35] *Agricultural Output of Ireland 1908*, p. 14; *Agricultural Output of Northern Ireland, 1925*, p. 32. Ó Gráda quotes figures of 2.3 to 2.5 gallons in a creamery and nearer to 3 gallons on the farm. C. Ó Gráda 'The beginnings of the Irish creamery system, 1880–1914', *Economic History Review*, 30 (1977), 302.

[36] Pringle, 'A review of Irish agriculture', 58.

[37] Solar, *Growth and Distribution*, p. 365; *Agricultural Output of Ireland 1908*, p. 14; *Agricultural Output of Northern Ireland, 1925*, p. 32. Free State figures from Ojala, *Agriculture and Economic Progress*, p. 204. On the basis of the Census of Agricultural

adopted the 1908 output estimate of 400 gallons per cow. She deducted 10 per cent to account for mortality, suckling and dry cows, then followed the 1908 estimates in assuming 35 gallons was consumed by each calf enumerated (that is for each animal of less than 1 year of age) and 5 gallons per head of each pig. Vaughan tackled yields in a different way and took a weight of 1 cwt. of butter per cow, and Ó Gráda took 350–385 gallons per cow for the two years 1854 and 1876 and made deductions in accordance with the 1908 estimates.[38] In their estimate of output for 1912/13 based on the twenty-six counties of modern Ireland O'Connor and Guiomard appeared to use a milk yield equivalent to 350 gallons per milch cow.[39] If Donnelly's assessment of contemporary observations is any guide then yields of 400–450 gallons per cow per annum were experienced in the 1880s, with exceptional farmers achieving 700+ gallons.[40] In 1905 James Wilson estimated that a dairy cow in the open pastures yielded 500 gallons while a cow receiving a cake or grain supplement in winter yielded 600 gallons per annum.[41] In 1872 Pringle estimated that the breed known as the 'Dutch' in the county of Cork yielded 24–30 quarts of milk daily. Other breeds in the same county were reported to yield 500–700 gallons per annum.[42]

In the end therefore there is a lot of variable advice as to how to treat milk output. My compromise is to employ lower and upper bounds of 350–400 gallons of 88 per cent of the enumerated milch cows at 3 gallons per 1 lb of butter.[43] The early supposed yield of 350 gallons is high when

Production of 1926–7, O'Donovan estimated that each milch cow produced 487 gallons of which 436 should be counted towards final output, *Economic History of Live Stock*, p. 352. In the whole of Ireland for the periods 1930–1 for Northern Ireland and 1928–9 for the Free State, milk production per cow per annum varied from highs of 593 and 576 gallons in the counties of Dublin and Tipperary to lows of 329, 358 and 360 gallons in the counties of Leitrim, Mayo and Donegal. Generally speaking, west of a line joining Belfast to the City of Limerick production was less than 500 gallons, O'Donovan, *Economic History of Live Stock*, pp. 351–2.

38 Solow, *The Land Question*, pp. 215–16; Vaughan, 'Agricultural output', p. 94; Ó Gráda, 'Irish agricultural output', 163.

39 Based on their output estimates related to the number of milch cows from the annual returns, R. O'Connor and C. Guiomard, 'Agricultural output in the Irish Free State area before and after Independence', *Irish Economic and Social History*, 12 (1985), 92–3.

40 Donnelly, *The Land and People of Cork*, p. 141.

41 Wilson, 'Tillage versus grazing', 224–5.

42 Pringle, 'A review of Irish agriculture', 6; Donnelly, *The Land and People of Cork*, p. 141 quotes a farmer in 1871 who produced 203 lbs of butter from 609 gallons of milk, or 3 gallons per lb.

43 This compromise lies between the estimates of Solow and Ó Gráda, and the likely proportion of production which formed output based on mid 1920s estimates from Northern Ireland and the Irish Free State. It is unlikely to be a generous estimate when compared with the rest of Britain where around 1908 output per head of cow was in excess of 400 gallons. See Ojala, *Agriculture and Economic Progress*, p. 204. More recent work reveals a best estimate for the 1860s and 1870s in England of 350 gallons, rising to

set against Solar's estimate of 304 gallons, and against Ó Gráda's estimate of 350 gallons from which he made a deduction for the recycling of milk to pigs and calves. However, my estimate excludes an allowance for the buttermilk which was produced as a by-product of butter production. Whilst some was used to feed animals, most was consumed by humans, at least in mid-century.[44]

Eggs

In 1908 it was assumed that ordinary fowl laid 100 eggs per bird net of eggs reserved for hatching, and in 1912–13 it looks as though about 86 eggs per bird was assumed.[45] In Northern Ireland in 1925 it was estimated that 95 per cent of the birds enumerated were egg layers (the other 5 per cent represented cock birds). They laid about 400 million eggs, or, *as reported*, 118 eggs per bird per year. A proportion would represent net additions to stock for hatching and rearing, but 394 million eggs would remain for consumption. The poultry population at the time was 5.741 million birds, which, *by my estimation*, suggests 68 eggs per hen for home consumption or export.[46] There is obviously a confusion here in the Northern Ireland estimate: was it 118 or 68 eggs per bird? 400 million eggs at 118 per bird suggests an egg-laying hen population of 3.39 million birds, not nearly 6 million. If my interpretation is at fault, and the true hen population was nearer to the smaller estimate, then it would appear that there were about 116 eggs per hen for export and home consumption (394/3.39). As recently as the 1950s the average number of eggs per bird was only 110 per annum.[47]

The Irish poultry population grew from just over 6 million birds at the end of the 1840s to 19 million in 1906. It jumped in an extraordinary fashion in 1907 to 25 million before rising steadily to 26 million on the eve of the Great War. The spurt in 1907 was certainly due to a change in the method of enumeration.[48] In addition to egg production there was also a trade in poultry meat and feathers. In 1912–13 the meat was

400+ gallons by the Great War. See D. Taylor, 'The English dairy industry, 1860–1930', *Economic History Review*, 29 (1976), 589. See also the discussion in T.W. Fletcher, 'The economic development of agriculture in East Lancashire, 1870–1939', M.Sc. Thesis, University of Leeds, 1954, appendix III, pp. xii–xvi.

[44] My thanks to Peter Solar for these ideas on the uses of buttermilk.

[45] *Agricultural Output of Ireland 1908*, p. 14 where the figure is actually stated; *Food Production, 1912–13*, p. 70, where it is stated that 9,342,000 'great hundred' eggs were derived from nearly 13 million birds, where a great hundred was equal to ten dozen.

[46] The *Agricultural Output of Northern Ireland, 1925*, pp. 31, 79.

[47] J. Bourke, 'Women and poultry in Ireland, 1891–1914', *Irish Historical Studies*, 25 (1987), 307–8.

[48] *Ibid.*, 307.

valued at £1.6 million with a further £67,000 for the feathers, or together nearly 3 per cent of the value of livestock output. In 1908 the equivalent figure was £1.2 million or again nearly 3 per cent of the final livestock output.[49] With the experience of other estimators available, as well as these contemporary estimates, I have used a productivity of 50 eggs per head of 95 per cent of the enumerated poultry population. Solar for 1856–60 used an estimate of 60 eggs per bird.[50] In view of the contemporary estimates my lower weight does not exaggerate the contribution of eggs, but to go as high as 100 eggs per bird would give them a prominence in value which might seem very high.

Other livestock

These include horses in the horse breeding and horse flesh trade;[51] the output from goats; the sale of poultry and feathers (discussed above), hides and other by-products from the cattle industry; the general slaughter of spent cows and other aged or otherwise unwanted animals; the value of honey; and finally the sale of turf.[52] Other estimators have used a mark-up of around 5 per cent over and above their estimates of output to account for these missing items. In 1908 the final output value of horses, goats and poultry was put at £2.64 million with a further £79,000 for hides and honey. Together these accounted for £2.7 million or 7 per cent of the final livestock output. In 1912–13 the final output value of horses, mules, goats and poultry was put at £3.128 million, and feathers, hides and honey at £182,000. Together they accounted for 7 per cent of final livestock output.[53] Therefore other estimators who adopt a mark-up of about 5 per cent have judged it very well, or certainly do not exaggerate the contribution of these peripheral products.

---

[49]   *Food Production, 1912–13*, p. 70; *Agricultural Output of Ireland 1908*, pp. 15–16, 18. Pringle, 'A review of Irish agriculture', 14, states that the value of fowl exported from the port of Waterford alone was over £1,000 each week. He was writing in 1872. From this port alone, therefore, there was an estimated export of fowl to the value of £50,000 per annum.

[50]   Solar, *Growth and Distribution*, p. 365.

[51]   The horses exported to Britain in the early twentieth century represented about 5 per cent of the total population of horses. In 1864 the British government price for a 4-year-old horse for cavalry purposes was £25. By my calculation, allowing for a 5 per cent mortality rate for horses over 2 years of age, in the 1860s perhaps 30,000 horses over 2 years of age disappeared other than through death. At a value of £25 each this would have amounted to £0.75 million. O'Donovan states that in the period 1904–21 the annual value of horses exported from Ireland was £1.5 million sterling. O'Donovan, *Economic History of Live Stock*, p. 283.

[52]   On this last item see O'Connor and Guiomard, 'Agricultural output', 93–4, where they valued turf at £3.3 million or nearly 7 per cent of final output.

[53]   *Agricultural Output in Ireland 1908*, p. 24; *Food Production, 1912–13*, p. 70.

Appendix 6 Richard Barrington of Fassaroe, County Wicklow

Labour costs at Fassaroe, 1837–1886

Richard Barrington's data on labour costs were derived from the meticulous farm accounts kept by his father and then himself over a period of fifty years on their tenant farm at Fassaroe in the parish of Kilmacanogue, in the barony of Rathdown, county of Wicklow.[1] They came to their tenancy in the 1830s. The annual average extent of the tillage cultivation over the 50 years to 1886 was 120 acres, with at times as much as 140 to 150 acres. These were Irish acres. On conversion into statute measure the average size was 194 acres, varying from 227 to 243 acres.[2] The crops grown included potatoes, turnips, mangolds, vetches, rape, wheat, oats, barley, clover and rye-grass. Of the main crops potatoes and turnips were grown in all years, wheat in all years except 1870, oats in all years except 1839, and barley was not grown in 1873, 1874, 1880 and 1882.[3] The usual rotation was in seven courses; oats, potatoes, wheat, turnips, barley, meadow and pasture. The most of any single crop grown in any year in Irish acres was, 22 acres of potatoes in 1842, 27 of turnips in 1851, 45 of wheat in 1841, 24 of oats in 1869, and 32 of barley in 1852.

Richard Barrington explains that his accounts were 'kept with a detail which is not usual with farmers'. These consisted of six main books.[4] They contained such details as weekly wages, the amount and type of work performed, and on which crops that work was performed, thus

[1] R.M. Barrington, 'The prices of some agricultural produce and the cost of farm labour for the last fifty years', *Journal of the Statistical and Social Inquiry Society of Ireland*, 9 (1887), 137–53. This appendix is based entirely on the accounts published in that paper. The family accounts have now been deposited in the Royal Dublin Society Library.

[2] The ratio of Irish acres to statute acres is 1:1.62. On the variation in units employed in mid-nineteenth-century Ireland see, P.M.A. Bourke, 'Notes on some agricultural units of measurement in use in pre-Famine Ireland', *Irish Historical Studies*, 14 (1965), 236–45.

[3] There is some ambiguity over wheat, although it does not figure in the manual labour costs for 1870 it does figure in the horse labour costs.

[4] Barrington, 'The prices of some agricultural produce', 137–8.

allowing the labour cost per crop to be deduced. There is also detail on the daily work sub-division for the horses on the farm, showing the physical time spent on the potatoes, the turnips, and so on. Horses inevitably have idle moments during the week, and yet they must be fed the whole time. To this extent there is some sensitivity about horse labour costs. Labourers were a cost measured only in terms of the time when there was work available, but in contrast, the horses were fed all of the time, whether working or not. When Barrington calculated his daily horse labour costs, therefore, he did so not by dividing 365 days into the cost of a year's feed, but instead he first deducted Sundays and those idle days when the horses did not work; 'Farmers who do not keep a horse account would be astonished if told what an idle day now and again amounts to at the end of the year.'[5]

Appendix 6, table A6.1 is Barrington's own estimate of the labour costs per Irish acre for five crops, and also the daily wage rate paid to his labourers.[6] This wage rate can be compared with national estimates (appendix 6, table A6.2), where Barrington's data are converted to an assumed five or six day working week.[7] Evidently he paid above the national norm. In Wicklow in 1902 farmers were the fifth highest payers of wages to ordinary labourers. These were defined as those regularly attached to the staff of a farm, though providing their own food.[8] Only in Dublin and in three Ulster counties were wages higher.

In appendix 6, table A6.1 everything has been changed to shillings, and a 'total' labour cost included based on cultivating an equal acreage of five crops (in terms of Irish acres). These labour costs included,

the expense of ploughing, etc., in preparing for the crop – putting it in, weeding, saving, and conveying to market. The heap of manure is supposed to be ready for putting in the ground in the case of potatoes and turnips. The previous collection and preparation of it is not included; but this labour appears in the Manure Account.[9]

This seems to have been the costs of production, but without the cost of seed, the costs of any storage, and the rent of the land. Although the cost of manure is not quoted, we do learn of the labour cost of using it; 'due to the higher wages given to labourers collecting and turning it and

[5]   *Ibid.*, 146.   [6]   *Ibid.*, 149.
[7]   *Second Report by Mr. Wilson Fox on the Wages, Earnings, and Conditions of Employment of Agricultural Labourers in the United Kingdom, B[ritish] P[arliamentary] P[apers]* [Cd. 2376], vol. xcvii, 1905, especially pp. 116–45, 220–5.
[8]   *Ibid.*, p. 122. A distinction is made here between the regular labourers (attached to the farm) and the casual labourers whose rates of cash wages were often higher than those of permanent staff. The conjecture is that the Barringtons most likely employed regular staff.
[9]   Barrington, 'The prices of some agricultural produce', 149.

Table A6.1   *Barrington's labour costs, 1837–1885*

	In shillings per Irish acre						Shillings
	Potatoes[a]	Turnips[a]	Wheat[a]	Oats[a]	Barley[a]	Total[b]	Daily wage rate[a]
1837	91.8	52.1	45.3	36.3	53.2	278.5	1.0
1838	78.7	35.1	53.3	34.2	38.9	240.2	1.2
1839	82.8	61.8	46.4		51.2	242.1	1.2
1840	82.8	55.7	46.0	33.7	45.5	263.6	1.2
1841	78.4	51.8	36.5	32.7	39.5	238.9	1.2
1842	91.3	33.2	37.9	35.7	41.6	239.6	1.2
1843	55.3	66.4	37.4	47.2	37.3	243.7	1.2
1844	66.5	41.7	36.8	30.9	37.8	213.7	1.2
1845	76.7	47.8	47.1	49.8	68.5	289.8	1.2
1846	69.4	56.0	46.0	43.1	27.1	241.6	1.2
1847	79.1	56.8	54.9	40.5	41.8	273.1	1.3
1848	98.8	43.8	40.5	40.7	40.9	264.6	1.3
1849	75.9	47.8	43.6	34.8	48.3	250.3	1.3
1850	62.1	47.0	42.6	38.0	43.3	232.9	1.3
1851	63.8	50.0	35.9	41.5	46.4	237.6	1.3
1852	64.2	41.7	44.7	45.6	52.7	248.8	1.3
1853	76.1	49.5	43.2	44.1	41.5	254.3	1.3
1854	59.8	48.0	50.0	43.3	28.3	229.4	1.3
1855	69.5	52.0	49.2	28.6	56.1	255.3	1.5
1856	94.2	41.2	52.2	46.7	51.3	285.5	1.5
1857	80.0	53.6	47.3	49.8	46.3	277.0	1.5
1858	85.3	55.1	39.1	44.8	43.8	268.0	1.5
1859	101.6	56.0	50.3	51.4	45.3	304.7	1.5
1860	93.3	76.1	37.0	55.1	45.4	306.9	1.5
1861	118.8	61.6	36.8	37.5	37.8	292.5	1.5
1862	108.9	66.9	52.2	39.9	27.0	294.9	1.5
1863	121.5	66.7	31.5	34.0	34.9	288.6	1.5
1864	84.4	56.7	31.5	25.3	32.8	230.8	1.5
1865	108.3	54.7	29.5	35.4	34.2	262.1	1.5
1866	109.7	58.0	24.7	29.4	29.6	251.3	1.5
1867	109.4	62.3	42.9	31.8	39.3	285.8	1.7
1868	108.5	77.7	46.1	49.6	41.3	323.2	1.7
1869	103.4	67.3	42.0	40.5	39.8	292.9	1.7
1870	105.7	62.8		35.3	35.9	239.6	1.7
1871	111.8	86.5	43.3	37.8	37.8	317.2	1.8
1872	112.7	85.2	46.6	58.8	57.3	360.5	2.0
1873	114.3	92.9	44.4	35.2		286.8	2.0
1874	142.8	82.6	53.6	48.8		327.7	2.0
1875	151.7	74.0	41.4	48.8	58.5	374.4	2.0
1876	138.2	105.4	54.8	54.8	51.4	404.7	2.0
1877	105.9	102.7	45.3	44.2	62.7	360.8	2.0
1878	136.5	93.7	53.6	71.3	54.8	409.8	2.0
1879	113.3	82.5	43.4	51.3	36.0	326.6	2.0
1880	141.0	71.2	53.2	50.3		315.7	2.0

In shillings per Irish acre

	Potatoes[a]	Turnips[a]	Wheat[a]	Oats[a]	Barley[a]	Total[b]	Shillings Daily wage rate[a]
1881	136.1	97.3	40.0	59.7	40.8	373.8	2.0
1882	130.4	84.5	46.3	61.4		322.7	2.0
1883	125.8	91.9	40.4	51.3	67.8	377.2	2.0
1884	111.9	74.3	35.4	42.7	57.7	322.0	2.0
1885	136.1	72.6	47.3	50.4	42.3	348.7	2.0

Notes: [a] Decimalised and rounded to one decimal place
[b] Subject to rounding errors
Source: See footnote 1.

Table A6.2 *Comparison of wage rates at Fassaroe with the national average* (in shillings/pence)

	At Fassaroe			Nationally – Cash wage per week for agricultural labourers		
	Per day	5 Day week	6 Day week			
1837	1/-	5/-	6/-			
1838–46	1/2	5/10	7/-			
1847–54	1/4	6/8	8/-	1850–4	5/10.5	
1855–66	1/6	7/6	9/-	1855–66	6/-	to 7/-
1867–70	1/8	8/4	10/-	1867–70	7/1	to 7/6
1871	1/10	9/2	11/-	1871	7/7	
1872–85	2/-	10/-	12/-	1872–85	7/9.5	to 8/11.5

Note: The dates are determined by the wage levels in operation at Fassaroe, hence the range of wages in the national data.

Sources: Appendix 6, Table A6.1; A Wilson Fox, *Second Report by Mr Wilson Fox on the Wages, Earnings, and Conditions of Employment of Agricultural Labourers in the United Kingdom, BPP,* Cd. 2376 (London, 1913), p. 137.

getting it ready to put out on the ground (a figure I have not as yet exactly determined) there will be a further increase in the cost of production of not less than 10 shillings per acre.'[10]

The cost of labour per unit changed over time, but the extra labour charge for manure seems to have been fixed at 10 shillings per Irish acre, which is equivalent to 6.2 shillings per statute acre. When the Barringtons

[10]   *Ibid.,* 151.

took up their tenancy the farm was very weedy. This may have had an effect on the labour input early on with a corresponding added labour cost. In 1860 'reaping was first done by machinery, and threshing by steam.' Whilst this may have been labour saving it was not necessarily a saving in costs. For example, 'In the case of the corn crops the cost has increased slightly notwithstanding machinery, the expense and wear and tear of which must be added. This is a larger item than is generally supposed.'[11]

From the initial data therefore it is possible to proceed from a measure of labour cost to one of labour input. This is quite simply the total cost per acre divided by the daily wage rate (and further divided by 1.62 to convert to statute acres), and expressed as the man labour days involved per acre of each crop. Appendix 6, table A6.3 reports the results from that calculation, but without the extra charge of bringing in the manure in readiness for putting on the turnip and potato fields. This amounted to 6.2 shillings per statute acre, and therefore, depending on the prevailing wage rate, implies an additional labour of 3.1 to 6.2 man days per statute acre. A half decade summary of annual average man labour days for five crops is given in appendix 6, table A6.4, less the extra labour charge for manure.

W.E. Vaughan first made this calculation when he produced a single annual average estimate for the five crops for the 1860s.[12] His estimates were, potatoes at 44 man days per acre, turnips 26, wheat 14, barley 14 and oats 20. My own arithmetic suggests 42.6, 25.8, 14.9, 14.1 and 15.1 man days per acre for the same decade. Whilst the precise numbers are not the same the only serious difference is the estimate for oats. According to Vaughan it took 118 man days to cultivate a sample 5 acres in the 1860s. My own calculation suggests 112.5 man days. The implications are that a single man with five such acres worked approximately only one-third of the year, or that such a person could manage a farm of 15 acres with his labour alone, or that a family farm with one full-time

---

[11] *Ibid.*, 149.

[12] W.E. Vaughan, 'Landlord and tenant relations between the Famine and the Land War, 1850–78', Ph.D. Thesis, University of Ireland, 1973, p. 37. I assume he made the same calculation, though his method is unclear. In addition, he made a national estimate of man labour based on the Barrington data for the two years 1854 and 1874 but which purports to include crops other than those for which Barrington gives information. This includes flax which was a crop not ever grown in and around Rathdown according to the annual agricultural statistics. Finally, Vaughan has used information from Sir Richard Griffith, see note 23 below, to include an estimate of livestock man labour days, but his method of estimation is obscure, W.E. Vaughan, 'Agricultural output, rents and wages in Ireland, 1850–1880', in L.M. Cullen and F. Furet (eds.), *Ireland and France 17th to 20th Centuries: Towards a Comparative Study of Rural History* (Ann Arbor, Michigan, and Paris, 1980), pp. 87, 91–2.

Table A6.3   *Man labour days for selected crops 1837–1885* (in standard man days per statute acre)

	Potatoes	Turnips	Wheat	Oats	Barley	Total
1837	56.6	32.2	27.9	22.4	32.8	171.9
1838	41.6	18.6	28.2	18.1	20.6	127.0
1839	43.8	32.7	24.6		27.1	128.1
1840	43.8	29.4	24.3	17.8	24.1	139.4
1841	41.5	27.4	19.3	17.3	20.9	126.4
1842	48.3	17.5	20.1	18.9	22.0	126.7
1843	29.3	35.1	19.8	24.9	19.7	128.9
1844	35.2	22.0	19.5	16.4	20.0	113.0
1845	40.6	25.3	24.9	26.3	36.2	153.3
1846	36.7	29.6	24.3	22.8	14.3	127.8
1847	36.6	26.3	25.4	18.8	19.4	126.5
1848	45.7	20.3	18.8	18.8	18.9	122.5
1849	35.2	22.1	20.2	16.1	22.4	115.9
1850	28.7	21.8	19.7	17.6	20.0	107.9
1851	29.5	23.2	16.6	19.2	21.5	110.0
1852	29.7	19.3	20.7	21.1	24.4	115.2
1853	35.2	22.9	20.0	20.4	19.2	117.8
1854	27.7	22.2	23.2	20.1	13.1	106.2
1855	28.6	21.4	20.2	11.8	23.1	105.1
1856	38.8	16.9	21.5	19.2	21.1	117.5
1857	32.9	22.1	19.5	20.5	19.1	114.0
1858	35.1	22.7	16.1	18.4	18.0	110.3
1859	41.8	23.0	20.7	21.2	18.7	125.4
1860	38.4	31.3	15.2	22.7	18.7	126.3
1861	48.9	25.3	15.2	15.4	15.6	120.4
1862	44.8	27.5	21.5	16.4	11.1	121.4
1863	50.0	27.4	13.0	14.0	14.4	118.8
1864	34.7	23.3	13.0	10.4	13.5	95.0
1865	44.6	22.5	12.1	14.6	14.1	107.9
1866	45.1	23.9	10.2	12.1	12.2	103.4
1867	40.5	23.1	15.9	11.8	14.6	105.8
1868	40.2	28.8	17.1	18.4	15.3	119.7
1869	38.3	24.9	15.6	15.0	14.7	108.5
1870	39.1	23.2		13.1	13.3	88.7
1871	37.7	29.1	14.6	12.7	12.7	106.8
1872	34.8	26.3	14.4	18.2	17.7	111.3
1873	35.3	28.7	13.7	10.9		88.5
1874	44.1	25.5	16.5	15.1		101.2
1875	46.8	22.8	12.8	15.1	18.1	115.6
1876	42.6	32.5	16.9	16.9	15.9	124.9
1877	32.7	31.7	14.0	13.6	19.3	111.3
1878	42.1	28.9	16.5	22.0	16.9	126.5
1879	35.0	25.5	13.4	15.8	11.1	100.8
1880	43.5	22.0	16.4	15.5		97.4
1881	42.0	30.0	12.3	18.4	12.6	115.4

	Potatoes	Turnips	Wheat	Oats	Barley	Total
1882	40.3	26.1	14.3	19.0		99.6
1883	38.8	28.4	12.5	15.8	20.9	116.4
1884	34.5	22.9	10.9	13.2	17.8	99.4
1885	42.0	22.4	14.6	15.6	13.0	107.6

Source: Derived from appendix 6, table A6.1, the labour charge divided by the daily wage rate and further divided by 1.62 to convert to Statute measure.

Table A6.4   *Labour inputs on the Barrington family farm, 1837–1885* (standard man days per statute acre)

	Potatoes	Turnips	Wheat	Oats	Barley	Total for 5 acres[a]
1837–39	47.3	27.8	26.9	20.2	26.8	149.1
1840–44	39.6	26.3	20.6	19.1	21.3	126.9
1845–49	39.0	24.7	22.7	20.6	22.3	129.2
1850–54	30.2	21.9	20.0	19.7	19.6	111.4
1855–59	35.4	21.2	19.6	18.2	20.0	114.4
1860–64	43.4	27.0	15.6	15.8	14.7	116.4
1865–69	41.7	24.6	14.2	14.4	14.2	109.1
1870–74	38.2	26.6	14.8	14.0	14.6	108.1
1875–70	39.9	28.3	14.7	16.7	16.3	115.8
1880–85	40.2	25.3	13.5	16.2	16.1	111.3
1837–85[b]	39.2	25.3	17.9	17.3	18.4	118.0

Note: [a] The 'Total' column is the sum across the page assuming that one acre of each crop was grown each year. An alternative calculation would be to take annual averages for each crop down the page, and then sum across the page. This gives a slightly different result. For example, the 49 year average becomes 115.8 standard man days.

[b] Estimates used in subsequent calculations are: potatoes at 40 man days per acre plus 4.5 for labour for manure; turnips at 25 man days per acre plus 4.5 for labour for manure; wheat at 18, oats at 17 and barley at 19 man days.

Source: Derived from appendix 6 table A6.3.

worker (the husband/father) plus the help on a part-time basis of the wife and the children might be able to tend a 30 acre farm.

This sets the labour norm of a family farm. It assumes that the working year could have been neatly divided into times when the potatoes could be tended, and different times for the turnips, the wheat and so on. This may have been possible with more labour units and job specialisation, but in reality some crops at some times of the year competed for labour. A family farm of somewhat less than 30 acres of tillage was therefore more like the maximum manageable without the necessary purchase of outside labour to any significant extent. J.S. Donnelly suggested that in Cork in or around 1891 the small farm of under 50 acres was the typical unit of production, and it was worked by the farmer and his family assisted by one or two hired labourers.[13]

These estimates imply an even distribution between crops. At the extremes, theoretically, all 5 acres could have been under potatoes at one time, but under barley at another. This would have required either 213 man days or 70.5 man days, or 58 or 19 per cent of the total year (not allowing for high days and holidays). Theoretically a potato farm approaching 10 acres could be managed by the farmer alone, or a barley farm of 25 acres. Using national data in conjunction with the same Barrington data, J.P. Huttman much more modestly estimated that 3 acres under traditional potato crops would have fully employed a farmer and his family.[14] In 1916 E. Barker reported that the minimum size of an economic holding in Ireland (not necessarily the same concept as employed here) was reckoned to be 15 acres, or at most perhaps between 20 and 40 acres.[15] Estimates presented here support that idea.

From data on the size of holdings and employment structure in 1912 it appeared that holdings in all size groups in Wicklow employed some hired labour.[16] Temporary employment was nearly always less than 20 per cent of total employment, but family labour was not significantly supplemented by other permanent or temporary labour until holding sizes reached about 50 acres (appendix 6, table A6.5, the reciprocals of column 3 and column 2). Aspects of this pattern more or less prevailed nationally, though as appendix 6, table A6.5 indicates, when Wicklow is compared with Ireland as a whole, for all holdings the proportion of non-family, and therefore of hired employment, was far larger in Wicklow (column 2). In this sense, in the country as a whole only 26 per cent of the agricultural labour force was wage earning compared with 45 per cent in Wicklow. In addition there were clear economies of scale:

---

[13] J.S. Donnelly, *The Land and the People of Nineteenth-Century Cork: The Rural Economy and the Land Question* (London, 1975), p. 133.

[14] J.P. Huttman, 'The impact of land reform on agricultural production in Ireland', *Agricultural History*, 46 (1972), 361, though his method of calculation is not explained.

[15] E. Barker, *Ireland in the Last Fifty Years, 1866–1916* (Oxford, 1916), p. 77.

[16] Saorstát Éireann, *Agricultural Statistics 1847–1926: Reports and Tables*, Department of Industry and Commerce (Dublin, 1930), p. 155 based on the 1912 annual statistics. As well as a sub-division in holding size groups there was further sub-division into males and females, and further still into those over and under 18 years of age. In only two years before 1930, in 1912 and 1927, were such employment statistics collected alongside the other farm data, *ibid.*, p. 1.

about five times as many persons worked per acre on farms of 15–30 acres as on the largest farms of over 200 acres, and about three times as many on farms of 30–50 acres (crudely calculated from columns 8 and 1 in appendix 6, table A6.5 with confirmation given from the untransformed data in the original source).

The small farms were more productive in terms of land productivity, but output per worker was greater on the larger farms. On holdings of all sizes, 74 per cent of the labour was provided by the farmers, their wives, sons and daughters, and other relatives, 15 per cent was provided by permanent wage earners and 11 per cent by temporary employees.[17] These are crude observations from gross summaries and while they may not define the limits of family farming, they may approach such a definition. Ernest Barker in 1916 may have been closer to the mark than anyone (quoted above).

On the Barrington farm, in the early days of the tenancy, the labour input for potatoes and turnips, and wheat and barley, was very high when set against the backcloth of the forty-nine-year period as a whole (tables A6.3 and A6.4). This was partly because of a problem with weeds, 'The cause of the high figure in the early decade is, so far as I have analysed the accounts, to be attributed to the extra labour spent in weeding, as the farm was taken up in very bad condition.'[18] The average labour input for potatoes in 1837–46 was 41.4 man days per statute acre. This would have been quite a bit higher but for one extraordinarily low year in 1843 of 29.3 man days. The annual average for 1847–56 was 33.6 man days. This is based on 'the normal wage for potatoes on fairly clean land at that time, and the second decade is a better starting point for comparing the labour spent on any of the crops.'[19] The labour input estimates for this decade were, potatoes 33.6 man days, turnips 21.6, wheat 20.6, barley 21.3 and oats 19.1. This produces a final total of 116.2 man days per 5 acres plus 3–6 additional labour man days for applying manure for every statute acre of potatoes and/or turnips; 200 acres of tillage (rounded for estimation purposes), evenly distributed between five crops at 40 acres each, produces a grand total of between 4,888 to 5,128 man days to look after Barrington's tillage crops.[20] On a 6-day week this would have provided work for about sixteen people. In reality the farming calendar would not have operated so efficiently as to employ all

[17] *Ibid.*, pp. l–liii, and notes on derivation on p. liv.
[18] Barrington, 'The prices of some agricultural produce', 148–50.
[19] *Ibid.*, 151.
[20] The simple arithmetic is 116.2*200/5 plus the additional manure labour for each acre of turnip and potatoes, but with lower and upper bounds of 3–6 man days per acre, that is a variation from 3*2*200/5 to 6*2*200/5.

Table A6.5   *Holding sizes and the employment of labour in Wicklow and Ireland, 1912*

1	2	3	4	5	6	7	8
A Wicklow							
< 1	76.4	83.4	0.2	neg	0.2	neg	0.2
1–5	69.3	86.3	0.6	0.1	0.6	0.1	0.7
5–15	75.7	88.7	0.7	0.1	0.9	0.1	1.0
15–30	74.6	89.8	1.1	0.2	1.3	0.2	1.5
30–50	69.8	88.2	1.4	0.4	1.8	0.2	2.0
50–100	56.7	83.0	1.5	0.7	2.1	0.4	2.6
100–200	43.7	79.7	1.6	1.3	3.0	0.8	3.7
200–500	22.4	76.5	1.2	2.9	4.0	1.2	5.3
> 500	7.3	87.0	0.8	8.3	9.1	1.4	10.4
Totals	55.2	84.1	0.9	0.5	1.4	0.3	1.6
B Ireland							
< 1	86.4	88.9	0.3	neg	0.3	neg	0.4
1–5	85.4	90.3	0.9	0.1	0.9	0.1	1.0
5–15	88.6	92.3	1.4	0.1	1.5	0.1	1.6
15–30	84.2	90.8	1.8	0.1	1.9	0.2	2.1
30–50	74.9	87.8	2.0	0.3	2.3	0.3	2.6
50–100	63.3	86.4	2.0	0.7	2.8	0.4	3.2
100–200	46.3	82.0	1.9	1.4	3.3	0.7	4.0
200–500	28.5	80.5	1.4	2.6	4.0	1.0	4.9
> 500	10.8	83.7	0.8	5.6	6.5	1.3	7.7
Totals	73.5	88.5	1.4	0.3	1.7	0.2	1.9

Notes: Subject to rounding errors.
1   Holdings size groups.
2   Family employment as a percentage of total employment.
3   Permanent employment as a percentage of total employment.
4   Number of family employees per holding.
5   Number of non-family but other permanent employees per holding.
6   Number of full-time employees per holding (4 plus 5).
7   Number of temporary employees per holding.
8   Total employment per holding (6 plus 7).

Sources: Derived from *Annual Agricultural Statistics*; Saorstát Éireann, *Agricultural Statistics 1847–1926: Reports and Tables*, Department of Industry and Commerce (Dublin, 1930), pp. 155, 160.

the workers on all six days of the week, and at other times there might have been work for more than sixteen. Nevertheless, perhaps these calculations are a starting point for an assessment of labour usage in post-Famine Ireland.

From a base of 1847–56 there was an increase in the labour input for both potatoes and turnips, but a considerable decrease for the three grains, especially for wheat (appendix 6, tables A6.3 and A6.4). The total wage bill per acre for these three crops increased from 1847, but this is accounted for by the increase in the daily wage rate. In fact the man days' requirement of these crops fell by large proportions, especially taking a new base of 1855–9, the last half decade before the introduction of machinery in 1860. There was a subsequent drop in the labour requirement per acre of the grains down to 1880–5, for wheat of over 30 per cent, for oats it was over 10 per cent, and for barley it was nearly 20 per cent (though for both oats and barley the man days' requirement increased dramatically in the half decade 1875–9).

Agricultural depression or expansion, or good or bad weather may have affected the amount of work required on the farm, but Barrington does not refer to cycles of economic activity or to environmental conditions. He does, though, have some interesting observations to make about the introduction of machinery. It looks as though its introduction in 1860, which was essentially assistance for reaping and threshing, was indeed labour saving. The man labour involved in cultivating the corn dropped dramatically and more or less immediately. Yet perversely, and surely hardly connected, the labour required per acre of turnips and potatoes increased dramatically. When the long-term variations are investigated the year 1860 itself does not appear as such an important watershed year. The upturn in labour input for potatoes and turnips occurred earlier, nearer to 1850 than 1860, and the downturn in labour input for the grains, and wheat in particular, was more or less continuous (with a tendency for barley and oats to turn upwards in the 1870s). The saving of labour by the introduction of machines was offset by the cost of the machines, their maintenance, and the increasing cost of labour per unit.[21] In other words, a simple amortization calculation based on the cost of a machine, the rate of application of labour in substitution for a machine, and the wage of that labour, cannot tell the whole story.[22]

The interesting question is why did labour costs increase? Simple correlation tests suggest only a weak relationship between average yields

---

[21] Barrington, 'The price of some agricultural products', 151–2.
[22] As in the calculation by J.W. Boyle, 'A marginal failure: the Irish rural laborer', in S. Clark and J.S. Donnelly (eds.), *Irish Peasants: Violence and Political Unrest 1780–1914* (Manchester, 1983), pp. 316–17.

in Wicklow as a whole and labour costs at Fassaroe. Therefore whatever caused those labour costs to rise it was not due to any extra work at a time of good harvest. The evidence points more to adjustments in the market price for labour (the wage rate) than it does to the extra work on the farm. It points to a labour shortage pushing up the price of labour. To this extent the introduction of machinery may not have been solely motivated by the need to shed labour because of rising costs, but rather it may have been a combination of labour scarcity and labour cost, which then had the effect of shedding labour.

The substitution of crops by the less labour intensive meadow and clover has yet to be considered. It can only be conjectured at how much less labour-intensive such a substitution was, though there is a clue from Sir Richard Griffith in his 'Instructions to Valuators' in 1853, in preparation of the national valuation of Ireland.[23] At one stage he advised his team of valuers that the cost of cutting and making hay on 3 Irish acres was £1.50. If this was entirely a labour charge it translates into 6.2 shillings per statute acre. With a wage rate at Fassaroe in 1853 of 1.333 shillings per day this suggests a modest 4.6 man days to produce an acre of hay. In addition, although Barrington gives no indication of his labour costs for livestock, and there is very little contemporary information to help, Griffith instructed his valuers that the expense for 'Dairy-maid, support and wages, for 6 cows' amounted to £8, which is 26.7 shillings per cow.[24] This was the labour charge, which at the going labourer's wage rate for 1853 of 1.333 shillings per day suggests 20 man days per cow. It follows that the introduction of each extra cow incurred a labour charge of 20 man days, but the shedding of arable incurred a labour saving. In addition, the extension of the meadow and clover (hay) incurred a modest labour charge per acre. With the various estimates of labour costs in place there are many possible combinations of labour shedding and labour gains to consider. The conversion of, say, 1 acre of turnips into hay would have incurred a saving of labour roughly equal to the extra cost per head of a cow; the conversion of potatoes to hay would also have resulted in a net saving; but the conversion of an acre of grain into hay and with the addition of a cow would have incurred a net addition to labour.

Compared with other livestock the man labour involved in rearing cows was quite high. There was daily attendance on the animals, and although pigs and poultry require daily feeding, this was spread over

---

[23] Sir Richard Griffith, *Instructions to the Valuators and Surveyors*, appointed under the 15th and 16th Vict, *c.* 63, for the Uniform Valuation of Lands and Tenements in Ireland (Dublin, 1853), p. 33.

[24] *Ibid.*, p. 33.

large pig litters or by the broadcast method of feeding. When store or stall feeding was prominent the labour charge rose (though the general literature on Irish agriculture tells us that though rainfall was high, the maritime climate was mild enough to allow outwintering of animals to dominate over stall feeding).

Bibliography

PARLIAMENTARY AND RELATED GOVERNMENT PAPERS

Note, while all of the annual *Agricultural Statistics* from 1847 to 1914 were consulted, only those specifically cited in footnotes have been included here.

The Agricultural Output of England and Wales, 1925 [Cmd. 2815], vol. xxv, 1927.

The Agricultural Output of Great Britain ... in connection with the Census of Production Act, 1906, but referring to 1908 [Cd. 6277], vol. x, 1912–13.

The Agricultural Output of Ireland 1908 (Department of Agriculture and Technical Instruction for Ireland, Dublin, 1912).

The Agricultural Output of Northern Ireland, 1925 [Cmd. 87] (Government of Northern Ireland, Ministry of Agriculture, Belfast, 1928).

Agricultural Statistics of Great Britain, 1907. Vol. XLII, Part III Prices and Supplies of Corn, Live Stock and other Agricultural Produce [Cd. 4264], vol. cxxi, 1908, where pp. 272–5 is 'Trade in Livestock with Ireland'.

Agricultural Statistics of Ireland ... 1861 [3156], vol. lxix, 1863.

Agricultural Statistics of Ireland ... 1871 [C. 762], vol. lxix, 1873.

Agricultural Statistics of Ireland ... 1878 [C. 2347], vol. lxxv, 1878/9.

Agricultural Statistics of Ireland ... 1891 [C. 6777], vol. lxxxviii, 1892.

Agricultural Statistics of Ireland ... 1901 [Cd. 1170], vol. cxvi, 1902.

Agricultural Statistics of Ireland ... 1902 [Cd. 1614], vol. lxxxii, 1903.

Agricultural Statistics of Ireland ... 1907 [Cd. 4352], vol. cxxi, 1908.

Agricultural Statistics of Ireland ... 1911 [Cd. 6377], vol. cvi, 1912–13.

Agricultural Statistics of Ireland ... 1914 [Cd. 8266], vol. xxxii, 1916.

Agricultural Statistics of Ireland ... 1916 [Cmd. 112], vol. li, 1919.

Agricultural Statistics, Ireland, 1907–8. Return of Prices, of Crops, Livestock and other Irish Agricultural Products [Cd. 4437], vol. cxxi, 1908.

Agricultural Statistics, Ireland, 1915. Return of Prices of Crops, Livestock and other Irish Agricultural Products [Cd. 8452], vol. xxxvi, 1917–18.

Census of Ireland for the Year 1841, vol. 24 for 1843, especially pp. xxix-xxx and 452–3 for land use, and pp. xxx-xxxii and 454–7 for live stock and land-holding sizes.

The Census of Ireland for the year 1851. Part II. Returns of Agricultural Produce in 1851 [1589], vol. xciii, 1852–3.

Cowper Commission, *Report of the Royal Commission on Land Law (Ireland)*

Act 1881 and the Purchase of Land (Ireland) Act, 1885, in three volumes [C. 4969, 4969 I, and 4969 II], vol. xxvi, 1887.

Devon Commission, *Report, Evidence, Appendix, and Index to the Minutes of Evidence taken before Her Majesty's Commissioners of Inquiry into the State of the Law and Practice in respect to the Occupation of Land in Ireland* [605, 606, 616, 657, 672, and 673], vols. xix, xx, xxi, and xxii, 1845, though only the *Appendix* [672], vol. xxii, was eventually used.

Saorstát Éireann, *Agricultural Statistics 1847–1926: Reports and Tables*, Department of Industry and Commerce (Dublin, 1930).

Food Production in Ireland (1912–13). Minutes of Evidence [Cd. 8158], vol. v, 1914–16.

A Wilson Fox, *Second Report by Mr. Wilson Fox on the Wages, Earnings, and Conditions of Employment of Agricultural Labourers in the United Kingdom* [Cd. 2376], vol. xcvii, 1905, for Irish data especially pp. 116–44, 220–5, 247–56 and 262–3.

Sir Richard Griffith, *Instructions to the Valuators and Surveyors, appointed under the 15th and 16th Vict, c. 63, for the Uniform Valuation of Lands and Tenements in Ireland* (Dublin, 1853).

Minutes of Evidence ... .Upon the Inland Transit of Cattle [C. 8929], vol. xxxiv, 1898, especially appendix III.

Report from the Committee ... .Transit of Animals by Sea and Land [C. 116], vol. lxi, 1870, especially appendix XXV, p. 110.

Report from the Select Committee of the House of Lords Appointed to Inquire into the Best Mode of Obtaining Accurate Agricultural Statistics from all Parts of the UK, vol. viii, 1854–5.

Report of Mr Chambers and Professor Ferguson on Pleuro-Pneumonia among Cattle recently imported from Ireland into Norfolk, vol. lx, 1875, pp. 621–30.

Report of the Irish Land Commission for the period 22 August 1890 to 22 August 1891 [C. 6233], vol. xxv, 1890–91.

Returns of Agricultural Produce in Ireland ... .1853 [1865], vol. xlvii, 1854/5.

Royal Commission of Her Majesty's Commissioners appointed to inquire into the Financial Relations between Great Britain and Ireland [C. 7720 and 7721], vol. xxxvi, 1895.

Table Showing the Estimated Average Produce of the Crops ... 1875 [C. 1407], vol. lxxviii, 1876.

BOOKS AND ARTICLES

Allen, R.G.D. *Statistics for Economists* (London, 1972).

Andrews, J.H. 'Limits of agricultural settlement in pre-Famine Ireland', in Cullen and Furet (eds.), *Ireland and France* (1980), pp. 47–57.

Armstrong, D.L. *An Economic History of Agriculture in Northern Ireland, 1850–1900* (Plunkett Foundation, Oxford, 1989).

Barker, E. *Ireland in the Last Fifty Years (1866–1916)* (London, 1916).

Barker, T.C., D.J. Oddy and J. Yudkin, *The Dietary Surveys of Dr Edward Smith, 1862–3* (Department of Nutrition, Queen Elizabeth College, University of London, Occasional Papers No. 1, 1970).

Barrington, R.M. 'The prices of some agricultural produce and the cost of farm labour for the last fifty years', *Journal of the Statistical and Social Inquiry Society of Ireland*, 9 (1887), 137–53.

'Notes on the prices of Irish agricultural produce illustrated by diagrams', *Journal of the Statistical and Social Inquiry Society of Ireland*, 9: 73 (1893), 679–91.

Barrington, T. 'The yields of Irish tillage crops since the year 1847', *Journal of the Department of Agriculture*, 21 (1921), in two parts, 205–29, and 289–305.

'A review of Irish agricultural prices', *Journal of the Statistical and Social Inquiry Society of Ireland*, 15 (1927), 249–80.

Bastable, C.F. 'Some features of the economic movement in Ireland, 1880–1900', *Economic Journal*, 11 (1901), 31–42.

Beilby, O.J. 'Changes in agricultural production in England and Wales', *Journal of the Royal Agricultural Society of England*, 100 (1939), 62–73.

Bekaert, G. 'Calorific consumption in industrializing Belgium', *Journal of Economic History*, 51 (1991), 633–55.

Bellerby, J.R. 'The distribution of farm income in the U.K. 1867–1938', reprinted and revised in W.E. Minchinton (ed.), *Essays in Agrarian History*, II (Newton Abbot, 1968), pp. 261–79.

Bender, A. 'The nutritional importance of fruit and vegetables' in Oddy and Miller (eds.), *Diet and Health* (1985), pp. 274–95.

Bew, P. *Land and the National Question in Ireland 1858–82* (Dublin, 1978).

Bew P. and F. Wright, 'The agrarian opposition in Ulster Politics, 1848–87', in Clark and Donnelly (eds.), *Irish Peasants* (1983), pp. 192–229.

Black, R.D. Collison *Economic Thought and the Irish Question 1817–1870* (Cambridge, 1960), pp. 47–8.

Bonn, M.J. 'The psychological aspect of land reform in Ireland', *Economic Journal*, 19 (1909), 374–94.

Bourke, J. 'Women and poultry in Ireland, 1891–1914', *Irish Historical Studies*, 25 (1987), 293–310.

Bourke, P.M.A. 'The extent of the potato crop in Ireland at the time of the Famine', *Journal of the Statistical and Social Inquiry Society of Ireland*, 20: 3 (1959–60), 1–35.

'Emergence of potato blight, 1843–6', *Nature*, 202 (22 August 1964), 805–8.

'Notes on some agricultural units of measurement in use in pre-Famine Ireland', *Irish Historical Studies*, 14 (1965), 236–45.

'The agricultural statistics of the 1841 Census of Ireland. A critical review', *Economic History Review*, 18 (1965), 376–91.

'The use of the potato crop in pre-Famine Ireland', *Journal of the Statistical and Social Inquiry Society of Ireland*, 21 (1967–8), 72–96.

'The average yields of food crops in Ireland on the eve of the Great Famine', *Journal of the Department of Agriculture*, 66 (1969), 26–39.

'The Irish grain trade, 1839–48', *Irish Historical Studies*, 20 (1976), 156–69.

Bowley, A.L. 'The statistics of wages in the United Kingdom during the last hundred years. (Part IV.) Agricultural wages – Concluded. Earnings and general averages', *Journal of the Royal Statistical Society*, 62: 3 (1899), 555–70.

Boyle, J.W. 'A marginal failure: the Irish rural laborer', in Clark and Donnelly (eds.), *Irish Peasants* (1983), pp. 311–38.

Campbell B.M.S. and M. Overton (eds.), *Land, Labour, and Livestock: Historical Studies in European Agricultural Productivity* (Manchester, 1991).

Cawley, M.E. 'Aspects of continuity and change in nineteenth-century rural settlement patterns: findings from county Roscommon', *Studia Hibernica*, 22:3 (1982–3), 106–27.

Chivers, K. 'The supply of horses in Great Britain in the nineteenth century', in F.M.L. Thompson (ed.), *Horses in European Economic History* (British Agricultural History Society, Reading, 1983), pp. 31–49.

Clark, C. *National Income and Outlay* (London, 1937).

Clark, S. 'The social composition of the Land League', *Irish Historical Studies*, 17 (1971), 447–69.

Social Origins of the Land War (Princeton, 1979).

Clark S. and J.S. Donnelly 'Introduction' to part III of Clark and Donnelly (eds.), *Irish Peasants* (1983), pp. 271–83.

Clark S. and J.S. Donnelly (eds.), *Irish Peasants: Violence and Political Unrest 1780–1914* (Manchester, 1983).

Collins, E.J.T. 'Harvest technology and labour supply in Britain, 1790–1870', *Economic History Review*, 22 (1969), 453–73.

Connell, K.H. 'The land legislation and Irish social life', *Economic History Review*, 11 (1958), 1–7.

Coppock, J. T. 'Crop, livestock, and enterprise combinations in England and Wales', *Economic Geography*, 40 (1964), 65–81.

An Agricultural Atlas of England and Wales (London, 1964 and 1976).

An Agricultural Geography of Great Britain (London, 1971).

An Agricultural Atlas of Scotland (Edinburgh, 1976).

'Mapping the agricultural returns: a neglected tool of historical geography', in M. Reed (ed.), *Discovering Past Landscapes* (London, 1984), pp. 8–55.

Cousens, S.H. 'Emigration and Demographic Change in Ireland', *Economic History Review*, 14 (1961), 275–88.

Craigie, P.G. 'Statistics of agricultural production', *Journal of the Royal Statistical Society*, 46 (1883), 1–47.

'On the production and consumption of meat in the United Kingdom', *Report of the British Association for the Advancement of Science* (1884), 841–7.

Crawford E.M. (ed.) *Famine the Irish Experience 900–1900: Subsistence Crises and Famines in Ireland* (Edinburgh, 1989).

Crotty, R.D. *Irish Agricultural Production: Its Volume and Structure* (Cork, 1966).

Crowe, W.R. *Index Numbers: Theory and Applications* (London, 1965).

Cullen L.M. and F. Furet (eds.) *Ireland and France 17th to 20th Centuries: Towards a Comparative Study of Rural History* (Ann Arbor, MI, and Paris, 1980).

Cullen L.M. and T.C. Smout (eds.) *Comparative Aspects of Scottish and Irish Economic and Social History 1600–1900* (Edinburgh, 1977).

Curtis, L.P. 'Incumbered wealth: landed indebtedness in post-Famine Ireland', *American Historical Review*, 85 (1980), 332–67.

Daly, M. *The Famine in Ireland* (Dublin, 1986).

Devine T.M. and D. Dickson (eds.), *Ireland and Scotland 1600–1850. Parallels and Contrasts in Economic and Social Development* (Edinburgh, 1983).

Donnelly, J.S. 'Cork Market: its role in the nineteenth-century Irish butter trade', *Studia Hibernica*, 11 (1971), 130–63.

The Land and the People of Nineteenth-Century Cork: The Rural Economy and the Land Question (London, 1975).

'The Irish agricultural depression of 1859–64', *Irish Economic and Social History*, 3 (1976), 33–54.

'Production, prices, and exports, 1846–51' in Vaughan (ed.), *A New History of Ireland* (1989), pp. 286–93.

'Landlords and tenants' in Vaughan (ed.), *A New History of Ireland* (1989), pp. 332–49.

Dovring, F. *Land and Labor in Europe 1900–1950* (The Hague, 1956).

Drescher, L. 'The development of agricultural production in Great Britain and Ireland from the early nineteenth century', *The Manchester School*, 23 (1955), 153–75, with a comment and critique by T.W. Fletcher, 176–83.

Edwards, C.J.W. 'Farm enterprise systems in east county Londonderry', *Irish Geography*, 7 (1974), 29–52.

'Changes in agricultural labour efficiency in Northern Ireland 1975–84', *Irish Geography*, 19 (1986), 74–82.

Engerman S.L. and R.E. Gallman (eds.) *Long-Term Factors in American Economic Growth* (Chicago, 1986).

Feingold, W.L. 'Land League power: The Tralee poor-law election of 1881', in Clark and Donnelly (eds.), *Irish Peasants* (1983), pp. 285–310.

Fitzpatrick, D. 'The disappearance of the Irish agricultural labourer, 1841–1912', *Irish Economic and Social History*, 7 (1980), 66–92.

Fletcher, T.W. 'The Great Depression of English agriculture, 1873–96', *Economic History Review*, 13 (1961), 417–32.

Floud R. and D. McCloskey (eds.), *The Economic History of Britain Since 1700: Part 2. 1860 to the 1970s* (Cambridge, 1981).

Fogel, R.W. 'Second thoughts on the European escape from hunger; famines, price elasticities, entitlements, chronic malnutrition, and mortality rates', *Working Paper Series on Historical Factors in Long Run Growth*, 1 (National Bureau of Economic Research, 1989).

'New sources and new techniques for the study of secular trends in nutritional status, health, mortality, and the process of aging', *Historical Methods: A Journal of Quantitative and Interdisciplinary History*, 26 (1993), 5–43.

Forsyth F.G. and R.F. Fowler, 'The theory and practice of chain price index numbers', *Journal of the Royal Statistical Society, Series A*, 144 (1981), 224–46.

Foster, R.F. *Modern Ireland 1600–1972* (London, 1988).

Freeman, T.W. 'Land and people, c. 1841' in Vaughan (ed.), *A New History of Ireland* (1989), pp. 242–71.

Geary, R.C. 'The future population of Saorstát Éireann and some observations on population statistics', *Journal of the Statistical and Social Inquiry Society of Ireland*, 15 (1935), 15–35.

Goldstrom, J.M. 'Irish agriculture and the Great Famine', in Goldstrom and Clarkson (eds.), *Irish Population, Economy and Society* (1981), 155–71.

Goldstrom J.M. and L.A. Clarkson (eds.) *Irish Population, Economy, and Society: Essays in Honour of the late K.H. Connell* (Oxford, 1981).

Gould, J.D. 'European international emigration: the role of "diffusion" and "feedback"', *Journal of European Economic History*, 9 (1980), 267–315.

Graham B.J. and L.J. Proudfoot (eds.) *An Historical Geography of Ireland* (London, 1993).

Grantham, G. 'Capital and agrarian structure in early nineteenth-century France', *Research in Economic History*, supplement 5 (1989), 137–59.

'Divisions of labour: agricultural productivity and occupational specialization in pre-industrial France', *Economic History Review*, 46 (1993), 478–502.

Grigg, D.B. *The Dynamics of Agricultural Change* (London, 1982).

Grimshaw, T.W. 'A statistical survey of Ireland, from 1840–1888', *Journal of the Statistical and Social Inquiry Society of Ireland*, 9 (1888), 321–61 and appendixes.

Guinnane, T. 'Economics, history, and the path of demographic adjustment: Ireland after the Famine', *Research in Economic History*, 13 (1991), 147–98.

Gwinnell, M. 'The Famine years in county Wexford', *Journal of the Wexford Historical Society*, 9 (1983), 36–54.

Hatton T. and J.G. Williamson, 'After the Famine: emigration from Ireland, 1850–1913', *Journal of Economic History*, 53 (1993), 575–600.

Holderness B.A. and M.E. Turner (eds.) *Land, Labour and Agriculture, 1700–1920: Essays for Gordon Mingay* (London, 1991).

Hoppen, K.T. *Elections, Politics, and Society in Ireland 1832–1885* (Oxford, 1984).

Ireland Since 1800: Conflict and Conformity (London, 1989).

Horner, A.A., J.A. Walsh and J.A. Williams, *Agriculture in Ireland A Census Atlas* (Dublin, 1984).

Hughes, T. Jones 'East Leinster in the mid-nineteenth century', *Irish Geography*, 3: 5 (1958), 227–41.

'Society and settlement in nineteenth-century Ireland', *Irish Geography*, 5: 2 (1965), 79–96.

Huttman, J.P. 'The impact of land reform on agricultural production in Ireland', *Agricultural History*, 46 (1972), 353–68.

Jensen, E. *Danish Agriculture: Its Economic Development* (Copenhagen, 1937).

Johnston, J. *Irish Agriculture in Transition* (Dublin and Oxford, 1951).

Jones, D.S. 'The cleavage between graziers and peasants in the land struggle, 1890–1910' in Clark and Donnelly (eds.), *Irish Peasants* (1983), pp. 374–417.

Kennedy, K.A., T. Giblin and D. McHugh, *The Economic Development of Ireland in the Twentieth Century* (London, 1988).

Kennedy, L. 'Retail markets in rural Ireland at the end of the nineteenth century', *Irish Economic and Social History*, 5 (1978), 46–61.

'Traders in the Irish rural economy, 1880–1914', *Economic History Review*, 32 (1979), 201–10.

'Regional specialization, railway development, and Irish agriculture in the nineteenth century', in Goldstrom and Clarkson (eds.), *Irish Population, Economy, and Society* (1981), pp. 173–93.

'Farmers, traders, and agricultural politics in pre-Independence Ireland', in Clark and Donnelly (eds.), *Irish Peasants* (1983), pp. 339–73.

'The rural economy, 1820–1914', in Kennedy and Ollerenshaw (eds.), *An Economic History of Ulster, 1820–1939* (1985), pp. 1–61.

Kennedy L. and P. Ollerenshaw (eds.) *An Economic History of Ulster, 1820–1939* (Manchester, 1985).

Kirkpatrick, R.W. 'Origin and development of the Land War in mid-Ulster, 1879–85', in Lyons and Hawkins (eds.), *Ireland under the Union* (1980), 201–35.

Klovland, J.T. 'Zooming in on Sauerbeck: monthly wholesale prices in Britain 1845–1890', *Explorations in Economic History*, 30 (1993), 195–228.

Lamartine Yates, P. *Food, Land and Manpower in Western Europe* (London, 1960).

Lawes J.B. and J.H. Gilbert, 'Home produce, imports, consumption and price of wheat, over forty harvest years, 1852–3 to 1891–2', *Journal of the Royal Agricultural Society of England*, 3rd series, 4 (1893), 77–133.

Leabo, D.A. *Basic Statistics* (5th edn, Homewood, IL, 1976).

Lee, J.J. 'Irish agriculture', *Agricultural History Review*, 17 (1969), 64–76.

The Modernisation of Irish Society 1848–1918 (Dublin, 1973).

'Patterns of rural unrest in nineteenth-century Ireland: a preliminary survey', in Cullen and Furet (eds.), *Ireland and France* (1980), 223–37.

Ireland 1912–1985 (Cambridge, 1989).

Lewis F., and R.M. McInnis, 'The efficiency of the French-Canadian farmer in the nineteenth century', *Journal of Economic History*, 40 (1980), 497–514.

Lyons F.S.L. and R.A.J. Hawkins (eds.), *Ireland under the Union: Varieties of Tension: Essays in Honour of T.W. Moody* (Oxford, 1980).

Macdonagh, O. 'Introduction: Ireland and the Union, 1801–70' in Vaughan (ed.), *A New History of Ireland* (1989), pp. xlvii–lxv.

McGregor, P. 'The impact of the blight upon the pre-Famine rural economy in Ireland', *The Economic and Social Review*, 15 (1984), 289–303.

McInnis, R.M. 'Output and productivity in Canadian agriculture, 1870–71 to 1926–27', in Engerman and Gallman (eds.) *Long-Term Factors in American Economic Growth*, pp. 737–78.

MAFF, *A Century of Agricultural Statistics: Great Britain 1866–1966* (London, 1968).

MAFF, *Farm Incomes in England and Wales, 1974–5*, Farm Incomes Series No. 28 (HMSO, February 1976).

Manchester University, *Farm Management Survey 1984/5* (University of Manchester, Department of Agricultural Economics, April 1986).

Miller, D.S. 'Man's demand for energy', in Oddy and Miller (eds.), *Diet and Health* (1985), pp. 274–95.

Minchinton W.E. (ed) *Essays in Agrarian History*, II (Newton Abbot, 1968).

Mitchell, B.R. *European Historical Statistics 1750–1975* (Cambridge, 1981 edn).

Abstract of British Historical Statistics (Cambridge, 1988 edn).

Mokyr, J. 'Malthusian models and Irish history', *Journal of Economic History*, 40 (1980), 159–66.

'Irish history with the potato', *Irish Economic and Social History*, 8 (1981), 8–29.

'Reply to Peter Solar', *Irish Economic and Social History*, 11 (1984).

Why Ireland Starved: A Quantitative and Analytical History of the Irish Economy, 1800–1850 (London, 1985 edn).

Mokyr J. and C. Ó Gráda, 'Emigration and poverty in prefamine Ireland', *Explorations in Economic History*, 19 (1982), 360–84.

'Poor and getting poorer?' Living standards in Ireland before the Famine', *Economic History Review*, 41 (1988), 209–35.

Moody, T.W. *Davitt and Irish Revolution 1846–82* (Oxford, 1981).

Morgan W.B. and R.J.C. Munton, *Agricultural Geography* (London, 1971).

Munting R. and B.A. Holderness, *Crisis, Recovery and War: An Economic History of Continental Europe, 1918–1945* (London, 1991).

Nicholas, S. and R.H. Steckel, 'Heights and living standards of English workers during the early years of industrialisation, 1770–1815', *Journal of Economic History*, 51: 4 (1991), 937–57.

'Tall but poor: nutrition, health and living standards in pre-Famine Ireland', *Working Paper Series on Historical Factors in Long Run Growth*, 39 (National Bureau of Economic Research, 1992).

O'Brien P.K. and Leandros Prados de la Escosura, 'Agricultural productivity and European industrialization, 1890–1980', *Economic History Review*, 45 (1992), 514–36.

O'Brien, P.K., D. Heath and C. Keyder, 'Agricultural efficiency in Britain and France, 1815–1914', *Journal of European Economic History*, 6 (1977), 339–91.

O'Connor R. and C. Guiomard, 'Agricultural output in the Irish Free State area before and after Independence', *Irish Economic and Social History*, 12 (1985), 89–97.

Oddy, D.J. 'Urban famine in nineteenth-century Britain: the effect of the Lancashire cotton famine on working-class diet and health', *Economic History Review*, 36 (1983), 68–86.

Oddy D.J. and D.S. Miller (eds.) *Diet and Health in Modern Britain* (London, 1985).

O'Donovan, J. *The Economic History of Live Stock in Ireland* (Cork, 1940).

Ó Gráda, C. 'Seasonal migration and post-famine adjustment in the West of Ireland', *Studia Hibernica*, 13 (1973), 48–76.

'Agricultural head rents, pre-Famine and post-Famine', *The Economic and Social Review*, 5 (1973–4), 385–92.

'Supply responsiveness in Irish agriculture during the nineteenth century', *Economic History Review*, 28 (1975), 312–17.

'The investment behaviour of Irish landlords 1850–75: some preliminary findings', *Agricultural History Review*, 23 (1975), 139–55.

'The beginnings of the Irish creamery system, 1880–1914', *Economic History Review*, 30 (1977), 284–305.

'Demographic adjustment and seasonal migration in nineteenth-century Ireland', in Cullen and Furet (eds.), *Ireland and France* (1980), pp. 181–93.

'Agricultural decline 1860–1914', in Floud and McCloskey (eds.), *The Economic History of Britain Since 1700: Part 2* (1981), pp. 175–97.

'Irish agricultural output before and after the Famine', *Journal of European Economic History*, 13 (1984), 149–65.

Ireland Before and After the Famine: Explorations in Economic History, 1800–1925 (Manchester, 1988).

The Great Irish Famine (London, 1989).

'Slices of Irish agricultural history: output and productivity pre-Famine and post-Famine', presented to a meeting of the Irish Agricultural Economics Society, May 1989.

'Poverty, population, and agriculture, 1801–45' in Vaughan (ed.), *A New History of Ireland* (1989), pp. 108–36.

'Irish agriculture north and south since 1900', in Campbell and Overton (eds.) *Land, Labour and Livestock* (1991), pp. 439–56.

Ojala, E.M. *Agriculture and Economic Progress* (Oxford, 1952).

Ollerenshaw, P. 'Industry, 1820–1914', in Kennedy and Ollerenshaw (eds.), *An Economic History of Ulster, 1820–1939* (1985), pp. 62–108.

Ollerenshaw, P. *Banking in Nineteenth-Century Ireland: The Belfast Banks, 1825–1914* (Manchester, 1987).

O'Neill, T.P. 'The food crisis of the 1890s', in Crawford (ed.), *Famine the Irish Experience 900–1900* (1989), pp. 176–97.

O'Rourke, K.H. 'Did the Great Irish Famine matter?', *Journal of Economic History*, 51 (1991), pp. 1–22.

'Rural depopulation in a small open economy: Ireland 1856–1876', *Explorations in Economic History*, 28 (1991), 409–32.

Orridge, A.W. 'Who supported the Land War? An aggregate-data analysis of Irish agrarian discontent, 1879–1882', *The Economic and Social Review*, 12 (1981), 203–33.

Orwin C.S. and E.H. Whetham, *History of British Agriculture 1846–1914* (Newton Abbot, 1971).

Perren, R. *The Meat Trade in Britain 1840–1914* (London, 1978).

Pollard, S. *Peaceful Conquest: The Industrialization of Europe 1760–1970* (Oxford, 1981).

Pringle, R.O. 'A review of Irish agriculture, chiefly with reference to the production of live stock', *Journal of the Royal Agricultural Society of England*, 2nd series, 8 (1872), 1–77.

Reed M. (ed.), *Discovering Past Landscapes* (London, 1984).

Rew, R.H. 'Production and consumption of meat and milk: second and third reports', *Journal of the Royal Statistical Society*, 67, 1904, 374–8, 385–93.

Rothenberg, W.B. 'A price index for rural Massachusetts, 1750–1855', *Journal of Economic History*, 39 (1979), 975–1001.

Shammas, C. 'The eighteenth-century English diet and economic change', *Explorations in Economic History*, 21 (1984), 254–69.

Smyth, W.J. 'Landholding changes, kinship networks and class transformation in rural Ireland: A case study from county Tipperary', *Irish Geography*, 16 (1983), 16–35.

Solar, P. 'The agricultural trade statistics in the Irish Railway Commissioners' Report', *Irish Economic and Social History*, 6 (1979), 24–40.

'Agricultural productivity and economic development in Ireland and Scotland in the early nineteenth century', in Devine and Dickson (eds.), *Ireland and Scotland 1600–1850* (1983), pp. 70–88.

'Why Ireland starved: A critical review of the econometric results', *Irish Economic and Social History*, 11 (1984), 107–15.

'The Great Famine was no ordinary subsistence crisis', in Crawford (ed.), *Famine the Irish Experience 900–1900* (1989), pp. 112–33.

'Harvest fluctuations in pre-Famine Ireland: Evidence from Belfast and Waterford Newspapers', *Agricultural History Review*, 37 (1989), 157–65.

'The Irish butter trade in the nineteenth century: new estimates and their implications', *Studia Hibernica*, 25 (1989–90), 134–161.

Solar P. and M. Goossens, 'Agricultural productivity in Belgium and Ireland in the early nineteenth century', in Campbell and Overton (eds.), *Land, Labour, and Livestock* (1991), pp. 364–84.

Solow, B. *The Land Question and the Irish Economy, 1870–1903* (Cambridge, MA, 1971).

Staehle, H. 'Statistical notes on the economic history of Irish agriculture, 1847–1913', *Journal of the Statistical and Social Inquiry Society of Ireland*, 18 (1950–1), 444–71.

Stamp, L.D. *An Agricultural Atlas of Ireland* (London, 1931).

Symons L. (ed.) *Land Use in Northern Ireland* (London, 1963).

Taylor, D. 'The English dairy industry, 1860–1930', *Economic History Review*, 29 (1976), 585–601.

Thompson F.M.L. (ed.) *Horses in European Economic History* (British Agricultural History Society, Reading, 1983).

'An anatomy of English agriculture, 1870–1914', in Holderness and Turner (eds.), *Land, Labour and Agriculture, 1700–1920* (London, 1991), pp. 211–40.

Toutain, J.C. *Le produit de l'agriculture française de 1700 à 1958* (Paris, 1961).

Tracy, M. *Agriculture in Western Europe* (London, 1982).

Turner, M.E. 'Livestock in the agrarian economy of counties Down and Antrim from 1803 to the Famine', *Irish Economic and Social History*, 11 (1984), 19–43.

'Towards an agricultural prices index for Ireland 1850–1914', *The Economic and Social Review*, 18 (1987), 123–36.

'Output and productivity in Irish agriculture from the Famine to the Great War', *Irish Economic and Social History*, 17 (1990), 62–78.

'Agricultural output and productivity in post-Famine Ireland', in Campbell and Overton (eds.), *Land, Labour, and Livestock* (1991), pp. 410–38.

'Output and prices in UK Agriculture, 1867–1914, and the Great Depression reconsidered', *Agricultural History Review*, 40 (1992), 38–51.

'Rural economies in post-Famine Ireland, c. 1850–1914', in Graham and Proudfoot (eds.), *An Historical Geography of Ireland* (London, 1993), pp. 293–337.

Van Zanden, J.L. 'The first green revolution: the growth of production and productivity in European agriculture, 1870–1914', *Economic History Review*, 44 (1991), 215–39.

Vaughan, W.E. 'Landlord and tenant relations in Ireland between the Famine and the Land War, 1850–1878' in Cullen and Smout (eds.), *Comparative Aspects of Scottish and Irish Economic and Social History* (1977), pp. 216–26.

'Agricultural output, rents and wages in Ireland, 1850–1880', in Cullen and Furet (eds.), *Ireland and France* (1980), pp. 85–97.

'An assessment of the economic performance of Irish landlords, 1851–81', in Lyons and Hawkins (eds.), *Ireland under the Union* (1980), pp. 173–99.

Appendixes in T.W. Moody, *Davitt and Irish Revolution 1846–82* (Oxford, 1981), pp. 569–70.

Landlords and Tenants in Ireland 1848–1904 (The Economic and Social History Society of Ireland, Dublin, 1984).

'Potatoes and agricultural output', *Irish Economic and Social History*, 17 (1990), 79–92.

Landlords and Tenants in Mid-Victorian Ireland (Oxford, 1994).

Vaughan W.E. (ed.) *A New History of Ireland V Ireland Under the Union, I 1801–70* (Oxford, 1989).

Warwick-Haller, S. *William O'Brien and the Irish Land War* (Dublin, 1990).

Weiss, T. 'Long-term changes in US agricultural output per worker, 1800–1900', *Economic History Review*, 46 (1993), 324–41.

Williamson, J.G. 'The evolution of global labor markets in the first and second world since 1830: background evidence and hypotheses', *Working Paper Series on Historical Factors in Long Run Growth*, 36 (National Bureau of Economic Research, 1992).

'Economic convergence: placing post-Famine Ireland in comparative perspective', *Irish Economic and Social History*, 21 (1994), 5–27.

Wilson, J. 'Tillage versus grazing', *Journal of the Department of Agriculture and Technical Instruction for Ireland*, 5 (1905), 217–35.

Winstanley, M.J. *Ireland and the Land Question 1800–1922* (London, 1984).

THESES AND DISSERTATIONS

Fletcher, T.W. 'The economic development of agriculture in East Lancashire, 1870–1939', M.Sc. Thesis, University of Leeds, 1954.

Huttman, J.P. 'Institutional Factors in the Development of Irish Agriculture, 1850–1915', Ph.D. Thesis, University of London, 1970.

Jordan D.E. Jnr, *Land and Politics in the West of Ireland: County Mayo, 1846–82*, Ph.D. Dissertation, University of California at Davis, 1982.

Kennedy, L. 'Agricultural Co-operation and Irish Rural Society, 1880–1914', D.Phil. Thesis, University of York, 1978.

Ó Gráda, C. *Post-Famine Adjustment, Essays in Nineteenth Century Irish Economic History*, Ph.D. Dissertation, University of Columbia, 1973.

O'Rourke, K.H. *Agricultural Change and Rural Depopulation: Ireland, 1845–1876*, Ph.D. Dissertation, Harvard, 1989.

Solar, P. *Growth and Distribution in Irish Agriculture Before the Famine*, Ph.D. Dissertation, University of Stanford, 1987.

Vaughan, W.E. 'A Study of Landlord and Tenant Relations between the Famine and the Land War, 1850–78', Ph.D. Thesis, University of Ireland, 1973.

Index

agricultural change 15–64
  and labour supply 164–70
  see also under individual crops and
    animals
agricultural depression 15, 25, 204, 216
  of 1859–64 30–2, 49, 54, 56–7, 75–6, 199,
    204
  and landholding 72, 83, 86
  and agricultural output 121–4
agricultural exports, *see* exports, *see also*
    trade
agricultural machinery 133–4
agricultural output 6, 12, 14–15, 29, 95–125,
    196–216
  as proportion of UK 127–8
  as share of England 154n
  as share of UK 154
  in constant volume 115–20
  distribution of 113–15
*Agricultural Output of England and Wales
    1925* 141
Agricultural Output of Ireland 1908 100,
    112, 257–8, 268–80
*Agricultural Output of Northern Ireland
    1925* 85, 223–4, 268–80
agricultural performance 12, 196–216
agricultural prices 255–67, *see also* under
    individual crops
agricultural productivity 6, 12
  see also land
  see also labour
agricultural prosperity 118
(The) *Agricultural Returns* 29, 220
agricultural specialisation 36, 60
agricultural statistics 216–54
Agricultural Statistics of Saorstát Éireann
    139–40, 223–4
America, N. 5, 10, 22
American Civil War 25, 32
  and cotton famine 113, 115, 121, 123
  and resurgence of flax 54, 123
  and wool prices 123

Antrim
  land use 38
  landholding changes 93
arable agriculture 15–47
  see also under individual crops
Ashbourne Acts (1885, 1888) 207, 210
asses, numbers 48, 235–8
Austria
  GDP 3
average size of holding 77–94
  county distributions 90
  provincial distribution 89
  see also under individual animals

Ballinasloe (Co. Galway) 259, 274–5
Baltic States, land/labour ratio 171
bank deposits 32, 202
Barker, E.
  on family farming 288–9
  on uneconomic holdings 187
barley 6, 20, 36
  acreages 227–8
  exports 18
  labour costs 281–92
  output 113, 116
  prices 257–60, 263–5
  seeding ratios 98
  standard man days 175, 178, 181
  weighting procedures 269–70
  yields 201, 244–5
Barrington, R. (of Fassaroe
Co. Wicklow) 174, 226, 281–93
Barrington, T.
  on prices 107, 200, 255–7, 259, 263–7
  on agricultural production 141
  on oversupply of labour 185
beans
  acreages 227–8
  yields 244–5
beef prices 263, 265–7
beets
  acreages 228–30